JONATHAN B. TUCKER

WAR OF NERVES

Jonathan B. Tucker received a B.S. in biology from Yale University and a Ph.D. in political science, specializing in defense and arms control studies, from the Massachusetts Institute of Technology. For the past ten years, he has been a chemical and biological weapons specialist at the Center for Nonproliferation Studies of the Monterey Institute of International Studies. Dr. Tucker previously worked as an arms control specialist for Congress and the State Department and as an editor at *Scientific American* and at *High Technology* magazine, where he wrote about biomedical research, biotechnology, and military technologies. He lives in Washington, D.C.

WAR OF NERVES

WAR OF NERVES

CHEMICAL WARFARE FROM WORLD WAR I TO AL-QAEDA

Jonathan B. Tucker

ANCHOR BOOKS
A DIVISION OF RANDOM HOUSE, INC.
NEW YORK

FIRST ANCHOR BOOKS EDITION, FEBRUARY 2007

The Library of Congress has cataloged the Pantheon edition as follows:
Tucker, Jonathan B.
War of nerves : chemical warfare from World War I to Al-Qaeda / Jonathan B. Tucker.
p. cm.
Includes bibliographical references and index.
1. Chemical warfare—History.   2. Chemical weapons—History.   I. Title.
UG447.T83 2006
358'.34'09—dc22 2005050053

Anchor ISBN: 978-1-4000-3233-4

Author photograph © Link Nicoll
Book design by Robert C. Olsson

www.anchorbooks.com

CONTENTS

Contents

ACKNOWLEDGMENTS

I am deeply grateful to the following individuals who reviewed part or all of the draft manuscript and provided valuable comments: Gordon Burck, Dawson Cagle, William C. Dee, Sigmund R. Eckhaus, John A. Gilbert, Olivier Lepick, David E. Kaplan, Ron G. Manley, Robert P. Mikulak, Vil Mirzayanov, Michael Moodie, John Ellis van Courtland Moon, Colonel Jonathan Newmark, Julian Perry Robinson, Barbara Seiders, Ralf Trapp, Mark L. Wheelis, and Jonathan Winer. Any errors of omission or commission that remain are clearly my responsibility.

Julian Perry Robinson deserves special thanks for making available his unparalleled archive of news clippings at the University of Sussex in Brighton, England.

Jeffrey K. Smart, the Chief Historian at Aberdeen Proving Ground in Maryland, was extremely helpful in identifying U.S. government documents relevant to the history of the U.S. nerve agent program. Deborah A. Dennis, the Freedom of Information Act (FOIA) Officer in the Office of the Chief Counsel at the U.S. Army Research, Development and Engineering Command, efficiently expedited the process of review, redaction, and public release of formerly classified U.S. government documents.

Sigmund Eckhaus generously granted me access to his collection of historical photographs, many of which are reproduced in this volume.

I am also grateful to the staffs of the U.S. National Archives and the Military History Institute at the U.S. Army War College in Carlisle, Pennsylvania, for help with archival research.

Finally, my sincere thanks for support and encouragement go to Victoria Wilson; her assistant, Zachary Wagman; Martha Kaplan, my literary agent; and my parents, Deborah and Leonard Tucker.

Jonathan B. Tucker
Washington, D.C.

LIST OF ILLUSTRATIONS

List of Illustrations

List of Illustrations

WAR OF NERVES

LIVE-AGENT TRAINING

The U.S. Army Chemical School at Fort Leonard Wood, near the edge of the Ozark Mountains in south-central Missouri, trains thousands of soldiers, sailors, and marines each year in the art and science of chemical warfare defense. Students enrolled in the Chemical Officer Basic Course learn to detect and identify the various types of chemical warfare agents, to don and seal a gas mask in seconds, to treat chemical casualties with injections of antidotes, and to decontaminate vehicles. The climax of the twenty-week course is a "live-agent" exercise in which a group of trainees, wearing full-body protective suits and masks, perform tasks inside a sealed chamber containing lethal concentrations of nerve agents, the deadliest class of chemical weapons.

Designed to kill, nerve agents such as Sarin and VX serve no peaceful purpose. They are colorless, odorless liquids that enter the body through the lungs or skin and attack the nervous system. Initial symptoms of nerve agent poisoning are runny nose, excess saliva, pinpoint pupils, and shortness of breath, followed by profuse sweating, stomach cramps, and involuntary muscle twitches. Finally the victim falls to the ground, convulses, and loses consciousness, after which inhibition of the breathing center of the brain and paralysis of the respiratory muscles cause death by asphyxiation within several minutes.

German chemists discovered nerve agents accidentally while doing industrial pesticide research in the mid-1930s. These compounds were then developed into weapons by the Nazi regime, which stockpiled but never used them during World War II. Although Adolf Hitler had at his disposal chemical weapons that were vastly more toxic than were any previous war gases, the Allied leaders had no inkling of this fact—a major failure of Western intelligence. Only after the war did the victorious Allies discover the secret of the German nerve agents and launch their own intensive development and production programs. During the 1950s and '60s, the United States and the Soviet Union manufactured vast quantities of nerve agents in a shadowy chemical arms race that paralleled the high-profile nuclear weapons competition.

For a brief period, nerve agents remained the exclusive province of the advanced industrial states. Beginning in the early 1960s, however, the technology and know-how to produce these weapons spread to about a dozen nations of the developing world. This process of chemical proliferation culminated in Iraq's large-scale use of nerve agents during the Iran-Iraq War of the 1980s and its brutal campaign of repression against the Iraqi Kurds.

Even more worrisome than the spread of nerve agents to so-called rogue states is the growing interest by terrorists in acquiring these weapons. Like some national leaders, terrorists might consider using lethal chemicals against civilians, who are far more vulnerable than troops wearing gas masks and protective suits. In a 1999 magazine interview, terrorist mastermind Osama bin Laden declared his intent to attack the United States and its allies using nonconventional means. "We don't consider it a crime if we [try] to have nuclear, chemical, biological weapons," he said. "Our holy land is occupied by Israeli and American forces. We have the right to defend ourselves and to liberate our holy land." Bin Laden added that any U.S. citizen who pays taxes is a legitimate target "because he is helping the American war machine against the Muslim nation." The dispersal of a volatile nerve agent such as Sarin in a crowded subway station, shopping mall, or sports arena could potentially claim hundreds or even thousands of victims.

Given the clear and present danger posed by chemical weapons, live-agent training at Fort Leonard Wood is not just a military rite of passage; it serves a vital purpose. Although soldiers can learn to use protective gear with nonlethal chemicals such as tear gas, only the experience of working with "live" nerve agents teaches them to handle the psychological stress of fighting on a battlefield that has been contaminated with invisible but deadly poisons. Unless troops gain confidence in the ability of their equipment to protect them, they could be paralyzed with fear when encountering nerve agents in combat for the first time.

LIVE-AGENT training at Fort Leonard Wood takes place inside a $27 million facility called the E. F. Bullene Chemical Defense Training Facility (CDTF), which opened in October 1999. From the outside, the modernistic, semicircular building with a domed roof bears a strong resemblance to a flying saucer. Built of reinforced concrete strong enough to withstand earthquakes and

A soldier from the 82nd Chemical Battalion during a chemical defense training exercise at the U.S. Army Chemical School.

tornadoes, the 72,600-square-foot structure contains classrooms, training and administrative areas, a medical clinic, and a chemical laboratory. The CDTF is one of two live-agent training facilities in the United States; the other is at the Center for Domestic Preparedness (formerly Fort McClellan) in Anniston, Alabama, where the Department of Homeland Security teaches city and state emergency personnel to respond to incidents of chemical terrorism.

At the core of the Fort Leonard Wood facility is a circular "hot zone," or containment area, which has been subdivided like a sliced pizza into eight bays for training exercises with lethal chemicals. A powerful ventilation system generates negative atmospheric pressure, so that if the containment area is breached the nerve agent vapors will remain inside rather than leaking into the environment. The building's energy and safety systems have multiple redundant backups, and contaminated air from the training area passes through eighteen high-efficiency filters before being released to the outside. Closed-circuit television cameras, air-sampling systems, and electronic alarms continually monitor the training bays. In the unlikely event of an accidental release of nerve agent, the facility has been sited so that the prevailing winds will carry the plume of lethal gas over unpopulated areas.

Prior to the live-agent exercise at the end of the Chemical Officer Basic Course, the trainees have their blood drawn to measure their baseline levels of cholinesterase, a key enzyme in the nervous system that is specifically targeted by nerve agents. Each student is assigned a buddy and spends a week in a training area designed to simulate the hot zone, learning to perform complex tasks while wearing a gas mask and a bulky protective suit that encapsulates the entire body and weighs fourteen pounds. Donning the chemical suit involves several steps. The soldier first puts on underwear, a cotton battle-dress uniform, white canvas sneakers covered with black vinyl booties, and a battle-dress overgarment (BDO): a set of camouflaged pants and jacket lined with a layer of activated charcoal that absorbs and neutralizes toxic agents. The trainee pulls the drawstrings of the BDO tight at the wrists and ankles to seal off leaks and then straps on a hood that protects the head, face, and neck, followed by white cotton gloves and heavy black rubber gloves. Finally, he dons the M40 gas mask, which contains a charcoal filter to protect the eyes and lungs, a voice transmitter, and a plastic tube that makes it possible to drink water without unmasking.

On the day of the live-agent exercise, the twelve masked and suited trainees, accompanied by their instructors, approach the containment area of the CDTF through a series of doors with electronic locks. Because the gas masks have oversized eye lenses and elongated snouts, the trainees resemble a swarm of giant insects. Before entering the hot zone, they move into a room where their gas masks are checked and rechecked. Even a two-day growth of beard is sufficient to break the seal around the face. To verify that the masks are airtight, the soldiers sit on chairs under clear plastic cowls and are sprayed with an acrid chemical called stannic chloride, which normally causes severe coughing. If the mask has been fitted properly, it screens out the toxic mist.

After passing this test, the twelve trainees file into the narrow corridor that encircles the building's core. The door to the training bays opens with a sucking sound, caused by the negative atmospheric pressure within. While a safety officer in another part of the building watches on closed-circuit television, the trainees gingerly enter the hot zone. By now, the air is thick with tension. Not only are the students anxious about the ordeal ahead, but their chemical suits are hot, awkward, and stressful. The BDO alone raises body temperature 10 degrees, and the gas mask has an unpleasant rubbery smell, restricts breathing, distorts vision, and dulls hearing.

Clammy with sweat, their hearts beating a rapid tattoo in their ears, the trainees experience an oppressive feeling of claustrophobia combined with a dread of poisoning. Having studied the toxicology of the nerve agents and seen videotapes of laboratory animals convulsing and dying, they understand all too well the consequences of exposure. The presence of two medics carrying antidote-filled syringes is only slightly reassuring. If antidote is injected into the thigh muscle within seconds, it can save the life of someone who has been "slimed." For one young soldier, the tension is too great; she breaks down in tears and has to be escorted out.

In the first bay, the trainees form a circle around a metal table in the center of the room. Attempting to overcome their fear with a show of bravado, they begin a muffled chant: "We want the nerve! We want the nerve!"

As if on cue, two agent handlers enter the room. Wearing heavy green rubber aprons over their protective suits, they walk with slow, measured steps. Each handler carries a small plastic tackle box containing a syringe filled with nerve agent, which has been synthesized in an on-site laboratory and stored in a guarded vault. Holding a bowl of decontaminating solution below the box to catch any stray drops, one handler carefully removes the syringe and uses it to deposit six drops of clear fluid at various points on the surface of the metal table. He then turns toward the video camera in the corner of the room and announces solemnly, "We've got the nerve."

The trainees now carry out their assigned task, which is to identify the toxic agent. Working awkwardly with thick rubber gloves, one member of each team removes a piece of M8 detection paper from a pocket-size detection kit and skims the paper over the surface of one of the drops, wetting a small portion. If the liquid contains a chemical warfare agent, the indicator paper will change color: red for a blister agent such as mustard, yellow for Sarin, green for VX. Holding the M8 paper at arm's length, the trainee watches it turn dark green, indicating that the clear fluid is VX. He then "kills" the paper by dropping it into a bucket of decontamination solution.

The teams now check for vapor hazard using an M256 chemical agent detector kit, a small card studded with transparent plastic bubbles containing liquid-filled glass ampoules. Crushing two of the ampoules with one's gloved fingers causes the chemicals to mix and react, yielding an indicator fluid that turns a test spot blue-green if no nerve agent is present. The spot remains colorless, confirming the presence of VX vapor. When the six teams have completed the exercise, the instructor decontaminates the metal table with a highly corrosive solution known as DS2. As the fluid reacts with and destroys the nerve agent, a column of white smoke rises eerily from the metal surface.

In the next two training bays, the trainees identify areas of VX contamination on items of military equipment, including a Humvee and the frame of a helicopter, and inject nerve agent antidote into a mannequin wearing a battle-dress uniform to simulate treatment of a chemical casualty. The soldiers also practice drinking from a canteen through a straw inserted through the tube in the gas mask.

After completing their assigned tasks, the students leave the hot zone and enter an air lock. They must now go through a lengthy, multistep process of decontaminating their protective garments and then doffing them one by one. This task requires a great deal of self-control, for by now the trainees are desperate to get out of the oppressive suits and masks. If a student breaks the seal on his gas mask prematurely or removes the elements of the protective ensemble in the wrong order, he is "red-tagged" and sent to the medics for a blood test. Finally, the trainees strip to their underwear, hold their breath, and pass through an ice-cold shower to close their pores.

Once the students have cleared the showers, the tension breaks and they discuss their experience in boisterous tones, laughing and joking. No one admits to having been afraid.

CHAPTER ONE

THE CHEMISTRY OF WAR

IN THE LATE AUTUMN OF 1914, the opposing armies on the western front huddled in their trenches near the Belgian town of Ypres, lobbing artillery shells at each other across a barren no-man's-land strewn with thickets of rusty barbed wire, craters, and splintered trees. Germany had launched the war on August 5 by carrying out the Schlieffen Plan, a massive surprise attack through neutral Belgium that sought to achieve the rapid conquest of France in the west, followed by a knockout blow to Russia in the east. The initial operations had gone according to plan, but when the kaiser's armies were thirty miles from Paris, a last-ditch counterattack by the French and British forces on September 6–12, the Battle of the Marne, had halted the German offensive. Then had come the "race to the sea" as each army tried to outflank the other, the German capture of Antwerp, and the First Battle of Ypres, where the British had blocked the German advance in Flanders. Seeking cover from the lethal hail of shrapnel and machine-gun fire, both sides had dug in, building labyrinthine trenches that would ultimately extend some four hundred miles across Belgium and France from the North Sea coast to the Swiss border.

With the onset of winter, the adversaries found themselves trapped in a bloody stalemate in which neither side was able to advance. Infantry offensives inevitably bogged down after taking negligible amounts of territory, at a heavy cost in lives. Seeking to break the deadlock and regain the offensive, the Germans began to consider the use of toxic chemicals delivered by artillery shells to force the enemy out of his trenches. This idea was not entirely new: in 1862, during the American Civil War, a New York City schoolteacher named John W. Doughty had written to the Secretary of War suggesting the use of poison gas shells against the Confederate forces. He had designed a 10-inch projectile in which one compartment was filled with a few quarts of liquid chlorine and the second with explosives; when the shell burst, the explosion would convert the chlorine into an asphyxiating gas. But the Union's chief of ordnance, Brigadier General James Ripley, had been resistant to new ideas and had rejected Doughty's invention.

Because Germany possessed the world's most advanced chemical industry, it enjoyed an inherent advantage in this type of warfare. The main obstacles were the existence of an international treaty specifically banning the use of shells to deliver asphyxiating gases and the deeply held belief that toxic weapons were illegitimate. This "chemical weapons taboo" appears to have originated in the innate human aversion to poisonous substances, as well as revulsion at the duplicitous use of poison by the weak (including women) to defeat the strong without a fair physical fight. Efforts to outlaw the use of poisons in war dated back to the classical Greek and Roman period. During the Middle Ages, German artillery gunners pledged not to use poisoned weapons, which were judged "unworthy of a man of heart and a real soldier." The first known international agreement banning chemical warfare, a Franco-German treaty prohibiting the use of poisoned bullets, was drawn up in Strasbourg in 1675.

Before the second half of the nineteenth century, numerous poisonous chemicals had been discovered, but they could not be produced on a large scale. The emergence of the European chemical industry, which was capable of manufacturing vast quantities of dyestuffs and other synthetic chemicals, gave rise to new concerns over the potential use of lethal gases on the battlefield. In 1863, the U.S. War Department issued the Lieber Code of Conduct, which prohibited "the use of poison in any manner, be it to poison wells, or food, or arms." Similarly, the 1874 Brussels Declaration on the laws and customs of war, signed by fourteen European countries but never ratified, banned the use of poison, poisonous gases, and weapons that caused unnecessary suffering.

At the 1899 International Peace Conference in The Hague, representatives of twenty-six countries, including Germany, signed the first Hague Convention Respecting the Laws and Customs of War on Land. Article 23(a) of this treaty prohibited "poison or poisoned weapons," including the deliberate tainting of arms, bullets, food, or wells. The contracting states also signed a separate document, the Hague Declaration Concerning Asphyxiating Gases, which specifically outlawed "the use of projectiles, the sole object of which is the diffusion of asphyxiating or deleterious gases." This treaty effectively banned the use of chemical shells even before they had been developed.

In late 1914, however, amid the futile slaughter of trench warfare, the traditional legal and moral restraints on the use of poison gas began to erode under the pressure of military necessity. From the outset, the German High Command had interpreted the Hague gas-projectile declaration as banning only the release of lethal gases from shells specifically designed for that purpose. The German military also considered tear gases and other nonlethal irritants to be equivalent to smoke, hence not covered by the legal ban. Indeed, the French had begun using tear gas grenades in August 1914, the first month of the war, albeit to little effect. Exploiting these loopholes, the Germans proceeded to develop a 105 mm artillery shell that was loaded with a lung irritant (dianisidine chlorosulfate) and was also designed to generate shrapnel, so that its "sole" purpose was not the delivery of a toxic gas. In October 1914, the Germans fired three thousand irritant shells at the British forces near Neuve-Chapelle, but because the high-explosive charge burned the chemical agent and neutralized its effects, the British remained unaware that they had been subjected to a chemical attack.

The Germans then developed a 150 mm howitzer shell containing seven pounds of another chemical irritant (xylyl bromide), once again combined with an explosive charge to disperse shrapnel. In January 1915, German troops fired more than 18,000 of these shells at the Russian positions near Bolimow, but the subfreezing temperatures prevented the liquid agent from vaporizing and rendered it harmless. The failure of these attacks with irritant gases, combined with a shortage of high explosives, led the German High Command to consider the use of shells containing lethal agents.

The individual who became responsible for developing chemical weapons for the German War Office was Professor Fritz Haber, a brilliant young chemist and ardent Prussian nationalist who directed the Kaiser Wilhelm Institute for Physical Chemistry in Berlin. Although born Jewish, Haber had converted to Christianity at the age of twenty-three. In 1909, he had invented a revolutionary method for the synthesis of ammonia from atmospheric nitrogen that was subsequently translated into a large-scale production process by the industrial chemist Carl Bosch. Ammonia was vital to the war effort because it was used to manufacture both ammonium sulfate fertilizer, required for intensive agriculture, and nitric acid, a key ingredient of the explosives nitroglycerine and TNT. The Haber-Bosch process for producing ammonia freed Germany from its previous dependence on the importation of Chilean nitrates by sea. That source had dried up shortly after the war began, when British warships had blocked German supply lines across the Atlantic. Without a synthetic source of ammonia, Germany would have quickly run out of food and ammunition, and Haber's essential invention made him a national hero.

In late 1914, Haber had the idea of loading artillery shells with chlorine, which the German chemical industry produced in large quantities for the production of dyestuffs. When a shortage of artillery shells ruled out this method of delivery, he proposed instead that chlorine be released directly from pressurized gas cylinders, allowing the wind to carry the poisonous cloud over the enemy's trenches. This tactic offered a number of potential advantages: chlorine released directly from cylinders would blanket a far larger area than could be achieved with projectiles, and the gas would dissipate rapidly, allowing the affected areas to be occupied by friendly troops.

In early January 1915, these arguments won over General Erich von Falkenhayn, the chief of the German General Staff, who considered poison gas "unchivalrous" but hoped that its use would result in a decisive military victory. As the site of the first chlorine attack, Falkenhayn selected the Allied-held town of Ypres in Flanders, Belgium. East of the town, the line of Allied trenches extended about four miles into German-controlled territory, forming a bulge called the Ypres Salient that was nine miles across at its widest point. Holding the line on the left side of the Salient, near the village of Langemarck, were the French 87th Territorial Division and the 45th Algerian Division, made up of French-Algerian soldiers known as Zouaves. Canadian and British units defended the center and right portions of the bulge, respectively.

In mid-January, Haber ordered the chemist Otto Hahn and several other colleagues to help prepare the chlorine attack. When Hahn objected that chemical warfare would violate the Hague Convention, Haber replied that the French had been the first to employ gas-filled munitions and that countless human lives would be saved if the effective use of chemical weapons brought the war to a rapid end. The German chemists helped to organize a special unit for gas warfare called Pioneer Regiment 35. These troops received training and equipment for handling chlorine, including the so-called Dräger self-preserver (Drägersche Selbstretter), which they would don for their protection when releasing the lethal gas.

On January 25, 1915, General von Falkenhayn ordered Infantry General Berthold von Deimling, who commanded the German XV Army Corps at Ypres, to report to the field headquarters at Mezières. According to Deimling's memoirs:

Falkenhayn revealed to us that a new weapon, poison gas, was to be used and that my corps area had been selected for the first attempt. The poison gas would be delivered in steel cylinders, which would be built into the trenches and opened when the winds were favorable. I must confess that the commission for poisoning the enemy, just as one poisons rats, struck me as it must any straightforward soldier: it was repulsive to me. If, however, the poison gas were to result in the fall of Ypres, we would win a victory that might decide the entire campaign. In view of this worthy goal, all personal reservations had to be silent. So onward, do what must be done! War is necessity and knows no exception.

Haber was dispatched to Flanders to organize and prepare the chemical attack. Under his direction, the German War Office shipped to the front 1,600 large and 4,130 small steel cylinders filled with pressurized liquid chlorine. On March 10, 1915, five hundred German troops from Pioneer Regiment 36 began to emplace the cylinders along a four-mile line opposing the French trenches, burying them vertically in slit trenches to prevent them from being ruptured or destroyed by enemy artillery fire.

After emplacing the cylinders, the Germans waited for the wind direction to change. For more than three weeks, the prevailing winds at Ypres blew from west to east, which would have carried the poisonous cloud back over the German lines. Finally, in the late afternoon of Thursday, April 22, the wind shifted and began to blow from the northeast. The velocity was sufficient to carry the chlorine gas away from the point of release, yet slow enough for the cloud to linger over the opposing trenches before dispersing.

AS THE LOWERING sun bathed the Ypres Salient in a warm, golden light, the French and Algerian troops rested in their trenches, preparing the evening meal and enjoying the cool breeze that had sprung up. Suddenly the Germans began a long-range artillery bombardment of the villages in front of Ypres. The thump of heavy shell fire from 17-inch howitzers echoed across no-man's-land, increasing rapidly in volume. Shell bursts flashed in the distance, spewing lethal fountains of dirt and shrapnel. Finally the bombardment stopped, and the evening was dead still. Then at exactly 5:00 p.m., a captive balloon rose above the German trenches and fired three flares, which sputtered brightly overhead.

At this signal, the German troops along the four-mile front simultaneously opened the cocks on the 5,730 buried cylinders. Pressurized streams of chlorine gas hissed from lead pipes extending out of the forward trenches and immediately turned white with the condensation of water vapor. A total of 168 metric tons of chlorine billowed out of the cylinders and merged into a vast, elongated cloud about five feet high. Heavier than air, the cloud drifted across no-man's-land toward the Allied trenches at a leisurely pace of about four miles per hour. Gradually the warmth of the ground caused the cloud to expand to a height of about thirty feet and assume a yellow-green color, darker near the ground and lighter on top.

Believing that the Germans were using smoke to mask an infantry assault, the French commanders ordered their men to mount the fire steps of their trenches and prepare to repel the enemy advance. But instead of the expected waves of German troops, the defenders saw only an endless bank of yellow-green fog, moving inexorably forward. The usual explosions and cries of battle had been replaced with an eerie silence.

As the wall of mist approached their lines, the French and Algerian troops smelled a pungent, acrid odor that tickled their throats, burned their eyes, and filled their mouths with a metallic taste. Moments later, the toxic cloud enveloped them, veiling the world in greenish murk as if they had suddenly been plunged several feet underwater. The chlorine seared their eyes and burned the lining of their bronchial tubes, causing blindness, coughing, violent nausea, splitting headache, and a stabbing pain in the chest. Hundreds of soldiers collapsed in agony, their silver badges and buckles instantly tarnished greenish black by the corrosive gas.

As the poisonous fog engulfed the Allied trenches, the Germans launched an artillery barrage. The remaining French and Algerian units fired wildly into the cloud and then broke and ran in terror, dropping their weapons and equipment. Within an hour, two divisions numbering some ten thousand men had collapsed in disarray, tearing a gap four miles wide in the Allied line. Six miles away, the British troops of the Queen Victoria Rifles saw the yellow-green cloud in the distance and began to murmur in confused speculation. A soldier named Anthony Hossack later wrote in his diary:

> Suddenly down the road from the Yser Canal came a galloping team of horses, the riders goading on their mounts in a frenzied way; then another and another, till the road became a seething mass with a pall of dust over all.
>
> Plainly something terrible was happening. What was it? Officers, and Staff officers too, stood gazing at the scene, awestruck and dumbfounded; for in the northerly breeze there came a pungent nauseating smell that tickled the throat and made our eyes smart.
>
> The horses and men were still pouring down the road, two or three men on a horse, I saw, while over the fields streamed mobs of infantry, the dusky warriors of French Africa; away went their rifles, equipment, even their tunics that they might run the faster.
>
> One man came stumbling through our lines. An officer of ours held him up with leveled revolver, "What's the matter, you bloody lot of cowards?" says he. The Zouave was frothing at the mouth, his eyes started from their sockets, and he fell writhing at the officer's feet.

More than six hundred French and Algerian troops lay blinded and dying in the wake of the poisonous cloud. Some of the victims managed to stagger to first-aid stations, where frantic doctors and nurses could do nothing to save them. Drowning on dry land as their lungs filled with fluid, the patients gasped painfully for air and coughed up a greenish froth flecked with blood. Gradually their faces changed from pallid white to grayish yellow, and their eyes assumed the glassy stare of death.

At 5:20 p.m., the German commander sent a few scouts wearing crude respirators into no-man's-land to determine if the air was now safe to breathe. After the scouts gave the all clear, the German troops moved forward cautiously, collecting the abandoned rifles that littered the field. Their advance was slowed by patches of chlorine gas that lingered in depressions in the ground, but within an hour the Germans had captured the villages of Langemarck and Pilckem, taken two thousand prisoners, and confiscated fifty-one artillery pieces.

Because the German High Command had been skeptical that the chlorine attack would produce significant results, it had not arranged for the reinforcements needed to exploit a possible breakthrough. By dusk, the German units had advanced less than three miles into the Ypres Salient. On reaching the Yser Canal, they dug in and awaited further orders. Meanwhile, fresh Canadian and British troops advanced from the opposite direction and re-formed a continuous defensive line about four miles behind the abandoned French trenches, so that the town of Ypres remained in Allied hands. Haber was furious that the German generals had failed to exploit the successful attack. Two days later, on April 24, the Germans again used chlorine gas against the Canadian troops deployed northeast of Ypres, gaining ground but failing to achieve a decisive breakthrough.

The German press hailed Haber's military innovation, but elsewhere in Europe the use of chlorine at Ypres provoked outrage and condemnation. The Allies claimed that the April 22 attack had killed 5,000 troops—a number grossly inflated for propaganda purposes—and branded it a flagrant violation of international law. Germany defended its actions in legalistic terms, arguing that the Hague gas-projectile declaration had banned the use of specialized chemical shells but not the release of poison gases from cylinders.

Although Haber's contribution to the German war effort enhanced his fame, it had tragic consequences for his personal life. His wife, Clara, was a talented chemist in her own right, having been the first woman to earn a doctorate in physical chemistry from the University of Breslau in 1900. But she had been denied a scientific career and forced into the traditional role of wife and mother, making her deeply unhappy and resentful. On May 1, 1915, a few days after Haber's return to Berlin from the Belgian front, Clara confronted her husband over the gas attack at Ypres. A passionate antimilitarist, she was horrified by his development of chemical weapons, which she considered a grotesque perversion of science. Haber responded angrily, and a bitter argument ensued. Late that night, overcome with despair, Clara shot herself through the heart with her husband's army pistol.

AFTER YPRES, THE military stalemate of trench warfare continued, but now the taboo against the use of poison gas had been broken. Seeking to avenge the German attacks, the British established special gas companies under the command of Lieutenant Colonel Charles Foulkes. On the evening of September 24, 1915, his men emplaced 5,500 cylinders of chlorine along a 25-mile front around the Belgian town of Loos. At 5:20 the next morning, the British released a mixture of chlorine gas and smoke from artificial smoke candles over the German lines for about forty minutes before commencing an infantry assault. The wind shifted unexpectedly, however, blowing the toxic cloud back toward the attackers and inflicting more British casualties than German ones. Because of the difficulty of controlling the poison gas released from cylinders, the British began using crude mortars called Livens projectors to deliver canisters filled with chlorine, reducing the warning of an attack to a few seconds. In 1916, both sides resorted to heavy artillery to deliver specialized chemical shells. By then, the legal constraints in the Hague gas-projectile declaration had been completely swept aside by "military necessity."

As the war ground on, the combatants developed defenses against chemical attack. The initial protective measures were improvised and ineffective. After the first German releases of chlorine, Allied troops were given motorcycle googles and cotton pads that they were told to soak in urine, which partially neutralized chlorine, and hold over the mouth and nose until they had escaped the poisonous cloud. A slightly improved method involved the use of handkerchiefs or flannel socks dampened with a solution of bicarbonate of soda. Later, troops were issued crude gas masks and small box respirators, in which air was filtered through a canister filled with charcoal and soda lime. Even so, soldiers who failed to don their masks in time were condemned to a cruel death, as the British poet Wilfred Owen vividly described in his classic poem "Dulce et Decorum Est":

> *Gas! GAS! Quick, boys!—An ecstasy of fumbling,*
> *Fitting the clumsy helmets just in time;*
> *But someone still was yelling out and stumbling,*
> *And flound'ring like a man in fire or lime . . .*
> *Dim, through the misty panes and thick green light,*
> *As under a green sea, I saw him drowning.*
>
> *In all my dreams, before my helpless sight,*
> *He plunges at me, guttering, choking, drowning.*

The development of improved chemical defenses was offset by the introduction of new poison gases of ever greater potency. In December 1915, the Germans fired shells containing phosgene, a gas used in the dye industry that was eighteen times more toxic than chlorine. Phosgene had a distinctive odor of new-mown hay, which soldiers quickly learned to recognize, but because it was less irritating than chlorine and caused severe lung damage only after a delay of a few hours, troops could unwittingly inhale a lethal dose before donning their gas masks. The French retaliated with phosgene in February 1916.

Soon after the United States intervened in Europe in April 1917, the American Expeditionary Force, led by General John J. Pershing, confronted the horrors of gas warfare. Totally unprepared for this new threat, U.S. soldiers had to be issued gas masks by their British and French allies. At Pershing's request, General Amos Fries organized the First Gas Regiment, which developed defenses against German chemical attacks and conducted offensive gas, smoke, and incendiary operations on the western front. Once the element of surprise had been lost and both sides were equipped with effective respirators, the number of chemical casualties declined sharply and battlefield deaths from chlorine or phosgene became relatively rare. Unprotected noncombatants were not as fortunate, however. Although civilians were not deliberately targeted by chemical attacks, the wind could carry the toxic clouds as far as twenty miles behind the front lines, killing and injuring humans, livestock, and wildlife alike.

IN AN EFFORT to circumvent the Allied use of protective masks and respirators, Haber and his colleagues developed a new chemical warfare agent that attacked the skin as well as the lungs. Called "mustard" because of its sharp, garlicky odor, it was an oily liquid (first synthesized in 1860) that was readily absorbed through the skin, giving rise after several hours to severe chemical burns and blisters. In July 1917, once again at Ypres, the Germans began firing mustard-filled shells containing an explosive burster charge that shattered the liquid agent into a fine mist that was colorless or light yellow. Like phosgene, the effects of mustard were insidious: symptoms developed only after a delay of three to twenty-four hours (with a mean of ten to twelve hours), so that troops often did not realize that they had been exposed to the agent until it was too late.

The first symptom of mustard exposure was that it caused areas of bare skin, such as the hands and neck, and sweaty regions, such as the groin and buttocks, to turn crimson and begin to itch and burn. A day later, the reddish patches turned into massive blisters a few inches across, filled with a watery fluid produced by the destruction of tissue. The weight of the fluid separated the skin from the underlying flesh, causing excruciating pain. When the blisters burst or were deliberately punctured to ease the torment, they could easily become infected, creating large suppurating wounds that required lengthy medical treatment. Although the inhalation of mustard vapor was relatively uncommon, it caused severe inflammation of the lungs and a slow, agonizing death by asphyxiation. Mustard soon became the most dreaded of chemical weapons and was dubbed "the king of the war gases." Not only was it highly persistent, clinging to clothing and equipment and contaminating the battlefield for days or even weeks, but its ability to penetrate the skin forced troops to augment their respirators with cumbersome oilskin capes, goggles, and leather or rubber garments. This protective gear could be worn only for short periods, however, because it caused heat stress and seriously impaired fighting efficiency.

Seeking the capability to retaliate with chemical weapons, the U.S. War Department launched a crash effort to develop phosgene, mustard, and other agents by employing large teams of chemists at the American University Experiment Station and the Catholic University in Washington, D.C. The Ordnance Department also built manufacturing and filling plants for mustard and phosgene at a military reservation on Gunpowder Neck, a secluded, wooded peninsula jutting into Chesapeake Bay some twenty miles northeast of Baltimore, Maryland. In May 1918, the laboratory at American University was closed and moved to Gunpowder Reservation, which was renamed Edgewood Arsenal and soon reported to the newly established Chemical Warfare Service, headed by General Fries. Within several months of the United States' intervention in Europe, about 10 percent of all U.S. artillery shells had chemical fills.

The Allies began to retaliate with mustard in June 1918. One of the German gas casualties was an obscure lance corporal by the name of Adolf Hitler, who served in the trenches as a messenger during the last major German offensive in the summer and fall of 1918. On the night of October 13–14, he was trapped on a hill near Werwick, just south of Ypres, by an artillery barrage that lasted for several hours and included the use of mustard-filled shells. Around midnight his eyes began to smart, and at seven the next morning he delivered his last message, staggering through the trenches as his vision dimmed. A few hours later, he recounted in *Mein Kampf,* "my eyes were transformed into glowing coals and the world had grown dark around me." Temporarily blinded, Hitler was evacuated to a military hospital in Pasewalk, near the city of Stettin in the eastern German province of Pomerania. He was still there, recovering from his eye injury, when the war ended with Germany's capitulation and the armistice of November 11, 1918. Although Hitler later recovered fully, he retained a deeply traumatic memory of the experience.

By the time the Great War came to an end, the major combatants had employed more than 124,000 metric tons of 21 different toxic agents, delivered primarily by some 66 million artillery shells. Chemical weapons had inflicted roughly one million casualties on all sides, of which an estimated 90,000 had been fatal, and many of the survivors had been left blind or chronically disabled. The Russian forces on the eastern front had suffered the greatest number of chemical casualties—approximately 425,000, of which about 56,000 were fatal—because of their lack of training and protective equipment. The American Expeditionary Force was also disproportionately affected. Chemical weapons accounted for 26.8 percent of the roughly 272,000 U.S. injuries and deaths, although only 2 percent of the chemical casualties were fatal.

IN 1919, IN the aftermath of the war, General Fries launched a public relations campaign to prevent the Chemical Warfare Service (CWS) from being disbanded. He and a colleague published a book in which they argued:

> [C]hemical warfare is an agency that must not only be reckoned with by every civilized nation in the future, but is one which civilized nations should not hesitate to use. When properly safe-guarded with masks and other safety devices, it gives to the most scientific and most ingenious people a great advantage over the less scientific and less ingenious. Then why should the United States or any other highly civilized country consider giving up chemical warfare? . . . It is just as sportsman-like to fight with chemical warfare materials as it is to fight with machine guns.

General Fries cultivated the support of chemical manufacturers, trade associations, and the American Chemical Society, and he gave speeches, wrote articles, and lobbied Congress to preserve the Chemical Warfare Service. The fruit of his tireless efforts was the National Defense Act of 1920, which made the CWS a specialized branch of the Army with its own mission and staffing level—although not a guaranteed budget. When the permanent service was activated on July 1, 1920, Fries became its first peacetime chief, and he remained in that position until his retirement in 1929.

Meanwhile, the horror and indiscriminate nature of gas warfare during the Great War inspired the international community to attempt to ban or control it. The 1919 Treaty of Versailles prohibited Germany from using toxic chemicals in war and from manufacturing or importing asphyxiating or poisonous gases and liquids. In 1921–22, representatives of the five major allied powers (Britain, France, Italy, Japan, and the United States) met in Washington, D.C., for a Conference on the Limitation of Armament. These states negotiated and signed a Treaty Relating to the Use of Submarines and Noxious Gases in Warfare, which outlawed "the use in war of asphyxiating, poisonous and other gases and all analogous liquids, materials or devices." Although the U.S. Senate ratified the treaty, French objections to the provisions on submarines prevented the agreement from entering into force.

The next opportunity to control chemical warfare arose in 1925 at the Conference for the Supervision of the International Trade in Arms and Ammunition and in Implements of War, convened by the League of Nations in Geneva. Although the United States proposed banning all trade in chemical weapons, other countries objected that such a ban would discriminate against states that did not manufacture chemical arms. Instead, the contracting parties agreed to outlaw the use in war of chemical (and bacteriological) weapons, but not their production and stockpiling. On June 17, 1925, the League of Nations adopted a treaty to this effect called the Protocol on the Prohibition of the Use in War of Asphyxiating, Poisonous or Other Gases, and of Bacteriological Methods of Warfare, better known as the Geneva Protocol.

Although the White House supported the Geneva Protocol and sought its ratification by the Senate, in 1926 General Fries organized a coalition of veterans' groups, chemical manufacturers, and the American Chemical

Society to lobby vigorously against the treaty. As a result, the Geneva Protocol remained bottled up in the Senate Foreign Relations Committee and was never released for a vote on the Senate floor. (Indeed, the United States would not ratify the treaty for another fifty years.) Over the next decade, however, the Geneva Protocol was ratified by some forty countries, including all of the great powers except Japan and the United States, establishing an important international legal norm against chemical and biological warfare.

Meanwhile, Germany was secretly pursuing a clandestine military buildup in violation of the Treaty of Versailles, including the restoration of a chemical warfare capability. Because World War I had left Germany humiliated and the Soviet Union isolated and weak, it was not surprising that the two countries would decide to collaborate in rebuilding their respective armed forces. Under the Rapallo Treaty of April 1922, the Reichswehr and the Red Army negotiated a military cooperation agreement that called for the establishment of German military bases on Soviet soil and the conduct of joint military exercises. Covert elements of the accord included research and development on armor, aviation, and chemical warfare.

In 1928, the Red Army and the German Reichswehr launched a top-secret program of chemical weapons development and testing at a site west of Volsk, in the Samara region of Russia. Code-named "Tomka," the proving ground was situated on a flat plain near the Volga River, surrounded by low mountain ranges. For the next five years, twenty-nine Germans (six chemists and engineers, plus pilots, mechanics, and laborers) worked alongside a larger number of Soviet staff, conducting open-air trials of mustard gas and other chemical agents in six-month campaigns, from May to October. In 1932, however, the Reichswehr decided that this would be their final cooperative venture with the Red Army. Not only was the joint testing program too expensive, but most of the activities carried out in Soviet Russia could now be done in Germany. In addition, after the appointment of Adolf Hitler, the leader of the National Socialist Workers Party, as chancellor of Germany on January 30, 1933, relations with Moscow rapidly soured. In September 1933, the German staff left Tomka for good and shipped their chemical testing equipment back into the Reich.

WITH HITLER'S RISE to power, Fritz Haber's life took another sad turn. Despite the distinction of winning the 1918 Nobel Prize in Chemistry (awarded in 1919) for the synthesis of ammonia, his work on chemical weapons had caused him to be shunned by the foreign scientific community. Now he faced a similar fate at the hands of his own countrymen. The Nazi regime moved quickly to purge Jews from the universities, and all Jewish scientists at the Kaiser Wilhelm Institute were forced to resign. Because Haber was a prominent figure and a war veteran, he was not immediately threatened, but he soon realized that he could not escape his ethnic heritage and would have to emigrate. Rejection and exile from the country he loved deeply left him a broken man. He accepted a position at the University of Cambridge, but the damp English climate depressed him and he developed a serious illness. He was en route to Switzerland to convalesce when he died suddenly in Basel on January 29, 1934, at the age of sixty-five.

Haber had argued that chemical warfare was more humane than blast or flame and would serve to shorten wars and save lives. Although the terrible suffering of gas victims had proved him wrong, some military strategists believed that chemical weapons had been tactically effective on the battlefield and might prove decisive in future conflicts. Accordingly, many countries that signed and ratified the Geneva Protocol reserved the right to use chemical weapons against states that were not among the contracting parties, or to retaliate in kind if an enemy used chemical weapons first. Once governments had claimed the option of retaliation, they found it necessary to continue research and development on chemical warfare agents and, in many cases, to produce and stockpile them as a deterrent, increasing the likelihood that the weapons would someday be used.

CHAPTER TWO

IG FARBEN

Wearing a stained white coat that hung below his knees, Dr. Gerhard Schrader surveyed the insecticide development laboratory, where his team of industrial chemists was performing the modern alchemy of organic synthesis. Sitting at benches with their hands inserted into fume hoods, they mixed and processed solutions in gleaming assemblies of blown glass. Their silent work was accompanied by the hiss of boiling water and the hum of exhaust fans venting noxious gases into the winter air.

The date was December 23, 1936. On the wall facing the laboratory benches was a large framed photograph of German Chancellor Adolf Hitler in heroic profile. The row of dusty windows looked out on the snow-covered brick buildings and fuming smokestacks of the Interessengemeinschaft (IG) Farben chemical complex in Leverkusen, north of Cologne. IG Farben was the world's largest corporation, having been created in 1925 from the merger of Germany's six largest chemical concerns, including BASF, Bayer, Hoechst, and Agfa. Headquartered in Frankfurt, it had a net worth of $2.5 billion, a workforce of about 200,000, and scores of research and production facilities across the Reich that were involved in every branch of industrial chemistry, from nitrogen fertilizers, gasoline, mineral oils, and dyestuffs to pharmaceuticals, photographic chemicals, and artificial fibers.

At Schrader's bench, a round-bottom flask immersed in a hot-water bath gave off a ribbon of steam into a condenser tube, which distilled the reaction product into drops of clear, colorless liquid. As always, he felt a pleasant tingle of anticipation as a new substance emerged from the synthetic process. Schrader, thirty-three, had a pale moon face, shrewd eyes framed by oversized glasses, dark hair slicked over a broad forehead, prominent cheeks, and a wide, amiable mouth. His personality was ideally suited to research, combining imagination and cleverness with an equal measure of determination and dogged persistence.

Schrader had grown up in a religious Protestant home and had enjoyed a sheltered and pleasant childhood. In October 1928, after completing his doctoral degree in chemical engineering at the University of Braunschweig, he had joined the research staff at the Bayer Company, a subsidiary of IG Farben. Although he had specialized in inorganic chemistry in graduate school, his work at Bayer focused exclusively on organic (carbon-based) compounds. Despite his young age, he was put in charge of a dyestuffs development laboratory at the company branch in Elberfeld, a town in the industrial Ruhr Valley. The year 1928 was eventful for Schrader in other ways as well. Around Christmastime, he became engaged to Gertrud Ahlers. They married in early 1929, and a year later his first daughter, Wiebke, was born.

After spending just two years at Elberfeld, Schrader was transferred to the main Bayer research laboratory in Leverkusen to work on naphthol dyes. In 1934, when the young Otto Bayer took over the leadership of the research department, he gave Schrader a new assignment, as head of the plant-protection group. Meanwhile, Schrader had purchased a house with a large garden in the village of Lützenkirchen. After a long day in the laboratory, he enjoyed relaxing after work in his rural idyll, surrounded by berries, fruit, vegetables, and free-roaming chickens. It was there, in April 1935, that his second daughter, Kristin, was born.

At Leverkusen, Schrader dove into the new field of synthetic pesticides with energy and enthusiasm. Beginning in 1933, the German Reich had sought to reduce its dependence on food imports, and the loss of the large territories in the East after World War I meant that the size of the grain harvest had to be expanded considerably. To improve crop yields, the German Reich had purchased 30 million marks' worth of pesticides from overseas, and it now sought industry's assistance in developing a cheaper domestic alternative. Otto Bayer gave Schrader's plant-protection group the task of developing a nonflammable fumigant that could destroy weevils in grain silos, as well as fleas in ships and living rooms. A huge potential market existed for such products because the main fumigants then in common use, ethylene oxide and methyl formate, could cause explosions in silos and other confined spaces.

Schrader was aware that organic compounds containing the element fluorine were generally toxic, making them good candidates for new insecticides. He therefore began to introduce fluorine into a wide variety of organic molecules. As Schrader and his team synthesized one new substance after another, he provided samples to Dr. Hans Kükenthal, a biologist at Leverkusen, who tested them for insecticidal activity.

The first set of compounds to emerge from Schrader's lab had a strong irritant effect on the eyes and the lungs, making them of no practical value as insecticides. He therefore moved on to organic compounds containing atoms of fluorine and sulfur. One such molecule appeared to be an effective fumigant against insect pests, but further testing showed that it was absorbed by the treated grain, rendering the food unfit for human consumption. When Schrader tried to exploit this defect by developing the toxic grain as a rat poison, he found that the absorbed chemical gradually evaporated, reducing the grain's toxicity over time. Once again, a promising line of research had led to a dead end.

Undaunted by these setbacks, Schrader and his coworkers continued their systematic synthesis of new carbon compounds containing sulfur as the central atom. Although many of these chemicals were toxic to insects, none met the standards of safety and stability required of a commercial insecticide. Schrader next decided to work on molecules containing phosphorus, the element next to sulfur in the periodic table. Because the two elements had similar chemical properties, he reasoned that compounds containing phosphorus might also be toxic to insects. In making this intuitive leap, Schrader also drew on the earlier work of chemist Willy Lange and his student Gerda von Krüger at the University of Berlin. In 1932, Lange and Krüger had synthesized some phosphorus-containing organic compounds that appeared to offer promise as insecticides. In 1935, however, Lange was forced to leave academia because he had a Jewish wife.

Schrader and his team proceeded to synthesize a series of organic molecules consisting of a central phosphorus atom with four bonds extending out from it like arms, each holding a different atom or cluster of atoms. One of these "organophosphate" compounds showed promising insecticidal activity: a water solution containing only 0.2 percent of the substance, sprayed on leaf lice, killed all of the insects on contact. IG Farben management considered the new compound sufficiently promising to patent it in Germany, the United States, England, and Switzerland.

Schrader and his team spent the next year searching for a more potent version of this molecule by synthesizing hundreds of structural variants, or "analogues," which were then screened for insecticidal activity. The researchers discarded all analogues that had low potency, were chemically unstable, gave poor synthetic yields, or required ingredients that were not available in sufficient quantities. This method of systematic trial and error was extremely labor-intensive.

Because cyanide—a carbon atom bound to a nitrogen atom—was a poison in its own right, Schrader decided to incorporate it into the structure of the phosphorus compound. After performing the initial synthesis in November 1936, he began to experience some highly unpleasant physiological effects, including headache, poor concentration, and shortness of breath. He also noticed a marked dimming of his visual field and difficulty with visual accommodation, the process of adjusting focus from a close object to a distant one. The visual impairment worsened until it became impossible for him to read under an electric lamp.

As Schrader drove home one evening to Lützenkirchen in his black-and-yellow Hanomag sedan, his vision had dimmed to the point that he could barely make out the road in front of him. His head throbbed and he felt painfully short of breath, with a feeling of pressure in his larynx. After reaching the house with great difficulty, he examined his eyes in a mirror and discovered that his pupils had constricted to pinpoints, giving him an eerie, zombielike appearance. Alarmed but intrigued by this phenomenon—the scientist in him was ever present—he discovered that his pupils failed to dilate in response to low light. Over the next few days, Schrader's symptoms worsened and he had to spend two weeks in the hospital before his vision recovered fully. After his release, he spent another eight days recuperating at his parents' home.

Returning to the laboratory shortly before Christmas, Schrader resumed work on the cyanide-containing compound. On December 23, 1936, the synthesis and purification process was nearing completion. Distillation of the final product yielded a clear, colorless liquid with a faint scent of apples, which Schrader termed Preparation 9/91. He gave a small sample of the substance to Dr. Kükenthal, who found that an extremely dilute solution—one part in 200,000—killed 100 percent of leaf lice on contact. Preparation 9/91 was a hundred times more potent than the original compound, and far more effective than anything Schrader's research group had developed before.

It also became clear that the highly unpleasant symptoms Schrader had experienced in November had been caused by exposure to the new substance. Although its mild, fruity odor made it seem benign, Schrader and his assistant Karl Küpper discovered upon further investigation (now carried out with extreme caution) that inhaling fumes from even a tiny drop, spilled by accident on the laboratory bench, gave rise in minutes to a cluster of striking physiological effects: strong irritation of the cornea, marked dimming of the visual field, and an oppressive feeling of tightness in the chest, as if a band were being constricted around it. Staying away from the lab for a few days and breathing fresh air caused most of the symptoms to vanish, although the impaired vision recovered only gradually.

As soon as Schrader and Küpper resumed work with Preparation 9/91 in January 1937, however, the unpleasant symptoms returned. Indeed, the two chemists had become hypersensitive, so that even the slightest whiff of the substance provoked the same array of symptoms. Küpper became agitated and feared he was losing his sight, and Schrader also worried that they were being slowly poisoned. This time they were forced to suspend their work in the laboratory for more than two weeks.

Intrigued by the powerful physiological effects of "this new and interesting substance," Schrader sent a letter on February 5, 1937, to Professor Eberhard Gross, the director of industrial hygiene at IG Elberfeld. At Gross's request, he sent a sample of Preparation 9/91 for toxicological testing. Meanwhile, Schrader continued to synthesize additional variants of the cyanide compound, and in March he and Kükenthal applied for a patent on this new class of insecticide.

In early May, Schrader received a lengthy report from Dr. Gross on the toxicity of Preparation 9/91 in mice, guinea pigs, rabbits, cats, dogs, and apes. Gross had renamed the compound Le-100, "Le" being an abbreviation for Leverkusen. In the experiments on apes, which are physiologically closest to humans, injecting as little as a tenth of a milligram of Le-100 per kilogram of body weight had given rise to dramatic toxic effects, including nausea, vomiting, constriction of the pupils and the bronchial tubes of the lungs, copious drooling and sweating, abdominal cramps, diarrhea, muscular twitching, gasping for air, violent convulsions, slowing of respiration and heartbeat, and finally paralysis of the breathing muscles, culminating in death.

Dr. Gross's laboratory also had a hundred-cubic-meter gas chamber suitable for inhalation experiments with large primates. It was made of glass and concrete bricks that were coated with rubber to permit a thorough cleaning. For security reasons, the gas chamber was accessible only through doors on the second and third floors of the laboratory, which were kept locked at all times. Gross had utilized the gas chamber to expose apes to Le-100 vapors at a concentration of 25 milligrams per cubic meter. After inhaling the vapor, all of the animals had convulsed and died in a time interval ranging from sixteen to twenty-five minutes.

Schrader was disappointed by the toxicology results because Le-100 was far too poisonous to warm-blooded animals to be marketed as a commercial insecticide. Nevertheless, IG Farben brought the new compound to the attention of the German government. According to an official Reich ordinance of 1935, all new discoveries and patents of potential military significance were to be reported to the War Office, which was empowered to classify any invention that might be useful for the nation's defense. Toxic industrial chemicals were of interest as chemical warfare agents, particularly since August 1936, when Hitler had ordered the armed forces (Wehrmacht) to prepare for war by 1940. German companies had already submitted more than a hundred compounds for evaluation.

A few influential figures within the Wehrmacht and the chemical industry viewed poison gas as a militarily "decisive" weapon. They noted that Benito Mussolini had employed chemical weapons extensively in 1935 and 1936 during the Italian conquest of Abyssinia (Ethiopia), using aircraft to drop mustard-filled bombs on Emperor Haile Selassie's army. Most of the Abyssinian fighters had been barefoot tribesmen lacking gas masks and protective clothing, making them extremely vulnerable to mustard. Although the chemical attacks had been a flagrant violation of the 1925 Geneva Protocol, to which Italy was a party, the League of Nations had done nothing to stop them.

Among the leading German proponents of chemical warfare was Dr. Heinrich Hörlein, the director of pharmaceutical research at IG Elberfeld. A physician by training, he had been involved in developing poison gases since 1933 and routinely advised the German Army on technical matters. After reading Dr. Gross's report on the mammalian toxicity of Le-100, Dr.

Hörlein forwarded a copy to the Army Ordnance Office (Heereswaffen-amt), which was responsible for the development, testing, production, and procurement of land weapons.

Within the Army Ordnance Office, the Weapons Development and Testing Department (Waffenprüfamt) was organized into several divisions. Division 9 (Wa Prüf 9), headquartered in the Charlottenburg district of Berlin and directed by Lieutenant-Colonel Dr. Kurt Rüdiger, specialized in the development and testing of chemical warfare agents, munitions, and protective equipment. In April 1937, Dr. Leopold von Sicherer, a senior official in Division 9, read Gross's report on Le-100 and, intrigued, requested a demonstration of the new compound. A few days later, Sicherer visited Gross's laboratory at Elberfeld, accompanied by Dr. Wolfgang Wirth, a specialist in pharmacology and toxicology at the Army Ordnance Office who had earlier worked at Tomka.

Dr. Gross demonstrated the effects of vaporized Le-100 on caged laboratory mice. Whereas standard chemical warfare agents such as phosgene and mustard took several hours to kill, exposure to small amounts of the new compound caused mice to go into convulsions and die within twenty minutes. Sicherer and Wirth concluded that Le-100 was a "remarkable compound" with great military potential. In early May, not long after the demonstration at Elberfeld, they invited Schrader to Berlin to demonstrate the synthesis of Le-100.

THE GERMAN ARMY'S Gas Protection Laboratory (Heeresgasschutzlaboratorium) was housed in Spandau Citadel, a brick fortress perched on a small island at the junction of the Havel and Spree Rivers in northwest Berlin. Although only the crenellated Julius Tower remained from the original structure, which had been built around 1200, the fortress had been greatly enlarged during the sixteenth century. Its layout was that of a square keep, with the four corners shielded by massive stone bastions in the form of arrowheads. One side of the citadel abutted the river, while the other three walls were surrounded by a moat covered with lily pads.

The name of the Army Gas Protection Laboratory was deliberately misleading. In fact, the roughly three hundred scientists and technicians worked not only on chemical warfare defense but also on the development of new agents and production methods. To accommodate the necessary facilities and equipment, numerous renovations had been made to the historical buildings inside the Citadel and four large new structures had been erected. The area beyond the entrance gate and the administration building was a military zone that was restricted and secured by an additional fence. Building 4, near the west curtain wall, contained a technical library and laboratories for the analysis and synthesis of chemical warfare agents; Building 6 was for work on chemical munitions and the development and testing of protective equipment; Building 8 contained pilot plants for producing up to 50 kilograms of agents for testing purposes; Building 14 did studies on the aerosolization of liquid agents; Building 15 was the human medical department and the staff clinic; and Building 15A housed the toxicological institute, which performed testing on a wide variety of experimental animals. The historical Armory (Zeughaus) was where manufacturing processes for chemical warfare agents were developed, and munitions testing took place in two explosive test chambers built of reinforced concrete next to the north curtain wall.

Schrader passed through a guard post and then crossed a bridge over the moat that led to the main gate. At the headquarters building, an officer checked his identity papers and escorted him to his appointment in Building 4. Much to Schrader's relief, the officers at Spandau wore army drab rather than the intimidating black uniforms of the Schutzstaffel (SS). Present at the meeting were Dr. Sicherer and Lieutenant-Colonel Rüdiger of Division 9 and Dr. Hans-Jurgen von der Linde, the chief of the Army Gas Protection Laboratory. After introductions had been made, Schrader described the properties of Le-100 and its remarkable physiological effects.

The Army scientists were deeply impressed by the potency of Le-100. One of them said that the new poison was "taboo" (*tabu* in German), meaning that it was too strong, and as a result it was named "Tabun." Schrader had discovered accidentally what many others had tried to develop deliberately. From 1925 to 1931, the Reichswehr had funded chemical weapons research at twelve German universities and institutes. Later, Army chemists at Spandau had searched the technical literature and pending industrial patents for references to highly poisonous compounds. Yet no substance had been identified that even approached the toxicity of Tabun.

Dr. Sicherer decided that the IG Farben patent for Tabun would hence-forth be classified top secret and that the Army would take charge of the compound's further development. In recognition of their work, Schrader and Gross received a reward of 50,000 marks. Schrader was requested to synthesize one kilogram of Tabun and send it to Spandau for preliminary testing. In the meantime, Dr. Rüdiger would arrange to remodel one of the chemical laboratories in the basement of the Citadel and install a modern apparatus for pilot-scale production.

IG Farben welcomed the Army's decision to assume responsibility for the further development of the new agent, for two reasons. First, the company could not manufacture highly toxic substances in its existing factories, all of which were located in densely populated areas. Second, it would be impossible for the firm to maintain a high level of secrecy about a chemical such as Tabun, tiny amounts of which produced striking physiological effects such as pinpoint pupils. In order to disguise the identity of the compound further, the Army developed a series of military code names for Tabun, including "Gelan," "Substance 83," and finally "Trilon 83" (or T-83), after a popular brand of laundry detergent manufactured by IG Farben.

The Army Gas Protection Laboratory subjected all candidate chemical warfare agents to a battery of tests to assess their toxicity in laboratory animals; the effects of temperature, humidity, and precipitation; and the feasibility of protecting friendly troops. Only about 2 percent of candidates survived the preliminary screening process and were sent for field testing. Those that proved effective in field trials had to overcome a final hurdle: the development of an economical process for industrial-scale production, including the availability of raw materials. By the end of the process, less than 1 percent of candidate chemical warfare agents were adopted by the Wehrmacht.

The testing and evaluation of Tabun, however, proceeded with remarkable speed. In late May 1937, field trials of the new agent began at the Army Proving Ground (Heeresversuchstelle) Raubkammer, north of the small town of Munster on the Lüneburger Heath. A fenced, roughly rectangular area covering seventy-six square miles of forest and scrub, Raubkammer (also known as Munster-Nord or Munsterlager) had been built originally as a troop exercise area and had later been expanded for the open-air testing of chemical warfare agents. The entrance gate was flanked by two giant pillars, each surmounted with a Nazi spread eagle and swastika.

The experimental station at Raubkammer had been designed with typical German thoroughness. Staffed by some five hundred scientists and technicians, it comprised a few dozen buildings in traditional German style, including well-equipped laboratories for chemical analysis and postmortem examination; animal facilities housing dogs, cats, guinea pigs, monkeys, apes, and horses; pilot production plants; and a large gas chamber in which animals could be exposed to toxic agents while scientists observed from a glassed-in balcony, as well as an administration building, barracks, and an officers' mess. For testing decontamination methods, a half-mile stretch of road had been paved with a variety of surfaces, including cement, granite blocks, asphalt, and gravel.

The field trials of Tabun at Raubkammer involved representatives from Division 9 and the Munitions Department of the Army Ordnance Office, the Army Gas Protection Laboratory, the Military Medical Academy, and the Air Ministry. Initial tests of Tabun-filled shells took place inside the Measurement House (Messhaus), a giant circular wooden chamber twenty meters high and thirty meters wide that was covered with scaffolding and had ventilator fans in the roof to remove the toxic gases. The inner walls of the building had numerous patches to repair holes made by flying shrapnel. During static firing experiments, an artillery shell charged with Tabun was detonated one and a half meters above the ground, and the concentration of the vapor cloud measured at various points within the chamber. The effects of the agent on tethered animals could be observed at the same time. These trials showed that the most effective way to disseminate Tabun was by using an explosive burster charge to break up the liquid agent into a fine mist of microscopic droplets, or aerosol, that would poison enemy troops by inhalation.

Raubkammer also had a vast outdoor testing area (Übungsplatz), including a range on which chemical artillery shells could be fired from a distance of up to ten kilometers at an instrumented target grid that contained a concentric array of sampling devices. Near the impact zone, three concrete bunkers linked by telephone lines shielded the test personnel as they controlled the artillery fire. Although trials conducted on the firing range were more realistic than those inside the Measurement House, the outdoor results were difficult to interpret because of the large number of variables that had to be taken into account, including wind direction and velocity, air and ground temperature, and weather conditions.

On October 27, 1937, Lieutenant Colonel Hermann Ochsner, the chief of the German Chemical Troops, prepared a memorandum for the Army General Staff in which he advocated the development of a new generation of chemical warfare agents. The first chlorine attack at Ypres in 1915 and Italy's use of mustard agent during the invasion of Abyssinia, he wrote, had demonstrated the devastating effects of poison gas against an unprepared or unprotected enemy. Now, however, Western armies were equipped with gas masks and other defenses that deprived the existing chemical warfare agents of their effectiveness. "The successful surprise that was possible during the last war because of the total novelty of the chemical weapon can no longer be achieved," Ochsner observed. "Back then, this means of warfare encountered a totally defenseless enemy. Today, the use of chemical warfare agents is generally known. Every modern army has gas masks that protect against all of the standard agents."

Another drawback of existing chemical warfare agents was that they all had distinctive odors that soldiers could be trained to recognize at low concentrations, enabling them to don their gas masks at the first whiff. Phosgene smelled like new-mown hay, mustard like garlic, lewisite like geraniums, and hydrogen cyanide like bitter almonds (although about 20 percent of people could not detect it). To achieve military surprise, Ochsner wrote, it would be necessary to develop new agents that had little or no odor, caused no sensory irritation, and were so toxic that one or two breaths could kill. In this context, he noted the recent discovery of a new substance that he called "Number 100," a probable reference to Le-100 or Tabun. "This agent has been produced only in the laboratory and has been demonstrated to have good—indeed remarkable—effects," he wrote. "An initial test in the open air failed, however. With respect to raw materials, the agent can be manufactured but requires as much chlorine as mustard gas."

Despite this mixed review, Tabun largely satisfied Ochsner's criteria for high toxicity and difficulty of detection. The compound had a faint fruity odor but did not cause noticeable irritation of the eyes or lungs. Although standard gas mask filters protected effectively against Tabun, the difficulty of detection meant that troops could be exposed by surprise before they had time to don their masks. The new agent could also penetrate the skin, although absorbing a lethal dose in this manner could take as long as an hour. Contamination of soil, clothing, and equipment with liquid Tabun posed an additional hazard because the agent evaporated slowly, giving off toxic fumes. Even if low-level exposures were not sufficient to kill, they could incapacitate soldiers by causing severe visual impairment and an asthmalike shortness of breath.

Field trials revealed certain limitations of Tabun as a war gas. Because the liquid agent was not particularly volatile, it was necessary to use a fairly large burster charge to transform it into a fine mist or vapor, yet the heat of the explosion destroyed much of the agent and reduced its effectiveness. The fact that Tabun vaporized more readily at higher temperatures and wind speeds made it more suitable for use during the summer months or in tropical climates.

Schrader's discovery of Tabun earned him some recognition from IG Farben. In November 1937, at Dr. Hörlein's request, he was transferred from Leverkusen to a new laboratory at IG Elberfeld, where he continued his research under conditions of tight secrecy. Although the German Army proposed to give Schrader a contract to develop an industrial-scale production process for Tabun, Dr. Hörlein turned down this offer. IG Farben management was reluctant to get involved in chemical warfare for several reasons: the company was fully occupied with the development and manufacture of civilian products, chemical weapons were unlikely to generate much profit, and they might stir up negative publicity that would harm foreign sales. Accordingly, Schrader was told to concentrate on the development of agricultural insecticides, although he was allowed to dabble in his spare time on an improved production process for Tabun.

In 1938, however, the IG Farben management had a change of heart about military work. Field Marshal Hermann Göring, the commandant of the Luftwaffe, asked Karl Krauch, the head of the company's board of directors, to prepare a detailed plan for German chemical rearmament. Krauch's report described poison gas in highly positive terms as "the weapon of superior intelligence and superior scientific-technical thinking. As such, it is called upon to be employed by Germany in a decisive manner, both on the front and against the enemy's hinterland."

These ideas won favor from Göring's air-warfare strategists, who considered chemical weapons to be potentially decisive because of their ability to elicit terror and confusion in the enemy population. On August 22, 1938,

Göring named Krauch his "Plenipotentiary for Special Questions of Chemical Production" under the Nazi Four-Year Plan. Because Germany had been forced to give up all of its chemical weapons factories after its defeat in World War I, they would have to be rebuilt from the ground up.

In order to test Tabun-filled munitions under more realistic conditions than static detonation, the Army Proving Ground Raubkammer developed a new short-range firing apparatus called the "Vz Tower," which began operation in November 1938. It consisted of a steel tower fifteen meters high, topped with a rotating platform on which were mounted a 105 mm light field howitzer and a 150 mm heavy howitzer. These guns fired chemical shells inside a circle with a radius of about fifty meters. To measure the concentration of the resulting gas clouds, a set of instruments was mounted on a miniature railroad car that ran along a track encircling the target area. Because of the configuration of the testing site, it was possible to take measurements regardless of the wind direction.

In close collaboration with the Air Ministry, Raubkammer also tested 250-kilogram aerial bombs containing 85 kilograms of Tabun. Dropped from aircraft, the bombs exploded on impact with an instantaneous or delayed fuse. These tests showed that about 25 percent of the agent remained in the bomb crater, while the rest was converted into a vapor cloud that traveled about 100 meters downwind, creating a lethal zone of 3,000 to 5,000 square meters. Although spraying Tabun from a low-flying aircraft was a more effective means of dissemination, the Luftwaffe rejected this approach because it would expose the aircraft and crew to hostile ground fire.

Field testing of Tabun was extremely hazardous and resulted in hundreds of injuries requiring medical attention. Even trace amounts of the agent, adhering to clothing, equipment, or the fur of dead animals, were sufficient to cause harmful exposures when masks and respirators were removed. Most cases of Tabun poisoning were mild, resulting in disturbances of vision and breathing that faded after a few days, but a few individuals were affected more severely.

Physiologists at Raubkammer also deliberately tested low doses of Tabun on human volunteers, mainly officers, clerks, employees, laborers, and students from the Army Gas Protection Laboratory, who received a small financial incentive to participate. By providing useful information about the physiological effects of the agent in humans, these experiments enabled the Military Medical Academy in Berlin to develop antidotes. To counteract the effects of Tabun poisoning, physicians administered injections of atropine and the related drug scopolamine, which was safer but worked more slowly. Atropine, extracted from the deadly nightshade plant (*Atropa belladonna*), produced physiological effects that were diametrically opposed to those of Tabun. Whereas the nerve agent slowed the heartbeat, constricted the pupils, and stimulated the salivary glands, atropine increased the heart rate, dilated the pupils, and dried out the mouth. Indeed, the herbal source of atropine was known as "belladonna" because women had used it for centuries as a beauty aid, to enlarge their pupils.

Meanwhile, working in his laboratory at Elberfeld, Schrader developed a new family of insecticides by replacing the cyanide group in Tabun with a fluorine atom and adding a phosphorus-methyl bond. On August 2, 1938, Schrader and Kükenthal filed a patent application for this class of compounds. (The patent was classified and was not published until September 1951.) Although all molecules of this type caused toxic effects in insects and animals, the various analogues differed considerably in potency. Lacking a theoretical explanation for why some structural variants were so much more potent than others, Schrader conducted his development work on a trial-and-error basis.

Toward the end of 1938, Schrader synthesized an organophosphorus compound containing fluorine whose toxicity against insects proved to be "astonishingly high." He gave a sample of the new substance to Dr. Gross for testing in a variety of warm-blooded animals. When the compound was injected into guinea pigs, a dosage of only 0.075 milligram per kilogram induced convulsions and rapid death. Inhalation tests also showed that the new substance was five to ten times more toxic than Tabun in dogs and twice as toxic in monkeys, ruling out its use as a commercial insecticide.

In early 1939, Dr. Gross sent his toxicology report to the German War Office, along with a sample of the fluorine-containing compound, which the Army code-named "Substance 146." Field testing at Raubkammer showed that the new agent was considerably more stable than Tabun and less likely to be destroyed by the explosion of an artillery shell or bomb. It was almost completely odorless, making it extremely difficult to detect, and it evaporated readily, increasing its potential utility on the battlefield. In June, Schrader traveled to Spandau Citadel in Berlin to deliver a presentation on Substance 146. The Army Gas Protection Laboratory assigned a large team of chemists to study the new agent and develop a simplified manufacturing process.

At the same time, other Army chemists at Spandau were synthesizing Tabun in small lots of about one kilogram while attempting to scale up to larger batches. After several false starts, these pilot studies led to marked improvements in the manufacturing process. In February 1939, the experimental production of Tabun in 30-kilogram lots resulted in good yields and a product that was 90 percent pure. The next step was to develop a pilot plant with a batch capacity of 400 kilograms. Because it was not feasible to build such a facility at Spandau, the Army decided to locate it in the experimental station at Raubkammer. Known as the "Vorwerk Heidkrug," the Tabun pilot plant was disguised as a government farm building.

In a memorandum to the Army General Staff dated June 28, 1939, Colonel Ochsner argued that the production of chemical weapons would conserve iron and other strategic materials, and that new agents such as Tabun were a major military asset. Chemical attacks, he wrote, should be carried out "on a very large scale against the enemy hinterland by air strikes, especially against industry concentrations and large cities." Such massive use would "overwhelm the enemy's medical facilities with a flood of sick and injured" and terrify the civilian population. "There is no doubt," Ochsner concluded, "that a city like London would be plunged into a state of unbearable turmoil that would bring enormous pressure to bear on the enemy government."

Despite the arguments by Ochsner and other advocates, Hitler showed little interest in chemical warfare and did not even visit the Raubkammer proving ground. Having been gassed during World War I, he had a strong aversion to such weapons and did not contemplate their use except for retaliation. Nevertheless, the Army Ordnance Office moved forward with preparations for the large-scale production of chemical agents. On August 5, 1939, General Walther von Brauchitsch, the commander in chief of the Army, approved the procurement of a stockpile of Tabun. This decision had been complicated by the fact that Substance 146 appeared to be significantly more effective than Tabun, but roughly two more years of development work would be required to bring the new agent to the point of large-scale production.

To construct the Tabun production facility, the Army decided to hire a commercial contractor, and the obvious choice was IG Farben. Not only had one of its scientists invented Tabun, but the company manufactured all of the necessary chemical ingredients and its depth of expertise was unparalleled. The Army therefore asked IG Farben executives to draw up preliminary estimates for a factory capable of producing 1,000 metric tons of Tabun per month.

ON SEPTEMBER 1, 1939, only a month after the German Army's decision to manufacture Tabun as a standard chemical warfare agent, the Wehrmacht invaded Poland, plunging Europe once again into the inferno of war. On September 7, the Army Ordnance Office summoned three members of the IG Farben board of directors—Heinrich Hörlein, Fritz ter Meer, and Otto Ambros—to Berlin for a meeting. The officials present included Division 9 chief Colonel Siegfried Schmidt and representatives from the Procurement Division, the Army High Command, and the Army Gas Protection Laboratory.

The two sides sat down across a conference table. Dr. Hörlein had sharp, rather sinister features and wore a pair of round glasses with black rims. Ter Meer, the chairman of IG Farben's Technical Committee, had a large rectangular face, with dark hair slicked back from his broad forehead and a stern gaze. Ambros, the youngest of the three executives, wore a finely tailored suit, and his lean features and intelligent eyes radiated confidence and authority. After earning a doctorate in chemistry in 1926, at the age of twenty-five, Ambros had joined the IG Farben plant in Ludwigshafen. Four years later, he had been sent to Sumatra for a year to study the chemistry of natural rubber, and by 1935 he had become IG Farben's leading expert on synthetic rubber, or Buna. Three years later, he had continued his meteoric rise through the company ranks by joining the board of directors, and he now managed eight of the company's chemical plants.

The Army officials opened the meeting by demanding IG Farben's full cooperation with the war effort. "The production of poison gases will be essential," Colonel Schmidt declared, "and IG Farben and other firms must do their part for the Fatherland." The three executives pledged to perform their patriotic duty. As the Army had requested, they provided cost estimates for the construction of two chemical weapons–manufacturing plants, one for mustard and the other for Tabun. The Tabun plant would have a production capacity of 1,000 metric tons per month and the potential to expand to 2,000 tons per month if necessary. At a follow-up meeting on November 7, 1939, the IG Farben executives accepted a preliminary set of instructions and signed an oath of secrecy. The production contract was finalized in early December, and two weeks later the Army Ordnance Office issued a preliminary "order to proceed."

To operate the chemical weapons plants, the IG Farben board of directors established a new subsidiary called Anorgana GmbH with 100,000 reichsmarks of working capital. They named Otto Ambros as the managing director. Anorgana was secretly financed and controlled by a Wehrmacht holding company called Montan Industriewerke, and Anorgana's board of directors consisted of three representatives from Montan and three from IG Farben. The reason for this byzantine organizational structure was to protect IG Farben's financial interests and conceal the company's involvement in chemical weapons production. Over the next few weeks, IG Farben's director of construction tried to find a suitable location for the Tabun plant in a remote portion of the Reich, far from populated areas. On December 30, 1939, after scouting several options, he recommended a site near Dyhernfurth, a small town and castle on the Oder River 40 kilometers northwest of Breslau, in the eastern province of Silesia.

Meanwhile, the development of Substance 146 continued in the basement of Spandau Citadel, where the Army Gas Protection Laboratory had constructed an apparatus to synthesize small amounts for testing purposes. Army officials named the new agent "Sarin," an acronym derived from letters in the names of the four key individuals involved in its development: Schrader and Ambros of IG Farben and Rüdiger and Linde of the Army Ordnance Office.

Although Schrader worked intermittently on a manufacturing process for Sarin, the Army expanded its technical staff and asserted full control over the development effort, limiting his involvement. Schrader resented being excluded and complained that the engineers at Spandau were mismanaging the process development effort and causing lengthy delays. He was also suspicious of the secrecy surrounding the physiological laboratory where Dr. Wirth and his colleagues were conducting experiments with Tabun and Sarin. When Schrader traveled to Spandau periodically to advise on technical issues, he was never allowed near the medical clinic and physiological laboratory in Building 15. Mystified, he suspected that some type of illicit activity was going on there, possibly experimentation on humans.

MEANWHILE, HITLER'S WAR was going well. The Wehrmacht's new blitzkrieg ("lightning war") tactic, involving rapid thrusts by mechanized tank columns supported by withering attacks from the air by Stuka close-support aircraft, had proved to be a dramatic success. The German Army had conquered Poland in a few weeks and achieved similar victories on the western front, rapidly overrunning Belgium and the Netherlands. Hitler's generals saw no reason to employ chemical weapons, which would only slow down the fast-moving campaign. But they worried that the Allies might resort to defensive chemical warfare tactics, such as using phosgene shells against armored columns or spraying mustard agent on the ground to contaminate the battlefield. The German generals were relieved when the feared attacks did not materialize.

On May 14, 1940, the Wehrmacht routed the French Ninth and Second Armies. The French government evacuated Paris on June 10, and four days later the German Eighteenth Army marched triumphantly down the Champs-Elysées and hoisted the swastika flag atop the Eiffel Tower. Britain was the next target in Hitler's sights. In a speech on June 18 to the House of Commons, Prime Minister Winston Churchill warned his countrymen of the severe trials that lay ahead. "What General Weygand called the Battle of France is over," he intoned. "I expect that the Battle of Britain is about to begin. Upon this battle depends the survival of Christian civilization. . . . The whole fury and might of the enemy must very soon be turned on us. Hitler knows that he will have to break us in this Island or lose the war."

If Britain did not prevail in the coming conflict, Churchill warned, "then the whole world, including the United States, including all that we have known and cared for, will sink into the abyss of a new Dark Age, made more sinister, and perhaps more protracted, by the lights of perverted science." Despite these vague premonitions, Churchill had no inkling that Germany had achieved a revolutionary advance in chemical weaponry— one for which the Allies were totally unprepared.

CHAPTER THREE

PERVERTED SCIENCE

ON JANUARY 29, 1940, the IG Farben board of directors founded a new subsidiary called Luranil (an abbreviation of Ludwigshafen Rhein Anilin) to build the nerve agent plant at Dyhernfurth and the mustard plant at Gendorf. Meanwhile, construction of the Tabun pilot plant at Raubkammer was delayed by shortages of materials and skilled labor, and corrosion problems forced a redesign of the apparatus. As a result, the pilot plant did not begin regular operation until July 1940. From then on, it manufactured a total of about fifty tons of Tabun for field trials, while providing valuable operating experience for the full-scale production facility at Dyhernfurth.

Because of harsh weather in Silesia during the winter months, the start of construction at Dyhernfurth was delayed until early spring 1941, when crews began to clear a dense tract of forest about one kilometer from the Oder River. Ninety technicians from Luranil and 120 prisoners of war worked to build the vast factory. Code-named "Hochwerk," it would eventually cover an area 1.5 kilometers long by 700 meters wide.

In August 1941, Otto Ambros summoned about a dozen young chemists and engineers from several IG Farben plants to a meeting in Ludwigshafen. He explained that they had been selected for a secret wartime assignment for the Reich and would be exempted from military service for the duration of the project. One of the chosen chemists, Dr. Wilhelm Kleinhans of the IG Farben laboratory in Mainkur, traveled to Elberfeld to work in Schrader's lab for several weeks. There he familiarized himself with the manufacturing process for Tabun before continuing on to Dyhernfurth.

Construction of the Tabun plant was slowed, however, by bureaucratic

Otto Ambros, chemist and industrialist, played a key role in the German nerve agent program. Ambros was a member of the Vorstand (managing board of directors) of IG Farben, chief of the Chemical Warfare Committee of Albert Speer's Ministry of Armaments and War Production, and manager of the Tabun and Sarin production plants at Dyhernfurth and Falkenhagen.

and logistical problems. In November 1941, for example, government officials ordered the entire workforce at Dyhernfurth transferred to the IG Farben plant in Heydebreck for the urgent production of fuel. Although Ambros managed to get the transfer order canceled, valuable time had been lost, making it impossible to complete the Tabun factory before the onset of winter.

IN FEBRUARY 1942, the Nazi regime undertook a sweeping reorganization of the weapons procurement bureaucracy previously headed by Fritz Todt, who had died in a plane crash. To replace the Todt organization, Hitler created a new Ministry for Armaments and War Production under the direction of his young protégé Albert Speer, who had previously served as his personal architect. An urbane man of high intelligence, ambition, and personal charm, Speer stood out among the group of crude, thuggish men who dominated the Nazi inner circle. Over the years, Speer had gained the confidence and affection of Hitler, who also fancied himself an architect and was fascinated with grandiose building projects.

Although the Army Ordnance Office survived the reorganization of the armaments bureaucracy, it increasingly came under the control of advisory committees created by Speer to oversee various aspects of weapons research, development, and production. One of these new bodies was Special Committee C, chaired by Otto Ambros of IG Farben, which managed the development and production of chemical weapons. To camouflage the true nature of its subject matter, the committee was denoted by the letter "C," for *Chemikalien* (chemicals), rather than by "K," for *Kampfstoffe* (chemical warfare agents).

On February 14, 1942, Colonel Schmidt of the Army Ordnance Office issued a top-secret report for senior government officials titled "Memorandum on a New War Gas, Trilon 83." In addition to describing the discovery and testing of Tabun, this memo discussed the possibility that Germany's enemies had developed similar agents. "[W]e have no evidence whatever that Trilon 83 or a similar compound is being made in foreign countries," the document concluded. "One must, however, . . . reckon with the fact that scientific research in other countries is sure to start sometime on the study of such compounds, for other great powers, especially England, America, and Russia, have been conducting an intensive search for new war gases for years."

ALTHOUGH HITLER had no plans to use chemical weapons against the Allied armies except in retaliation for an attack, the SS began to employ a different poison gas to murder millions of Jews and other defenseless civilians in the extermination camps. The compound selected for this purpose was hydrogen cyanide, also known as prussic acid. A potent, fast-acting poison, cyanide blocks the ability of cells to utilize oxygen, starving the brain and other vital organs and resulting in dizziness, vomiting, unconsciousness, and death. Before the war, a formulation of hydrogen cyanide known as Zyklon B had been developed to exterminate vermin in ships, buildings, and factories. This product was manufactured by the Frankfurt firm Dagesch (an abbreviation of Deutsche Gesellschaft für Schädlingsbekämpf-ung, or German Society for Insecticide Research) under license from IG Farben, which held the patent. Zyklon B consisted of pea-sized, gray-blue pellets of diatomaceous earth that had been impregnated with a mixture of hydrogen cyanide, a stabilizer, and a warning chemical with an unpleasant odor. Once the pellets were removed from their sealed metal container and exposed to air, they began to give off the lethal gas.

Zyklon B was brought to the main Auschwitz concentration camp (Auschwitz I) in the summer of 1941 for the delousing of prisoners. In September, however, the SS conducted experiments to test the suitability of the poison for the mass killing of inmates in gas chambers. When Zyklon B proved effective for this purpose, the Nazis ordered Dagesch to manufacture the pellets without the warning chemical, a violation of German law. The Hamburg firm of Tesch & Stabenow supplied the modified product to the concentration camps at Auschwitz, Maidanek, Sachsenhausen, Ravens-brück, Stutthof, and Neuengamme. In 1942 and 1943, nineteen metric tons of Zyklon B were delivered to the Auschwitz-Birkenau extermination camp, three kilometers northwest of Auschwitz I, where most of the mass gassings took place. During the single night of March 13, 1943, for example, the SS used six kilograms of Zyklon B to murder 1,492 Jewish women, children, and old people from the Kraków ghetto in the gas chambers.

BECAUSE OF SHORTAGES of key equipment and manpower, it took two years and an expenditure of 120 million reichsmarks to complete the Hochwerk plant at Dyhernfurth, which the Anorgana company headed by Ambros began to operate in the spring of 1942. The sprawling production complex included buildings for manufacturing basic chemical ingredients, intermediates, and final products; numerous warehouses and storage tanks; a bombproof bunker that could hold 1,000 tons of bulk agent; filling lines for loading Tabun into artillery shells and aerial bombs; a well-equipped medical clinic with a staff of trained physicians; and barracks for the plant workers. To reduce exposure to air raids, the main production facility was built partially underground and camouflaged with trees planted on the roof.

Although the Dyhernfurth plant had managed to manufacture a few hundred tons of chemical intermediates, in April the large-scale production of Tabun finally got under way. At that time, the Hochwerk complex employed about ninety scientists, technicians, and other white-collar staff, along with 560 German workers. It was a major challenge to obtain sufficient quantities of the basic ingredients needed for Tabun production, such as elemental phosphorus. Because no reserves of phosphate ore existed in the German Reich or the newly occupied territories, the mineral had to be imported from mines in North Africa. A single factory at Piesteritz in central Germany processed raw phosphate ore into elemental phosphorus, with an output of 1,300 tons per month. In addition to being used for the production of incendiary grenades and smoke bombs, phosphorus was combined with chlorine to yield phosphorus oxychloride ($POCl_3$), the starting material for Tabun production.

The Tabun factory contained twelve separate but parallel production units, each of which was theoretically capable of producing a metric ton of agent every twenty-four hours. A production unit consisted of a large iron reaction kettle with a volume of 1,500 gallons, lined with a special corrosion-resistant iron alloy called Remanit. Chemical ingredients were introduced into the kettle through a long pipe that penetrated the vapor-tight lid, and the reaction products were removed through the long pipe by injecting pressurized air into a short pipe that ended above the surface of the mixture. The rate of the chemical reaction could be increased by heating the vessel with hot water that circulated through an external steel jacket, and slowed by cooling the solution inside the kettle with a set of immersion coils containing a chilled solution of calcium chloride.

The process for manufacturing Tabun was essentially the same as that developed by Schrader, but scaled up to industrial volume. First the kettle was filled with the two starting materials, which took about thirty minutes. Then hot water was allowed to circulate through the metal jacket, heating the mixture inside the kettle and causing the two chemicals to react. After an hour and forty minutes, the reaction reached completion, yielding an intermediate called Product 39 that was highly irritating to the eyes. In the second step, Product 39 was mixed with two additional chemicals for a period of two hours. Because this reaction generated heat, the cooling coils were used to keep the mixture at a constant temperature. Finally, the end product was drawn from the kettle into a holding tank. The raw Tabun that emerged from the kettle was an oily liquid with a dark reddish brown hue that, when filtered to remove solid precipitates, became clear and colorless. Whereas small amounts of pure Tabun gave off a faint aroma of ripe fruit, large quantities had a fishy odor.

Initially the end product was prepared in a form called Tabun A, containing 5 percent chlorobenzene, the solvent used in its preparation. Due to the presence of impurities left over from the production process, however, Tabun A was unstable and had a limited shelf life: its toxicity declined by 5 percent after six months and 20 percent after three years. Beginning in mid-1944, Dyhernfurth began to produce a new formulation called Tabun B that contained 80 percent Tabun and 20 percent chlorobenzene. This mixture was more stable, had a longer shelf life, and evaporated more readily.

Because of Tabun's extreme toxicity, the design of the Hochwerk facility included special features to protect the plant workers against exposure. Each kettle was housed in an enclosed operating chamber formed of two spaced glass walls. Between the glass walls, a ventilation system produced greater than atmospheric pressure, so that the flow of air was always toward the reaction kettle. Inside the operating chamber, the air above the kettle was continuously changed by means of a separate ventilating duct, creating negative pressure. This pressure differential meant that the air contaminated with Tabun fumes was retained inside the operating chamber. All pipes used to transfer solutions containing Tabun were double-walled, and their outer surfaces were sprayed frequently with a weak solution of ammonia and water to neutralize minor leaks. After each production run, the kettles were decontaminated with steam and ammonia.

No technicians were allowed to enter the operating chambers while the production of Tabun was under way. Instead, the operators opened and closed valves with long-handled mechanical levers that penetrated the double glass walls through rubber-sealed gaskets. This system enabled them to control the flow of chemical ingredients to and from the reaction kettles without being exposed to the deadly fumes. Because the rubber seals were not perfectly airtight, however, trace amounts of Tabun managed to leak out. As a result, the plant workers at Dyhernfurth were never free of the symptoms of low-level Tabun poisoning.

Since the harmful effects of the nerve agent were cumulative, repeated low-level exposures over a period of several days could be fatal. Accordingly, every few weeks, scientists and technicians at Dyhernfurth were ordered to remain outside the production area for two or three days to allow their bodies to recover. IG Farben workers also received extra rations of high-fat foods, such as milk and cheese. This apparent act of generosity had a utilitarian purpose: consuming a high-fat diet was known to increase resistance to Tabun poisoning.

Although the standard German Army gas mask protected against breathing contaminated air, Tabun could also be absorbed through the skin. For this reason, all mechanics who entered the sealed production chambers to perform repairs and maintenance wore not only a respirator but a protective suit, cap, boots, and gloves, encapsulating the entire body. The suit consisted of two layers of rubber separated by a layer of cloth, making it cumbersome and unbearably hot in summer. Despite these precautions, about a dozen fatal accidents occurred during the two and a half years of Tabun production, most of them affecting mechanics performing overhauls of the plant. In one incident, seven pipe fitters were struck in the face by a pressurized stream of liquid Tabun that forced itself between their respirators and rubberized suits. The victims became giddy, vomited, and removed their masks, causing them to inhale more of the deadly fumes. They then collapsed and went into convulsions. According to a report by the chief medical officer at Dyhernfurth, "On examination they were all unconscious . . . , had a feeble pulse, marked nasal discharge, contracted pupils, asthmatic type of breathing, and smelled strongly of flowers. Involuntary [urination] and diarrhea occurred."

All seven victims were given intramuscular injections of atropine and a new drug called Sympotal, but five did not respond to the antidotes and died. When the two survivors regained consciousness in the clinic, they were overexcited and continued to have minor convulsions. To counteract these symptoms, the doctors injected them with a sedative called sodium evipan that put them to sleep for eight to ten hours, after which they awoke fully recovered. A pathologist from the Military Medical Academy in Berlin autopsied the five deceased. The only abnormality he could observe with the naked eye was congestion of the lungs and brain, but he removed the major organs for detailed examination.

ALL ASPECTS OF life at Dyhernfurth were overshadowed by elaborate security measures. Access to the site was strictly controlled and required passing through a series of heavily guarded perimeters and checkpoints. In addition, the technical details of the Tabun manufacturing process were classified and the "need to know" principle was strictly enforced: factory personnel were informed only about those operations in which they were directly involved. Although IG Farben chemists and engineers were naturally curious about other aspects of the production process, they did not ask their colleagues too many questions for fear of being informed on or suspected of espionage, which could result in interrogation and torture by agents of the Geheimstaatspolizei (abbreviated Gestapo), or secret state police. Accordingly, the scientists and technicians at Dyhernfurth wore psychological "blinders" and kept their attention tightly focused on their narrow roles in the production process.

German counterintelligence officials also developed elaborate methods to conceal the nerve agent program from foreign intelligence services. Tabun was given a variety of cover names, including "Gelan I" and "Substance 83," although the preferred designation was "Trilon 83." The ethyl analogue of Tabun was called "Gelan II" or "Trilon 32," while Sarin was referred to as "Gelan III" or "Trilon 46." Chemical ingredients used in the manufacture of Tabun were also designated with code names to make it harder for enemy spies to track shipments. These codes were kept in a secret "black book" and deciphered with the aid of an index. For example, raw materials were coded as follows: ethanol (A4), chlorine (A5), phosphorus (A6), sodium hydroxide (A9), and sodium (A17). Whenever an ingredient for Tabun arrived at Dyhernfurth, it was assigned another local code name, making correct identification nearly impossible if the plant and its records were to fall into enemy hands. The code-name system also had the effect of keeping most of the technical staff in the dark about the precise chemical reactions involved in the manufacture of Tabun.

Because of the elaborate counterintelligence measures designed to protect the secrecy of Tabun and Sarin, the Allies remained unaware of these dramatic developments. A U.S. intelligence report in July 1942 titled "New German Poison Gas" read as follows: "Disclosures relative to so-called 'Blau Gas' have occurred numerous times in the past and . . . are no longer seriously regarded. Intelligence reports lend considerable weight that new German agents are not of the nature of so-called nerve gases." In hindsight, this assessment could not have been more wrong.

DYHERNFURTH HAD its own munition-loading facility, which was built underground and equipped with ventilation shafts. Steel artillery shells and bomb casings were manufactured in a separate building and placed on conveyor belts that transported them to the filling plant. Liquid Tabun pumped from underground storage tanks was loaded into empty 105 mm and 150 mm artillery shells, 250-kilogram aerial bombs, and artillery rockets. Whereas a 105 mm shell contained about a kilogram of liquid agent, a 250-kilogram bomb contained 80 to 85 kilograms. To compensate for Tabun's lack of volatility, the bombs contained a central "burster" tube filled with a high explosive that, detonated on impact with the ground, would shatter the liquid agent into a mist of tiny droplets, poisoning enemy soldiers through inhalation and skin contact.

Once an aerial bomb had been loaded with Tabun, the filling port was closed with a plug that incorporated a tightening pin. Using a wrench, a technician applied seating pressure to the pin, causing it to shear off and leave the plug in a sealed position, flush with the surface of the weapon. The sealing plug and adjacent surface were then coated with a slow-drying pink lacquer that would turn a deep carmine if Tabun leaked through the plug. Near the base of the bomb or shell, workers painted three green rings around the munition to indicate its contents, along with stenciled numbers providing the date of manufacture and a code letter indicating the ratio of Tabun to chlorobenzene. In September 1942, the first 138 metric tons of Tabun-filled shells and bombs produced at Dyhernfurth were delivered to the Wehrmacht. Packed into crates, the munitions, fuses, and other components were loaded onto trucks and railway freight cars for transport to storage depots controlled by the Luftwaffe and the Army.

When serious shortages of raw materials prevented Anorgana from meeting its manufacturing targets for mustard agent at Gendorf, the Speer ministry decided to give priority to the production of Tabun, ensuring qualitative if not quantitative superiority over the Allies. To increase output at Dyhernfurth, the Nazi regime decided to employ forced prison labor and built a satellite of the nearby Gross-Rosen concentration camp there in early 1943. Known as Dyhernfurth I, the labor camp (Arbeitslager) initially housed some two hundred prisoners, mostly Poles, Russians, Germans, and Czechs.

The forced laborers were assigned the most menial, backbreaking, and dangerous tasks at the Tabun plant, including construction, maintenance, and loading munitions with the liquid agent. On the filling line, they wore protective clothing similar to a deep-sea diving suit, with a helmetlike mask covering the entire head and a hose providing a supply of fresh air. Because of the short length of the hose, the suit permitted only limited movement. Whenever a worker needed to drink or go to the bathroom, he had to remove the mask, exposing himself to toxic fumes. As a result, the forced laborers suffered continually from the symptoms of low-level Tabun poisoning. Those exposed accidentally to a lethal dose were denied medical treatment and left to die.

Prisoners at Dyhernfurth were also exploited for medical experiments involving deliberate exposure to nerve agents. Early in the war, guinea pigs and white rats were found to be inadequate for testing Tabun and Sarin, and apes were used instead because their physiological reactions were closer to those of humans. The Speer ministry purchased a colony of apes from Spain at a cost of 200,000 Swiss francs and transported them to Germany by train, but many of the animals died in transit. Given the difficulty and high cost of procuring nonhuman primates, it was decided to experiment on concentration camp inmates. At Dyhernfurth, about twenty prisoners were exposed to nerve agents for varying lengths of time in a sealed glass chamber and then examined; about a quarter suffered painful deaths during the trials. Prisoners were also misused as human "canaries" by being locked up for long periods without a gas mask in train cars or munitions depots loaded with Tabun-filled bombs or shells.

MEANWHILE, German military scientists continued to search for more lethal and effective nerve agents. Because the mechanism of action of Tabun and Sarin was poorly understood, the research-and-development process was based largely on trial and error. In early 1943, Colonel Schmidt of the Army Ordnance Office asked Professor Richard Kuhn, the director of the Institute of Chemistry at the Kaiser Wilhelm Institute for Medical Research in Heidelberg, to analyze the effects of nerve agents on the central and peripheral nervous systems. As this task had been assigned a high priority, the members of Kuhn's research team were exempted from military service.

Kuhn, forty-two, was one of Germany's most eminent organic chemists, having been awarded the Nobel Prize in Chemistry in 1938 for his work on the synthesis of carotenoids and B vitamins (although he had been prevented from accepting the prize). Before the war he had taught for a year at the University of Pennsylvania and had been a prominent member of the international scientific community. After Hitler's rise to power, however, Kuhn had remained president of the German Chemical Society and served as a consultant to other Nazi Party organizations. He seemed to be an enthusiastic supporter of the regime, giving the Hitler salute at the beginning of his classes and shouting "*Sieg heil!*" with apparent gusto. Kuhn's close friends later claimed that he had only feigned support for Hitler to shield academic science from political interference, and that he had retained the presidency of the German Chemical Society to prevent it from being taken over by a Nazi hack. In any event, Kuhn willingly accepted the Army assignment to study the physiological action of the nerve agents.

In conducting this investigation, Kuhn drew on some recent discoveries about the role in the nervous system of a natural chemical substance called acetylcholine. In 1914, Henry Dale, a physiologist at the National Institute for Medical Research in London, had described the physiological effects of acetylcholine on various organs. Then, in 1921, Otto Loewi, a German-born professor of pharmacology living in Graz, Austria, had provided the first proof that chemical messenger substances are involved in the transmission of nerve impulses from one nerve cell to another and from a nerve cell to a responsive organ. Loewi focused on the function of the autonomic nervous system, which governs the activity of involuntary smooth muscles (such as those of the pupil, the heart, and the gastrointestinal tract) and secretory organs (such as the salivary, sweat, and adrenal glands). The autonomic nervous system is in turn divided into two parts, "sympathetic" and "parasympathetic," with opposing physiological effects. For example, the sympathetic system increases the heart rate, whereas the parasympathetic system slows it.

In a landmark experiment, Loewi and his colleagues found that by electrically stimulating the vagus nerve (part of the parasympathetic nervous system) enervating the isolated heart of a frog, they could slow the heart's rate of beating. The investigators then took the saline solution perfusing the frog heart and used it to perfuse a second isolated frog heart in which the vagus nerve had not been stimulated. Surprisingly, the rate of the second heart also slowed, indicating that the nerve ending had released a chemical substance that mediated its physiological effect on the heart muscle. This substance, which Loewi termed *Vagusstoff*, was later shown to be acetyl-choline. In 1926, Loewi and his colleague Ernst Navratil demonstrated that acetylcholine is broken down in the body by a specific enzyme, which they named cholinesterase; and in 1929, Henry Dale and Harold Dudley isolated acetylcholine from animal tissue. For their important discoveries, Loewi and Dale shared the Nobel Prize for Physiology or Medicine in 1936.

When Kuhn and his colleagues began their research in 1943 on the mechanism of action of nerve agents, they knew from the work of Loewi and Dale that acetylcholine plays a key role in the parasympathetic part of the autonomic nervous system and in the peripheral nervous system, which provides voluntary control over the skeletal muscles. The arrival of a nerve impulse at the junction between a nerve and a muscle cell induces the release from the nerve ending of molecules of acetylcholine, which diffuse across a narrow gap called the synapse and stimulate receptors on the surface of the muscle cell, triggering a series of biochemical events that cause the muscle fibers to contract. Under normal conditions, cholinesterase enzymes in the synapse immediately break down the acetylcholine and halt the stimulation of the receptors, allowing the muscle fibers to relax. In this way, acetylcholine and cholinesterase operate as a biochemical on-off switch: the messenger substance activates the circuit, and the enzyme breaks it.

Kuhn found that exposing laboratory animals to Tabun strongly inhibited the action of cholinesterase, an effect that he hypothesized was key to the toxic effects of nerve agents. By preventing cholinesterase from destroying acetylcholine, nerve agents freeze the biochemical on-off circuit in the "open" position, allowing the messenger substance to build up to toxic levels. Because acetylcholine plays multiple roles in the peripheral, autonomic, and central nervous systems, excessive amounts give rise to diverse physiological effects. In the peripheral nervous system, a surfeit of acetylcholine causes the skeletal muscles to go into violent, uncontrolled spasm, followed by a state of vibration and then paralysis. In the autonomic nervous system, too much acetylcholine affects the smooth muscles and glands involved in digestion, excretion, and respiration, resulting in pinpoint pupils, excessive salivation, intestinal cramps, vomiting, and constriction of the bronchial tubes. In the central nervous system, acetylcholine overstimulates groups of nerve cells in the brain, causing seizures. Nerve agents can induce death by asphyxiation through three different mechanisms: constriction of the bronchial tubes, suppression of the respiratory center of the brain, and paralysis of the breathing muscles. These diverse effects of excess acetyl-choline are collectively known as a "cholinergic crisis."

Based on these insights, Kuhn's research team developed a standardized assay that measured the ability of nerve agents to inhibit purified cholin-esterase enzymes in the test tube. Over the next two years, they used this assay to screen a variety of candidate nerve agents, some of their own inven-tion and others synthesized by the German Army chemists at Spandau Citadel.

Despite Germany's invention of the nerve agents, Hitler held back from unleashing this secret weapon. At the same time, he encouraged the Ger-man Army to proceed with production and testing so as to ensure a position of military superiority should the Allies decide to initiate chemical warfare. By the end of 1944, the production of Tabun was slated to rise from 1,000 to 2,000 tons per month.

Meanwhile, the Anorgana company headed by Ambros moved forward with plans for the industrial production of Sarin, which was militarily more effective than Tabun but more difficult to manufacture. Construction of a Sarin pilot plant at Dyhernfurth had been authorized in late 1942, and full-scale production was scheduled to begin in March 1945. Two competing manufacturing processes for Sarin, one developed by Schrader and the other by an army chemist named Reetz, were tested at pilot plants con-structed at Spandau, Raubkammer, and Dyhernfurth. Schrader's process involved a series of five reactions, two of which were highly corrosive and required the use of reactors lined with silver, glass, or fused quartz.

The German Army proposed building a full-scale Sarin production plant with a capacity of 500 tons per month alongside the Tabun factory at Dyhernfurth, taking advantage of common elements in the manufacturing processes of the two agents. But the Luftwaffe objected strongly to this plan on security grounds: if the enemy ever discovered the location of Dyhern-furth, a single air raid could deprive the Reich of its two most effective war gases. Given the desirability of dispersing military production, the Luft-waffe insisted that the full-scale Sarin plant be built at a separate location and operated independently.

Anorgana eventually identified a suitable site in the forest of Falken-hagen near the town of Frankfurt-on-the-Oder, about a hundred kilometers east of Berlin. Before the war, this area had been developed as a proving ground and had a vacuum tunnel for testing ballistic missiles. Falkenhagen offered several logistical advantages: it was connected by road and rail to the nearby town of Brisen, fifteen kilometers away, and had worker housing, a high-capacity waterworks, and a power plant with transformer stations and cables that provided ample electricity. In May 1943, Otto Ambros visited Falkenhagen and approved it as the site of the Sarin facility. Given the code name "Seewerk," it would be built largely underground and equipped with the most modern production equipment. Whereas the Sarin plant at Dyhernfurth would manufacture 100 metric tons of agent per month with Schrader's process, the full-scale facility at Falkenhagen would produce 500 tons per month with Reetz's process.

On May 11, 1943, the British captured a German Army officer in Tunisia. Under interrogation, he revealed that he was a chemist who had done chemical weapons research at Spandau Citadel in Berlin. He described the development of a new warfare agent that was colorless, had little odor, and possessed "astounding properties." Minute doses made the pupils shrink to pinheads and constricted the bronchial tubes, causing an asthmalike shortness of breath, and higher doses were lethal within fifteen minutes. The informant knew the substance only by the code name "Trilon 83."

The British interrogators judged the prisoner's information to be reliable and wrote a ten-page secret report that was sent on July 3, 1943, to Military Intelligence in London and the Chemical Warfare Experimental Establishment at Porton Down. (Founded in 1916 on 7,000 acres of rolling English countryside in Wiltshire, Porton Down was the British government's primary center for chemical weapons research and development.) Because of a lack of corroborating evidence, however, British officials had doubts about the veracity of the intelligence report and decided to take no action.

ON THE EASTERN front, Hitler's campaign to conquer the Soviet Union was reaching a fateful turning point. In the fall of 1942, the German Sixth Army had attacked Stalingrad, and during the ensuing weeks, street fighting had raged throughout the city. At dawn on November 19, the Soviets launched a major counteroffensive in which armored spearheads drove in a pincer movement from the north and the south, cutting off Stalingrad and forcing the Sixth Army to retreat to the west or be surrounded. After Hitler refused to authorize a retreat, twenty German divisions were encircled by the Soviet forces. On January 8, 1943, the Soviet commander gave the doomed Sixth Army a final chance to surrender. When Hitler again refused, the Soviets began a massive artillery bombardment with 5,000 guns. Over six days of bitter fighting, the German pocket was reduced by half, and on February 2, 1943, the battered remnant finally surrendered. Of the more than 250,000 German soldiers who fought at Stalingrad, 70,000 were killed and 91,000 captured, including 24 generals. Half starved and frostbitten, the POWs were sent to camps in Siberia, and only about 5,000 survived the war. This terrible defeat halted the German advance into southern Russia and provoked much soul-searching in Berlin.

After the debacle at Stalingrad, the proponents of gas warfare in the Nazi inner circle believed that their time had come. Martin Bormann, the head of the Party Chancellery and private secretary to the Führer; Joseph Goebbels, the Reich Minister for Popular Education and Propaganda; and Robert Ley, the leader of the German Workers' Front, all argued for unleashing Tabun against the Red Army. Hitler was prepared to consider the use of chemical weapons against the Russians, whom he despised and considered subhuman. In February 1943, he ordered preparations for a chemical attack on the eastern front, setting a deadline of April 20. But when that day arrived, he continued to equivocate.

On May 15, 1943, the Führer called a war conference of his closest advisers at his new military headquarters near Rastenburg in East Prussia known as "Wolf's Lair" (Wolfsschanze). Situated in dense forest, the compound consisted of three concentric circles, each protected by minefields, pillboxes, and an electrified barbed-wire fence that was continually patrolled by SS guards. To enter the innermost zone where Hitler lived and worked, even the most senior officials had to obtain a special onetime pass and be personally inspected by the SS chief of security or one of his deputies.

Attending the May 15 conclave at Wolf's Lair were Albert Speer, the Minister of Armament and War Production; Field Marshal Wilhelm Keitel, the supreme commander of the Wehrmacht; Brigadier General Walther Schieber, who oversaw the chemical industry for the Speer ministry; and other senior military officials and directors. Otto Ambros, IG Farben's leading expert on poison gases, was also present. He had received a telegram from Speer a few days earlier ordering him to come to Berlin, whence he had flown in an official airplane to Rastenburg.

The last item on the agenda of the daylong war conference was a one-hour discussion of the situation in the chemical weapons field. Speer and Schieber began by describing Germany's readiness for waging gas warfare and the likelihood that the Allies would resort to such weapons. Then Ambros took the floor and reported objectively on the production of war gases, referring to a table that described the Wehrmacht's military requirements for the various agents and the existing stockpiles. Ambros noted that, thanks to the outstanding work of IG Farben scientists, Germany had developed a new class of war gases that targeted the nervous system and were best described as "nerve agents." The first such compound, Tabun, could kill in minute doses, while a second agent called Sarin was six times more potent. Because of their extraordinary lethality, the use of these gases would have a severely demoralizing effect on the enemy.

As of May 1, 1943, Ambros said, Germany had produced a total of 44,764 metric tons of chemical warfare agents, including 1,500 tons of Tabun. Moving from pilot- to industrial-scale production of Tabun had been extremely challenging because the manufacturing process involved highly toxic and corrosive materials. "Nevertheless," Ambros added, "in the past few months, remarkable progress has been achieved." Although the level of Tabun production at Dyhernfurth at that time was 350 metric tons per month, Anorgana expected to reach the full capacity of 1,000 tons per month in early 1944. The company also planned to construct a Sarin pilot plant at Dyhernfurth with a capacity of 100 tons per month. Ambros concluded by urging Hitler to allocate more resources to the chemical weapons sector, including manpower for construction and operations, materials for buildings and installations, and air defenses to protect the major storage depots.

Hitler was clearly disappointed by this report, noting that for most types of chemical warfare agents, not even half the requirements of the General Staff had been achieved. He then asked about the enemy's chemical warfare capabilities and Germany's relative strength in this area. Ambros replied that although the Wehrmacht possessed all of the major choking and blister agents, the enemy had larger stockpiles and production capacity. Any industrial power that could manufacture petrochemicals such as ethylene oxide was capable of mass-producing mustard. "I believe," Ambros added, "that the enemy, because of his greater supplies of ethylene, probably has the capacity to produce larger quantities of mustard than does Germany."

Hitler's toothbrush mustache bristled with irritation. "I understand that countries that have oil are in a better position to make mustard," he snapped, "but what about the special gas Tabun? I have been told that in this area Germany has a monopoly. Do you believe that the enemy has also developed nerve agents?"

Aware of Hitler's dangerous temper, Ambros chose his words carefully. Whether Germany had a monopoly in the nerve agents could not be judged with any certainty, he said. German scientific papers and patents on related compounds had been published in the open scientific literature as early as 1902, and only much later had these materials been classified and withdrawn from commercial development. It was therefore possible that the enemy had developed nerve agents like Tabun. "I am also convinced," Ambros added, "that in the event that Germany were to use this special gas, other countries would not only be able to imitate it quickly but could produce it in considerably larger quantities."

Visibly distressed by this remark, Hitler turned abruptly on his heels and strode out of the room.

AMBROS'S BELIEF that the Allies had independently discovered Tabun or related compounds was based largely on inference. He was aware that German intelligence had surveyed the U.S. chemical literature before the war and found published papers on compounds whose chemical structure was distantly related to Tabun. Since the war began, all such information had disappeared from U.S. scientific journals, suggesting that the research had become classified. German intelligence was also familiar with the famous Soviet school of organophosphorus chemistry led by Professor Alexander E. Arbusov in the Russian city of Kazan. Schrader had used a reaction sequence developed by Arbusov to synthesize an intermediate in the production of Sarin.

In fact, Ambros had overestimated the extent of the Allies' progress. Although American, British, and Soviet chemists were studying various organophosphate agents and assessing their military potential, they had not independently discovered Tabun or Sarin. Beginning in 1941, a chemistry professor at the University of Cambridge named Bernard Charles Saunders had synthesized several fluoride-containing organophosphate compounds, of which the most promising was diisopropyl fluorophosphate, or DFP.

(Saunders was unaware that the German chemist Willy Lange had previously synthesized DFP in 1932.) In addition to being quite toxic, DFP had desirable physical properties and was cheap and easy to manufacture. Saunders conducted a series of risky experiments on himself and his colleagues in which he used a bicycle pump attached to a round-bottom flask to disperse low concentrations of DFP in a sealed room. The scientists then entered the room and cautiously sniffed the air to assess the agent's physiological effects.

On December 11, 1941, Saunders reported to the British Ministry of Supply, which was responsible for chemical weapons development, that high levels of DFP had a rapid "knockout" action comparable to that produced by hydrogen cyanide. At much lower doses, DFP constricted the pupils of the eyes, resulting in a marked dimming and impairment of vision that could put enemy soldiers out of action for an extended period. British officials considered the new agent promising enough to commission the Chemical Warfare Establishment at Sutton Oak, England, to develop a small-batch production plant. It gradually became clear, however, that DFP did not offer a significant improvement in toxicity over standard agents such as mustard or phosgene, making it chiefly of interest as a harassing agent.

The British shared their findings on DFP with American military chemists working in Division 9 of the National Defense Research Committee (NDRC), a wartime agency reporting to the U.S. Office of Scientific Research and Development led by Dr. Vannevar Bush. From December 1942 to the end of 1945, the NDRC issued contracts to academic chemists at the University of Illinois, the University of Chicago, and the California Institute of Technology to synthesize and evaluate roughly two hundred different organophosphorus compounds, many of them containing fluorine, yet none approached the toxicity of Tabun or Sarin. The U.S. Chemical Warfare Service was sufficiently interested in DFP (which the Americans called PF-3) to contract with the Monsanto Corporation's Phosphate Division to build a pilot production plant and to conduct some tests with the agent on soldiers at Edgewood Arsenal. But although test runs of the pilot plant produced 535 pounds of DFP for testing purposes, the Army decided not to stockpile it as a standard agent. Other U.S. wartime chemical research focused on the development of insecticides to combat malaria mosquitoes and other vectors of infectious disease. (A common misconception holds that DDT, a powerful insecticide discovered by the Swiss chemist Paul Hermann Müller in 1939, was related to the German nerve agents. In fact, DDT is a chlorinated hydrocarbon whose chemical structure has nothing in common with Tabun or Sarin.)

In a broader sense, however, Ambros's calculation was correct. If the Germans had employed Tabun or Sarin during the war and the Allies had obtained samples of the new agent, ongoing British and American research on DFP and related substances would have enabled those countries to identify the structure of Tabun and replicate its synthesis in a fairly short time. The Allies would then have launched a wartime crash program to mass-produce the new agent. Scaling up from the laboratory bench to industrial production would have been difficult and time-consuming, but still feasible with a sufficient investment of money and effort. Thus, any German military advantage arising from the first use of nerve agents would have been short-lived, and the consequences for the German population severe, given the inadequate gas protection of most civilians and the Allies' growing air superiority.

Hitler also knew that even if the Allies did not possess nerve agents, they did have vast stockpiles of aerial bombs filled with phosgene and mustard agent with which to retaliate against German cities. Indeed, Allied leaders made explicit threats to deter the German use of chemical weapons. In June 1943, President Franklin D. Roosevelt declared that the United States would under no circumstances resort to chemical weapons unless they were used by the enemy first. But he then warned, "Any use of gas by any Axis power will immediately be followed by the fullest possible retaliation upon munition centers, seaports, and other military installations through the whole extent of the territory of such Axis country."

To ensure German superiority in the event that chemical weapons were used, Ambros recommended the immediate expansion of production capacity for both Tabun and Sarin. This argument fell on receptive ears, and on May 26, 1943, only ten days after the meeting at Wolf's Lair, Hitler ordered the doubling of Tabun production at Dyhernfurth from 1,000 to 2,000 metric tons per month by the end of 1944, and an increase in Sarin output from 100 to 500 tons per month. Hitler and Speer also approved Anorgana's request for more resources for nerve agent manufacture, including 88 million reichsmarks, 55,000 tons of steel, and 6,900 additional workers.

Despite continued pressure from Bormann, Goebbels, and Ley, however, Hitler showed no inclination to initiate gas warfare against the Soviet Union. Fearing that even a limited use of chemical weapons might trigger massive Allied retaliation, he ordered that no chemical munitions were to be transported outside the Reich, including Bohemia and parts of Poland, or deployed to the Russian front. Tight control over the chemical stockpile would reduce the risk of unauthorized use and avoid the capture of the weapons by enemy forces.

In early November 1943, Ambros informed the Army Ordnance Office that because of shortages of raw materials such as phosphorus, chlorine, and sodium cyanide, it would not be possible to double the production of Tabun at Dyhernfurth by the end of 1944 as planned. In fact, production of Tabun never approached the ambitious goal of 2,000 tons per month; the maximum output was about 800 tons per month of the 80 percent formulation. In order to increase the rate of Tabun production, Anorgana expanded the number of forced laborers at Dyhernfurth. A second satellite labor camp for up to 3,000 prisoners was completed in the fall of 1943, with plans to increase the total number to 9,700. The inmates were housed in thirty drafty barracks, poorly clothed and fed, and forced to work twelve hours a day. Between twenty and thirty prisoners died each week from malnutrition, beatings, exhaustion, and presumably Tabun exposure. To preserve the secrecy surrounding nerve agent production at Dyhernfurth, forced laborers who tried to escape were summarily executed. In any event, few inmates could expect to live long enough to tell the outside world about their experiences.

Because of the operational drawbacks of Tabun, which decomposed during storage and did not vaporize at low temperatures, Ambros wished to give priority to the manufacture of Sarin, which was both more toxic and more volatile. But construction of the Sarin production facility at Dyhernfurth was far behind schedule. Although the Luranil construction company had broken ground for the second Sarin plant at Falkenhagen in September 1943, difficulties in obtaining building materials and skilled labor meant that production was unlikely to start until the middle of 1945. Ambros had assigned his protégé, a young IG Farben chemist and SS officer named Jürgen von Klenck, to be the future director of the Falkenhagen plant, despite his lack of experience.

IN APRIL 1944, Hitler made his personal surgeon, Dr. Karl Brandt, responsible for protecting the German civilian population against an Allied chemical attack. Brandt had enjoyed a remarkable career under the Nazi regime. His wife had been Hitler's swimming instructress, and while on holiday in Bavaria in 1932, the couple had visited the Nazi leader's country retreat, a rustic villa called Berghof on the Obersalzberg mountain near Berchtesgaden. During the visit, Hitler's entourage was involved in an automobile accident, injuring his adjutant and three relatives. Brandt skillfully treated their injuries and gained Hitler's gratitude and affection shortly before the Nazi seizure of power in January 1933. The following year, at the age of twenty-eight, Brandt became Hitler's personal surgeon and accompanied him to Venice for a summit meeting with Mussolini. Later that year Brandt joined the SS, and in the summer of 1942 he was appointed General Commissioner for Reich Medical Services with the rank of colonel.

In his new position as Special Commissioner for Gas Defense, Brandt was responsible for shielding the German population against chemical attack. Of the 68 million inhabitants of the German Reich, few were equipped with effective chemical protective gear. Beginning in 1938, some 30 million respirators had been produced and issued, but only about half were still in good working order. Brandt planned to implement a program for the manufacture of 45 million "people's gas masks" (Volksgasmasken), incorporating a filter made of activated charcoal. Although a plan for mass production of the masks was drawn up, German industry was already working at full capacity and key materials such as rubber were in short supply. The resulting delay in production meant that the populations of major German cities remained vulnerable to Allied chemical attack.

Also during the spring of 1944, Richard Kuhn, working at the Kaiser Wilhelm Institute in Heidelberg, made a striking discovery. He was continuing his research for the German Army by screening a wide variety of organophosphorus compounds—some of which he had synthesized himself—for the ability to inhibit cholinesterase. Because of the Nazi obsession with secrecy, his research was "compartmented": he was not put into contact with other scientists in the nerve agent field and was completely unaware of Schrader's work. When Kuhn replaced the isopropyl alcohol used to make Sarin with a more complex alcohol known as pinacolyl, the resulting substance (which he called Compound 25075) had a camphorlike odor and was roughly twice as potent as Sarin in inhibiting cholinesterase.

The War Office code-named this new compound "Soman," and Kuhn synthesized small amounts in the laboratory.

Toxicological testing of Soman in animals by Dr. Gross at IG Elberfeld revealed that the new agent was twice as toxic as Sarin by inhalation, readily penetrated the skin, and passed rapidly from the bloodstream into the brain, enhancing its lethal effects. Even more striking, Soman inactivated cholinesterase irreversibly within two minutes, severely limiting the effectiveness of atropine as an antidote. Over the next several months, Kuhn and his colleagues tested about fifty analogues of Sarin and Soman for their ability to inhibit cholinesterase in his test-tube system. The most promising compounds were then tested on dogs and apes.

Meanwhile, the war was reaching a major turning point—the Allied invasion of German-occupied France—in which a decision by Hitler to employ nerve agents could have a decisive impact on the outcome.

CHAPTER FOUR

TWILIGHT OF
THE GODS

SHORTLY AFTER DAWN on June 6, 1944 (D-Day), a huge armada of Allied warships approached the rainy, windswept coast of France and began disembarking thousands of landing craft filled with American, British, and Canadian soldiers. It was the start of Operation Overlord, the Allied invasion of German-occupied Europe. Under the cover of heavy fire from the battleships' big guns, the troops crossed the choppy waves toward the Normandy beaches, facing a withering hail of machine-gun fire from German pillboxes on the cliffs above. Although the first waves of infantry suffered heavy losses, by afternoon the Americans had seized a portion of two beaches, and the British three.

Commanding all American, British, Canadian, and French forces involved in Operation Overlord was the Supreme Headquarters Allied Expeditionary Forces (SHAEF), which had been established in February 1944 under Major General Dwight D. Eisenhower. Before the Normandy landings, SHAEF military planners had worried that Hitler might employ chemical weapons in a desperate attempt to repel the invasion. Allied beachhead operations would be concentrated in a relatively small area, providing an ideal target for chemical attack, and many of the invading troops did not even carry gas masks. As General Omar Bradley noted in his 1951 memoir, *A Soldier's Story,* "While planning the Normandy invasion, we had weighed the possibility of an enemy gas attack and for the first time speculated on the probability of his resorting to it. . . . I reasoned that Hitler, in his determination to resist to the end, might risk gas in a gamble for survival."

To deter the German use of chemical weapons, Roosevelt and Churchill warned of severe retaliation in kind and prepared to follow through with this threat. The two leaders ordered the stockpiling of a sixty-day supply of chemical bombs at depots in England and the training of air crews to deliver the weapons. In the event that the Germans unleashed a chemical attack, two Allied retaliatory operations could be mounted within forty-eight hours, each involving four-hundred-bomber formations that would deliver hundreds of tons of mustard and phosgene against German cities. Fortunately, the feared attack did not materialize. As General Bradley wrote in his memoir, "When D-Day finally ended without a whiff of mustard, I was vastly relieved. For even a light sprinkling of persistent gas on Omaha Beach would have cost us our foothold there."

Although the rainy weather on June 6 was far from optimal for chemical warfare, the effects would still have been disastrous. German use of mustard or Tabun against the Normandy beachhead might have repelled the Allied invasion of France and delayed another attempt for six months, possibly necessitating landings at a new location. In the event, however, the threat of massive retaliatory strikes successfully deterred any German use of chemical weapons on D-Day. Because the Allies had achieved air superiority, cities and factories throughout the Reich were now exposed to aerial attack. According to a report written after the war by General Ochsner, chief of the German Chemical Troops, "the initiation of gas warfare by us might have had incalculable consequences for our homeland if the enemy had decided to bomb our factories and communications facilities with gas, thus compelling us to carry out extensive decontamination work, not to mention the detrimental effect gas would have had on the morale of the population of big cities already severely stricken." Furthermore, the Wehrmacht faced logistical constraints on its ability to launch chemical attacks. Ochsner wrote, "It also had to be taken into account that supply transportation for the Atlantic front, which already was not functioning too well on account of enemy air superiority, would not have been able to cope with the additional load of material for chemical warfare."

Despite Hitler's forbearance at Normandy, British civil defense authorities believed that he would eventually resort to poison gas attacks against London and other cities in southeast England. Although the Germans no longer had enough aircraft to deliver chemical weapons across the English Channel, they had developed long-range "vengeance" weapons with which to terrorize the British capital. On June 13, 1944, one week after the Normandy landings, the Germans began to launch hundreds of V-1 flying "buzz bombs" at London and southeast England; those attacks were supplemented on September 6 with V-2 ballistic missiles, which were impossible to shoot down. The British authorities feared the potential use of the V-1 and V-2 to deliver chemical agents, creating a formidable weapon of terror. Responding to this threat, they distributed 30 million gas masks to civilians of all ages.

German scientists had indeed developed a proximity fuse for the V missiles that could detonate chemical warheads a few hundred feet above the ground, creating drifting clouds of lethal vapor. But although Wehrmacht planners considered delivering Tabun with the V-1 or V-2, they calculated that because of the missiles' limited payload, high-explosive warheads would cause more casualties than chemical ones. A liquid-filled warhead would also adversely affect the V-2's ballistics.

On July 20, 1944, a group of disgruntled German military officers led by Colonel Claus von Stauffenberg tried to assassinate Hitler with a time bomb as he met with his senior advisers in an aboveground conference room at "Wolf's Lair" in East Prussia. The explosive device was concealed inside a briefcase that, at the last minute, was moved by an aide behind a heavy support for the map table, deflecting the blast. Four people died in the explosion but Hitler was relatively unscathed, although it left him with two ruptured eardrums and persistent dizzy spells. Stauffenberg and the other ringleaders were arrested and summarily executed by firing squad in the courtyard of the Army headquarters in Berlin.

Although Hitler had survived the attempt on his life, the military situation was becoming increasingly dire. Half a million German soldiers on the western front had been killed, injured, or taken prisoner, and the remaining units had lost most of their armored vehicles and artillery. In the east, the German Army was also in retreat. On August 24, the Army General Staff proposed to the Supreme Command of the Wehrmacht that poison gas be used to halt the advance of the Red Army, with an emphasis on "those chemical agents that are based on entirely new German developments and hence are probably unknown to the enemy."

In September 1944, Robert Ley, a chemist by profession, tried to persuade Speer that the large-scale use of chemical weapons could stave off defeat and strike a decisive blow against the Soviet enemy. His proposal was to create a *cordon sanitaire* along the 750-kilometer German-Soviet front by contaminating the terrain with persistent mustard and Tabun, blocking the Red Army's advance into central Europe. A notorious drunk, Ley raised his idea over glasses of strong wine. "His increased stammering betrayed his agitation," Speer later wrote in his memoir. " 'You know we have this poison gas. I've heard about it. The Führer must do it. He must use it. Now he has to do it! When else? This is the last moment. You, too, must make him realize that it's time.' "

Appalled by Ley's drunken tirade, Speer remained silent. But Ley made a similar appeal to Goebbels, who in turn persuaded Hitler to hold a high-level conclave to discuss the possible use of poison gas. Despite his vacillations about chemical weapons, the grave military setbacks forced the Führer to reconsider their use. During the meeting, he speculated that because the British and the Americans had an interest in slowing the Soviet advance toward Berlin, the Western powers might tolerate German chemical attacks against the Red Army. Militating against this course of action, however, was the risk of Allied retaliatory strikes and the fact that limited stocks of Tabun were available. Production of the nerve agent had been considerably lower than expected because of severe shortages of raw materials, such as phosphorus and sodium cyanide. Furthermore, construction of the Sarin plant at Dyhernfurth was only about 70 percent complete. The manufacturing units involved in Steps I and II were operational, but Steps III through V were not. As a stop-gap measure, the Step II product was shipped in lead-lined iron tanks to Spandau, where the remaining steps were performed on a small scale. As a result, less than ten metric tons of Sarin were produced during the war.

The second Sarin production facility at Falkenhagen, which had been slated to go on line in mid-1945, was also far behind schedule. One reason for the delay was a conflict between the Anorgana officials who managed the Sarin plant and the Technical Office of the SS over an adjacent facility for the production of "N-Stoff" (chlorine trifluoride), an incendiary chemical. Although all three branches of the Wehrmacht had rejected N-Stoff as a useless weapon, Hitler had ordered the SS to reassess the chemical's utility as a filling for antiaircraft shells. The SS had reached a favorable verdict, and

Luranil had proceeded to build an N-Stoff plant at Falkenhagen with a capacity of fifty tons per month. In August 1944, SS chief Heinrich Himmler told his officers to seize control of the N-Stoff plant, leading to a confrontation with Anorgana. Ambros complained to Speer, who ordered that the plant be returned. The SS then retaliated by interfering repeatedly with the construction of the Sarin plant.

Meanwhile, the push by Ley and Goebbels to unleash Tabun against the Red Army encountered strong opposition from Speer. Once again, Hitler asked his armaments minister if he thought the Allies possessed stocks of nerve agents. Speer checked with Ambros, who said that his earlier assessment remained unchanged. The Wehrmacht General Staff also opposed any resort to chemical warfare. From the start of the conflict, the German military had resisted assimilating chemical arms into its doctrine, training, and logistics, creating major impediments to the use of such weapons on the battlefield. No Luftwaffe personnel, for example, had been trained to handle or deliver Tabun-filled bombs. According to Speer's testimony at the Nuremberg War Crimes Tribunal, "In military circles, there was certainly no one in favor of gas warfare. All sensible Army people turned gas warfare down as being utterly insane since, in view of [Allied] superiority in the air, it would not be long before it would bring the most terrible catastrophe upon German cities, which were completely unprotected." Even if gas masks had been distributed widely, they would have provided limited protection against skin-penetrating agents such as mustard.

Determined to end the debate over chemical warfare once and for all, Speer decided to take matters into his own hands. He had no intention of devoting scarce materials and skilled labor to the production of weapons whose use was uncertain and probably undesirable. On October 11, 1944, he drafted a message to Field Marshal Keitel describing the adverse effects of Allied bombing on German armaments production. "Due to the extraordinarily effective enemy attacks on our raw materials industry," Speer wrote, "a situation has arisen that, taking account of the current demands from the fronts, requires sharply cutting production of the most important chemical agents—Tabun and mustard—to the benefit of powder and explosives." He directed that Tabun production be cut back to 100 tons in October and halted entirely on November 1 unless an improvement in the supply of cyanide was achieved.

On November 2, SS Brigadier General Walther Schieber, the head of the

Armaments Supply Office in the Speer Ministry, convened a meeting with Dr. Karl Brandt and other senior Nazi officials to discuss the Speer directive. Brandt was strongly opposed to halting production of nerve agents and noted the standing order from Hitler that the manufacture of poison gas was not to be compromised under any circumstances. "We have in Tabun a new type of chemical agent that alone is capable, in the event of a massive enemy use of chemical agents, to provide an effective countermeasure that possibly could compel the adversary to halt his use of poison gas," Brandt said. "This applies in even greater measure to the more effective agent Sarin, whose more rapid production could help decide the outcome of the war and must be promoted with all available means. With respect to raw materials, Sarin is preferable to Tabun, because Sarin avoids the current constraints on the availability of cyanide."

With Brandt's encouragement, Schieber decided to resist the Speer directive. A few days later, however, Speer fired Schieber and replaced him with a more compliant bureaucrat who went on to halt the production of chemical intermediates for nerve agents. By the end of 1944, the Tabun plant at Dyhernfurth had ceased operation, and all artillery shells and bombs were henceforth filled with conventional explosives.

At the same time, the Nazi regime launched a systematic effort to cover up the nerve agent development and production program. Thousands of secret research documents and testing protocols were shredded, starting with those at the highest security classification, and scientists were ordered to burn their laboratory notebooks. All sensitive items were removed from Spandau Citadel, and the Tabun and Sarin pilot plants there were disassembled and shipped west to Raubkammer. IG Farben also undertook the massive shredding and burning of files, totaling some 15 tons of paper.

ON NOVEMBER 20, 1944, as the Red Army advanced on a broad front toward the eastern German provinces of East Prussia and Silesia, Hitler left his Wolf's Lair headquarters and returned to Berlin. In a desperate move, the Nazi regime had drafted all able-bodied men between the ages of sixteen and sixty who did not already serve in the armed forces into a homeland defense force called the People's Army (Volkssturm), but the poorly trained and equipped militia was little more than cannon fodder. By early January 1945, Soviet forces were approaching the Oder River, putting Dyhernfurth at risk. In two and a half years of operation, the factory had produced nearly 12,000 metric tons of Tabun, of which 10,000 tons had been loaded into 250-kilogram aerial bombs for the Luftwaffe and 2,000 tons into artillery shells for the Army.

In late January, the approach of Soviet troops triggered frenetic activity at Dyhernfurth, as if an ant colony had been breached. All remaining Tabun-filled munitions were loaded onto trucks and freight trains and shipped west to depots deep inside the Reich. Most of the bulk Tabun remaining in two underground storage tanks was pumped into the Oder River, and stocks of Product 39, the main intermediate used in Tabun production, were also destroyed. The entire factory was then prepared for demolition: a Pioneer commando brought in dozens of explosive charges that were laid at key points and wired to detonators. At the last minute, however, an urgent message arrived from the Army High Command in Berlin rescinding the demolition order. Hitler had changed his mind, apparently believing that German forces could recapture the factory intact and resume nerve agent production at a later date.

On January 24, shortly before the first vanguard of Soviet troops reached the Oder River, the Dyhernfurth director, Dr. Albert Palm, gave the order to evacuate all staff members and the 3,000 inmates of the two satellite labor camps. Over the previous two years, hundreds of forced laborers at the Tabun factory had died of exhaustion, malnutrition, disease, and toxic exposure. Now the survivors, in dirty striped uniforms, were compelled to march from the Dyhernfurth subcamp to the main concentration camp at Gross-Rosen. Emaciated to the point of resembling "walking corpses," they were driven and beaten by the SS guards, and those who collapsed by the side of the road were summarily shot.

As the ragged column passed through the town of Neumarkt, the sight of thousands of skeletal, foul-smelling prisoners, coatless in the bitter cold, aroused disgust and pity in the town's citizens. The physician Hildegard Staar and her husband asked one of the SS guards if they could give the inmates food, clothes, and medicine, but they were harshly rebuffed. Two residents of Neumarkt who defied the SS order not to provide assistance were arrested and later executed by local Nazi officials. By the time the forced laborers reached the Gross-Rosen concentration camp, two thirds of the original 3,000 had died or been killed. On February 11, the SS transferred the survivors from Gross-Rosen to Mauthausen concentration camp. The weak, ill-clothed prisoners were transported in open vegetable wagons in the subfreezing weather, causing scores to succumb to pneumonia. Because the Nazis wanted to eliminate all outside witnesses of nerve gas production at Dyhernfurth, the Gestapo tracked down the survivors at Mauthausen and murdered them.

IN EARLY FEBRUARY 1945, the Red Army occupied the village of Dyhernfurth and the nearby castle but halted for a few days on the banks of the Oder River because of overextended supply lines, indiscipline and drunkenness, and threats to their flanks. The Russians were unaware of the great prize that lay only a few kilometers from their garrison. At the abandoned Tabun factory, the manufacturing and filling lines were still intact, along with stocks of such raw materials as white phosphorus. Moreover, several gallons of Tabun remained in the production kettles and the two underground storage tanks.

Fearing that the Soviets would discover the factory, take samples of Tabun away for analysis, and thereby learn the secret of the nerve agents, the Nazi leadership decided to send in a special raiding party to clean and decontaminate the kettles and storage tanks before the Red Army arrived. To lead this sensitive mission, the Supreme Command of the Wehrmacht selected one of Germany's youngest general officers, Major General Max Sachsenheimer, thirty-two, the commander of the 17th Infantry Division. Under his control were several hundred infantry, supply troops, and a light Pioneer assault boat company with eighty-one motorized boats and three antiaircraft batteries.

The operation began in the morning darkness of February 5, 1945. Although the railroad bridge over the Oder had been partially destroyed by bombing, enough of the superstructure remained to be usable by the raiding party. After crossing the bridge, Sachsenheimer's troops followed the railroad tracks to the Tabun plant, a kilometer inland from the river. They were accompanied by two army chemists who had worked at the factory and were familiar with its layout, as well as eighty former plant technicians from Anorgana. After a tense 65 minutes, the commandos reached the Tabun factory and secured it. Then the Anorgana technicians, wearing gas masks and protective rubber suits, pumped the remaining Tabun into the Oder and scrubbed the storage tanks and kettles to remove any telltale residues of nerve agent.

Meanwhile, the rest of Sachsenheimer's troops went into action. A half hour after the first group of commandos had crossed the railroad bridge, the Pioneer company launched its assault boats across the Oder 2.5 kilometers downstream and seized the village of Dyhernfurth. This attack was a diversionary maneuver designed to draw the approaching Soviet forces away from the factory site. German infantry armed with panzerfaust antitank weapons then held off a series of uncoordinated Soviet counterattacks, including one at about 1:00 p.m. involving eighteen tanks.

By late afternoon, the decontamination work at the Tabun plant had been completed, and the raiding team withdrew. A few days later, the Red Army occupied the area. Although the Luftwaffe later tried to bomb the factory from the air, they were unsuccessful and the Soviets captured both the full-scale Tabun plant and the pilot Sarin plant intact.

UNABLE TO HALT the advance of the American and British forces in the west and the Soviets in the east, the Nazi leadership faced the dilemma of what to do with the thousands of tons of superlethal poisons it had manufactured and stockpiled. On February 2, 1945, Hitler ordered that "chemical agents and chemical munitions must not fall into enemy hands." Destruction of chemical weapons was permitted only when it could be carried out in an inconspicuous manner because of the risk that the Allies might perceive it as a deliberate attack.

Two days later, at Hitler's direction, Field Marshal Keitel issued an order to the general quartermasters of the armed forces stating that all chemical munitions stored at nine depots near Berlin, in the direct path of the Red Army, should immediately be transported west into the state of Lower Saxony. Priority was to be given to nerve agents, which "under no circumstances" were to be captured by the enemy. To accomplish this task, a massive transport operation was put into motion. Although trucks and fuel were in extremely short supply, both were made freely available for this purpose.

As the transfer of chemical weapons was getting under way, a final debate took place within the Nazi inner circle over whether to initiate their use. After the horrific firebombing of Dresden by British and American bombers on February 13–14, 1945, an outraged Goebbels, supported by Ley and Bormann, demanded that the Luftwaffe retaliate by drenching British cities with Tabun. On February 19 and 20, Hitler convened meetings with his generals to discuss whether Germany should respond to the burning of Dresden by formally withdrawing from the 1925 Geneva Protocol. This action would remove any legal constraints on the first use of chemical weapons and "demonstrate to the enemy that we are determined to use all means to fight for our existence." But the Navy's commander in chief, Karl Dönitz, and other senior officers warned that the risks of this action would outweigh the benefits and that resorting to chemical warfare would not significantly delay the end of the war. Once again, Hitler put off making a final decision.

By late February, the military situation had grown so desperate that the Nazi leadership was more preoccupied with safeguarding its arsenal of nerve agent weapons than with planning for their use. Pressure was building on the eastern front as the Red Army surged forward. General Heinz Guderian, the Army chief of staff, told Foreign Minister Joachim von Ribbentrop that the war was lost. Informed by Ribbentrop of Guderian's remarks, Hitler summoned the Army general and warned him angrily that such statements were tantamount to treason.

At the end of February, the Red Army was eight kilometers from the unfinished Sarin plant at Falkenhagen when director Klenck ordered a general evacuation. The production line was partially dismantled and corrosion-resistant pieces of apparatus that were silver-lined or made of solid silver were transferred to the Sarin pilot plant at Raubkammer. Beyond the considerable intrinsic value of this equipment, it may have been salvaged with the possible intent of resuming production of nerve agents after the official end of hostilities, perhaps by guerrilla bands such as the Freikorps Adolf Hitler, founded by Ley, or the Werwolf, under Bormann. For Nazi partisans, even small amounts of Sarin produced in a pilot plant could be employed for terrorist attacks against the occupying armies. Due to the bombardment of Berlin, Division 9 and the Army Gas Protection Laboratory also moved to Munster for the remainder of the war.

One of the last messages to reach Falkenhagen before the plant was evacuated was an order from Ambros to destroy all classified materials. In fact, Klenck did not follow this order completely. He took several thousand pages of secret documents with him when he fled to Heidelberg in March 1945, possibly with the intent of using them as bargaining chips with the Allies. The classified materials in his possession included several laboratory notebooks, five packets of contracts between IG Farben and the Army High Command, information on Dyhernfurth, and production sheets for Tabun and other gases.

Klenck went to the Villa Kohlhof, the IG Farben guesthouse outside Heidelberg, where he met up with Ambros. On March 26, the two men traveled to the Bavarian town of Gendorf, where Anorgana had built a multipurpose chemical plant that produced mustard agent as well as paints, cleaning powder, and other commercial products. Klenck and Ambros supervised the conversion of the plant from war to peace production, including the removal of specialized equipment for the manufacture of mustard. Meanwhile, without informing Ambros, Klenck placed the secret documents from Falkenhagen in a steel barrel, which he told the chief of the Anorgana-Gendorf fire brigade to bury on an isolated farm six miles outside of town. (The documents were later discovered by Allied intelligence.)

WITH THE SPECTER of defeat looming, Hitler and Goebbels began to advocate a "scorched earth" policy under which all military, industrial, transportation, and communications facilities throughout the Reich would be destroyed to prevent them from falling intact into enemy hands. Hitler no longer cared about the welfare of the German people, whom he believed had not fought bravely enough and were undeserving of his leadership. Instead, he would transform Germany into a desert to deprive the Allies of the spoils of conquest.

Speer, who would soon turn forty, faced an acute dilemma. For eleven years, he had grown rich and powerful under Hitler's wing while exploiting hundreds of thousands of slave laborers in his armament factories. But even he could not fail to grasp the wanton destructiveness of the dictator's "scorched earth" proposal. Hoping to avoid a further waste of lives and hasten the war's inevitable end, Speer conceived the idea of using Tabun to assassinate Hitler and the other top Nazi leaders. Whenever Hitler was in Berlin, he routinely held nighttime military conferences in his spacious underground bunker fifty feet beneath the marble halls of the Reich Chancellery building, which Speer had designed. The meetings in the Führerbunker were attended not only by the generals of the Supreme Command but also by the Nazi inner circle, including Göring, Himmler, Goebbels, and Ley.

Ever since the failed attempt on Hitler's life on July 20, 1944, no one was allowed to approach the underground entrance of the bunker without being searched for weapons and explosives. To get around this problem, Speer devised a plan to introduce Tabun into the bunker's external air intake, which projected into the garden of the Chancellery building. During walks in the garden, he found the opening of the ventilation shaft, which was level with the ground, covered by a thin iron grate, and hidden by a small shrub. Speer believed that if a fine mist of nerve agent could be introduced into the external air intake at the time of one of Hitler's meetings, the lethal vapor would spread rapidly through the bunker's ventilation system, killing all those inside.

On February 20, 1945, Berlin experienced a major air raid, and Speer waited out the attack in a starkly furnished concrete bomb shelter together with Dieter Stahl, the head of munitions production in the Ministry of Armament. The two men had recently become close after Stahl had been overheard making a defeatist remark about the impending end of the war and had been interrogated by the Gestapo. Fearing the worst, he had appealed for help to Speer, who had managed to quash the investigation. The two men now held a frank conversation about the folly of Hitler's policies, which they agreed were leading the nation to disaster. Stahl gripped Speer's arm and murmured, "It's going to be frightful, frightful!"

At this point, Speer inquired discreetly about the poison gas Tabun and whether Stahl might be able to obtain a small supply of it. After a pause, Speer blurted out, "It is the only way to bring the war to an end. I want to try to inject the gas into the Chancellery bunker." Although Speer was shocked by his own frankness, Stahl seemed unperturbed and promised soberly to investigate ways to obtain the agent. Several days later, in early March, Stahl told Speer that he had consulted with the head of munitions at the Army Ordnance Office, who had told him that any midlevel employee at a weapons depot would have access to Tabun-filled artillery shells. Stahl had learned, however, that Tabun was effective only when vaporized by an explosion. This property made it impractical for Speer's assassination scheme because the blast would shatter the thin-walled ventilation duct.

Despite this setback, Speer decided to pursue the plan using a more traditional agent, such as phosgene. He met with Johannes Hentschel, the chief engineer at the Chancellery, and told him that Hitler had complained about the poor air quality in the bunker. At Speer's request, the air filters were removed for replacement, leaving the bunker temporarily unprotected. A few days later, however, Speer returned to the Chancellery garden and was stunned to discover that a series of new security measures had been put into effect at Hitler's personal order. Armed SS guards stood on the roofs of the buildings, searchlights had been installed, and the air intake for the underground bunker had been covered by a cylindrical metal chimney more than ten feet high.

It was now impossible for Speer to inject poison gas into the bunker without being detected by the guards patrolling the garden. Hitler had ordered the installation of the chimney not because he suspected an assassination plot but because he feared that Red Army troops attacking Berlin would fire chemical shells at the Führerbunker. His worst nightmare was that the Soviets had developed a knockout gas that would render him unconscious, allowing him to be captured alive "like a stunned animal in the zoo." Poison gas, being heavier than air, would not penetrate the ten-foot chimney.

Speer was actually relieved that the assassination plot had been thwarted. For several weeks he had lived in a state of constant tension, fearing that the plan would be exposed and that he, his wife, and his six children would suffer terrible consequences. Years later, he admitted in his memoirs that his strong personal feelings for Hitler would have prevented him from carrying out the attack. "The whole idea of assassination vanished from my considerations as quickly as it had come," he wrote. "I no longer considered it my mission to eliminate Hitler but to frustrate his orders for destruction. That, too, relieved me, for all my feelings still existed side by side: attachment, rebellion, loyalty, outrage. Quite aside from all question of fear, I never could have confronted Hitler pistol in hand. Face to face, his magnetic power over me was too great up to the very last day."

On March 15, Speer prepared a memorandum opposing the "scorched earth" plan, which he delivered to the Führer in person on the evening of March 18. Unmoved by Speer's plea, Hitler said coldly, "If the war is lost, the nation shall also perish." The next day he issued a directive for the systematic demolition of German towns and factories, dams and bridges, food and clothing stores, railways, ships, and trains. Speer resolved to do everything in his power to countermand Hitler's orders.

ON MARCH 28, the British armies, under the command of Field Marshal Bernard Montgomery, crossed the lower Rhine River and headed northeast toward Bremen, Hamburg, and the Baltic Sea. Montgomery was relieved that the German defenders did not resort to chemical warfare, as it was their last tactical opportunity to do so. Even so, he feared that Hitler might still unleash poison gas in a final act of desperation.

The next day, Field Marshal Keitel gave the order to remove Germany's most modern chemical munitions—those filled with nerve agents—from depots threatened by the enemy to more secure locations. Keitel's order read as follows:

I. The components of the Armed Forces must ensure that as the enemy advances, the "special" chemical agents (Sarin and Tabun) . . . , which are presumably not known to the enemy, must under no circumstances fall into his hands, but must be removed from storage facilities before his arrival. Older chemical agents known to the enemy are to be removed only as a secondary effort. When, in exceptional cases, this is not possible, they are to be abandoned as necessary.

II. Never is chemical agent to be identified as such; any markings to this effect must be removed so as not to call the enemy's attention to them.

III. The storage space needed for transportation of the munitions must be made available immediately by the Army Transport Chief and the Army components, as long as it is not required for operational purposes.

IV. Destruction of chemical agents and munitions should be undertaken in such a way as not to give the enemy an excuse to claim that Germany has initiated chemical warfare.

v. Chemical production facilities and storage bunkers are to be destroyed when threatened by the enemy. In so doing, it is essential that even after destruction, the enemy is not able to obtain information on the type and composition of the chemical agents and munitions produced and stored there.

The rapid advance of Allied forces created numerous problems with the evacuation of nerve agent munitions. In addition to the lack of cargo capacity, the bombing of railway lines hampered transportation. Damaged tracks forced the unloading of several trains in the middle of the stretch, and others had to be abandoned on secondary rail lines. Meanwhile, American forces were approaching the industrial Ruhr Valley in a pincer movement, with the Ninth Army to the north and the First Army to the south. On April 1, the two armies converged, trapping twenty-one German divisions in the Ruhr and tearing a 200-mile gap in the German front, through which the Allies advanced toward central Germany. General Eisenhower planned for the American forces to meet up with the Soviets on the Elbe River south of Berlin, cutting Germany in two.

By April 2, it was clear that Field Marshal Keitel's order to evacuate all nerve agent munitions to secure depots was no longer practicable. Not only were the necessary means of transport lacking, but Germany no longer controlled weapons depots that were inaccessible to the enemy. Sinking munitions in rivers or lakes had been ruled out because of the risk of contamination or discovery. The only alternative was to move the weapons by barge. At Hitler's direction, Keitel ordered the evacuation of Tabun-filled bombs and shells stockpiled in Silesia and their loading onto barges on the Elbe and Danube Rivers.

Despite the desperate military situation, the Wehrmacht managed to organize a major riverine transport operation for the chemical munitions. The chief of the Transport Department requisitioned hundreds of freight barges, while the Reich transport minister and the general inspector for water and energy arranged for suitable anchorages. Officers from the SS, Army, Luftwaffe, and Navy took extraordinary measures to carry out the evacuation order, including the provision of air cover and military police to guard the loaded barges.

One of the Luftwaffe's largest munitions depots was near the town of Lossa in the eastern German state of Thuringia. This depot contained several thousand aerial bombs filled with Tabun. On April 5, 1945, shortly after the entry of American forces into Thuringia and their expected advance along the Eisenach-Erfurt-Jena highway, a top secret "flash" telegraph signed by Hitler arrived at Luftwaffe headquarters. The coded message read, "The Führer has ordered the immediate evacuation of K-Muna [chemical munitions depot] Lossa, north of Kolleda, on the Kolleda rail line. All stocks of III-Green aerial bombs are to be removed immediately." The designation "III-Green" referred to the three green rings painted on the bombs to indicate that they were filled with Tabun.

A total of eleven trainloads would be needed to evacuate the special chemical munitions from Lossa. According to the plan, two trains per day would transfer the weapons to the outskirts of Torgau, a major transportation hub on the Elbe, where the munitions would be stockpiled in the open air until the requisitioned barges had arrived at the docks. Shortly after receiving Hitler's order, the general quartermaster of the Luftwaffe, Lieutenant General Dietrich Georg von Crigern, contacted Field Marshal Keitel to express his deep concern about the plan. In view of the acute threat of enemy air raids, he warned, the large-scale transport of chemical munitions posed an unacceptable risk to the civilian population. But over Crigern's objections, the operation went ahead as planned.

On April 8, 1945, as Tabun-filled bombs were being loaded into a freight train at Lossa station, a pair of U.S. Thunderbolt fighter-bombers swooped out of the sky and flew low over the town with a deafening roar. The planes strafed the station platform with heavy machine-gun fire and then dropped high-explosive bombs onto the train cars and the exposed crates of munitions, pulling up sharply at the end of the run. Moments later, orange fireballs blossomed over the station with a series of thunderclaps, and the destroyed train burned fiercely. Much to the pilots' surprise, the munitions did not erupt with large secondary explosions. Instead, liquid Tabun streamed from the ruptured bombs, forming a shallow lake that spread over the train-station landing. Four town residents who were near the station at the time of the attack were overcome by the lethal vapors and died in convulsions within minutes. Thus, in a dark irony, the first combat deaths from Tabun were German civilians.

The consequences of the Lossa attack were so grave that General Crigern reported the incident immediately to Field Marshal Keitel, despite the fact that hundreds of more important strategic targets were also being hit by Allied bombers. German Army chemical troops wearing gas masks and rubberized suits cordoned off Lossa station and evacuated the entire population from a radius of twenty kilometers. The troops spent the next twenty-four hours decontaminating the area.

Even in the face of the Lossa disaster, the Nazi leadership was determined to continue loading the Tabun-filled bombs onto river barges. The special munitions were diverted to a train station in the nearby town of Bill-roda, from which the transport to Torgau continued. One modification was made in the weapons evacuation plan, however. In addition to Torgau, Hamburg harbor had been designated as a hub for loading nerve agent munitions from depots in Lübbecke and Walsrode onto barges on the Elbe. After the Lossa disaster, the Reich defense commissar in Hamburg expressed strong objections to this part of the plan, and Field Marshal Keitel agreed to move the loading zone outside the city limits.

ON APRIL 9, the Wehrmacht Supreme Command sent a certain Captain Hemmen from the Army quartermaster's office on a one-week inspection tour of the northern region of Germany, which remained under Nazi control. Hemmen's mission was to verify that Hitler's orders for the evacuation of nerve gas weapons were being properly executed. On April 10, when the captain arrived at the Army Munitions Depot in Walsrode, the facility had been taken over by a German Army group. Logistics officers had arranged for ten railroad cars to transport Tabun-filled munitions to Nordenham and had requisitioned another twenty-nine cars. At Nordenham, the bombs would be off-loaded onto barges, which would be tugged to the mouth of the Weser River and brought into a holding area in the vicinity of the Elbe.

On April 11, Hemmen reported to Berlin that fighting had disrupted key rail and road links, making the chemical weapons transport operation increasingly chaotic. Breakdown of the telephone network was also impeding the delivery of orders. According to one report, a train loaded with Tabun-filled weapons from the Army depot at Lübbecke in Westphalia had fallen into Allied hands. Some German commanders abandoned their depots while others buried munitions, left transport trains standing in mid-stretch, or sank containers of bulk agent in rivers.

On April 15, the general quartermaster of the German Army, Major General Alfred Toppe, reported that no more barges were available for the transport of chemical weapons. Instead, the munitions would have to be stored on trains in inconspicuous locations or towed out to sea and dumped to create more space on the existing barges. The next day, Hitler told Field Marshal Keitel that nerve gas weapons must not be disposed of under any circumstances. If the barges could not be towed downstream, they should be moored outside cities and harbors, away from the likely targets of enemy air raids. Once again, this order reflected Hitler's determination to retain control over the weapons to the bitter end.

Hitler planned to leave Berlin on April 20, 1945, his fifty-sixth birthday, and travel to the Berghof, his mountain retreat on the Obersalzberg above the town of Berchtesgaden, in the southeast corner of Bavaria. Long convoys of trucks loaded with state papers and anxious officials had already fled south, and Hitler had sent most of his household staff to the Berghof to await his arrival. During the early years of Nazi rule, the area around Hitler's villa had been transformed into a highly secure zone called the Leader's Territory (Führergebiet), which contained summer homes for members of the Nazi inner circle, administration buildings, SS guard barracks, a hotel for visiting dignitaries, worker housing, and tunnel and bunker complexes. In addition, a small concrete and granite building known as the Kehlsteinhaus ("Eagle's Nest") perched atop a nearby peak. It was widely believed that the Nazi regime would make its last stand in this "Alpine Redoubt." Rumors were rife that thousands of trained German troops, Werwolf partisans from the Hitler Youth, weapons stockpiles, and even armaments factories had been hidden in the mountains of southern Germany and Austria to defend the Führer. Goebbels also warned that the Nazi last stand would involve the use of a "wonder weapon."

As the planned April 20 departure date approached, however, Goebbels urged the Führer to remain in Berlin. Hitler agreed, believing against all evidence that the Soviets would suffer a bloody defeat. His generals were under no such illusions, however, and during the regular military conference in the Führerbunker after Hitler's birthday party, they urged him to leave Berlin immediately for the south, warning that the Red Army would cut off the last escape route within days. Although Himmler and Göring decided belatedly to flee, Hitler hesitated, unwilling to admit that the capital of the Thousand-Year Reich was about to fall. His one concession was to establish separate northern and southern military commands in case the Allies cut Germany in half.

Meanwhile, thousands of nerve agent weapons stacked on barges were traveling down the Danube to Passau in southeastern Bavaria, from whence they were towed down the Inn River and the Alz River to Lake Chiemsee, west of Salzburg. In one incident during the final days of the war, American artillery units began to shell four barges on the Danube and were stunned when the German officers on board quickly raised the white flag of surrender. After being taken prisoner, the Germans explained that the barges held chemical agents that might be released by the shelling, with disastrous consequences for the surrounding area. In this way, the cargo of Tabun-filled bombs fell into American hands.

The final destination of the chemical transports was a large munitions depot outside the village of Sankt Georgen, about ten kilometers east of Lake Chiemsee and close to Hitler's retreat in the Bavarian Alps. This depot comprised hundreds of sod-covered bunkers camouflaged to blend in with the adjacent pines and filled with orderly stacks of chemical bombs and shells.

On the afternoon of April 25, 1945, the American and Soviet armies met at the Elbe, trapping the remaining Nazi leaders in Berlin. Five days later, determined not to be captured alive, Hitler shot himself in the Reich Chancellery bunker at 3:30 p.m. His suicide precluded a Nazi last stand at the "Alpine Redoubt" that might have involved the use of chemical weapons, including Tabun, causing thousands of additional deaths. In the words of the historian William L. Shirer, the German dictator had planned a Wagnerian finale in which he would go down "like Wotan at Valhalla, in a holocaust of blood—not only the enemy's but that of his own people."

Whether this scenario was realistic or an elaborate bluff remains a matter of historical debate. The Führergebiet on the Obersalzberg was not a well-prepared defensive complex, and toward the end of the war, German resistance was crumbling so rapidly that it is doubtful that a last-ditch use of nerve gas would have posed a major threat to the Allied armies and later to the occupation troops. In the last days of the Third Reich, Hitler was living in a fantasy world, directing from his bunker a series of phantom divisions that had already been decimated in the Soviet drive to Berlin.

At midnight on May 8, 1945, the guns finally fell silent. During the five years and eight months since Hitler had ordered the invasion of Poland, tens of millions of soldiers and civilians had been slaughtered. Yet even as the victorious Allies celebrated the end of the long, bloody conflict, they began to compete among themselves for Germany's precious trove of military secrets—including those of the nerve agents.

CHAPTER FIVE

FIGHT FOR
THE SPOILS

As the Allied forces advanced into German-occupied France after the June 1944 landings on the Normandy beaches, British and American technical intelligence teams followed close behind them, scouring the newly liberated areas for arms caches, military laboratories, and information on German breakthroughs in various fields of science, industry, and weapons development. U.S. technical experts were attached to military intelligence units called "T Forces" and assigned targets for collection. The Army Counter-Intelligence Corps, for example, deployed thirty-five mobile microfilm teams to photograph captured German documents. Another U.S. technical intelligence unit called the ALSOS Mission had the special task of investigating Germany's unconventional weapons programs: nuclear, chemical, and biological.

Working in parallel with the Americans, British experts collected technical intelligence under the auspices of a separate organization, the British Intelligence Objectives Sub-Committee (BIOS). The poor coordination between the U.S. and British collection efforts often resulted in confusion and redundancy. In an attempt to address this problem, on August 21, 1944, SHAEF established the Combined Intelligence Objectives Subcommittee (CIOS), which was based in London and staffed jointly by American and British officers. Extending through the various Army groups and combat elements of SHAEF, the CIOS administrative organization compiled the latest information on potential intelligence targets and dispatched field teams and investigators to capture and interrogate German scientists, discover and microfilm important technical documents, and confiscate useful equipment from German laboratories and factories. As its peak, CIOS coordinated the work of more than 10,000 U.S. and British intelligence personnel scattered over France, Belgium, the Netherlands, Luxembourg, and Nazi Germany. Although CIOS existed for only eleven months, it investigated thirty-three different industrial fields and compiled and published more than 1,200 reports.

On August 27, 1944, two days after the U.S. Army liberated Paris, the Chemical Warfare Service (CWS) technical exploitation team arrived at the French capital to begin its work. During the weeks leading up to D-Day, CWS intelligence officers had been dispatched to London, where they were joined by technical specialists from the Navy and the Army Air Forces and civilians from the Office of Strategic Services (the predecessor of the CIA). Heading the group of fifty chemical weapons specialists was Lieutenant Colonel Paul R. Tarr, the head of the CWS Intelligence Division for Europe. Attached to the Seventh Army, the CWS intelligence team had orders to search for German chemical warfare scientists, weapons stocks, and production facilities. Tarr and his colleagues traveled across liberated France in armored personnel carriers, following closely on the heels of the advancing U.S. combat troops. They drove through bomb-damaged towns and villages, over pontoon bridges, and past endless columns of refugees, finally crossing the Rhine at the end of March 1945, only hours before the final collapse of the German defenses.

Tarr's top priority was to arrest and interrogate German military scientists and officials who had participated in the chemical weapons program. The team inspected the IG Farben factories at Elberfeld and Leverkusen and then traveled south to the plant at Ludwigshafen, which had suffered heavy bomb damage. At each site, Tarr and Edmund Tilley, an Army Air Forces investigator who spoke fluent German, ordered cowed IG Farben officials to tell them where the company's top scientists were hiding. Anyone who refused to cooperate was arrested and jailed.

ON APRIL 16, 1945, British soldiers from Field Marshal Montgomery's 21st Army Group occupied the chemical warfare experimental station at Raubkammer, which the Germans had abandoned nearly intact. There the British found a wealth of information. One month earlier, files and equipment from the German Army's Gas Protection Laboratory at Spandau

Citadel in Berlin, including a pilot plant with silver-lined reactors and pipes, had been transferred to Raubkammer to avoid capture by the Red Army.

About four kilometers from Raubkammer, British troops discovered the German Army Munitions Depot at Munster-East. This installation consisted of dozens of wooden buildings and concrete bunkers filled with more than 100,000 chemical-filled artillery and mortar shells. Fuses and other components had been carefully packed into boxes, crates, and wicker baskets with leather straps and handles. About the same distance from Raubkammer in another direction was the Luftwaffe Munitions Depot at Oerrel, comprising 175 concrete bunkers concealed in a pine forest that held aerial bombs containing toxic agents. The bunkers were so well protected that only a direct hit with a 1,000-pound bomb would have done much damage. Many of the weapons stored at Munster-East and Oerrel had been painted with a single yellow ring, the standard marking for munitions filled with mustard agent. But there were also stacks of 105 mm shells and 250-kilogram bombs that were painted with three green rings—a novel marking that mystified the British chemical officers.

On April 23, a CIOS chemical weapons intelligence team consisting of ten British and nine American and Canadian specialists, led by Commander A. K. Mills of the British Ministry of Aircraft Production, examined the unidentified German munitions. With the cooperation of German chemists from Raubkammer who had been taken into custody, British experts from the Chemical Defence Experimental Establishment at Porton Down analyzed the contents of one of the German 250-kilogram bombs. The munition was 64.5 inches long and had a set of three green rings painted on the nose, and another set around the middle. Stenciled on the body of the bomb was the code number KC 250 III Gr. The central burster tube contained fourteen pellets of TNT, but the impact fuse was missing; the fuse hole, covered by a Bakelite disk, was filled with cardboard and paper packing.

The Porton chemists opened the charging hole on the side of the bomb and carefully extracted a sample of the chemical fill, a dark brown liquid containing 20 percent chlorobenzene. Initially the British experts suspected that the substance, which they designated "T-2104," was a new blister agent similar to mustard gas. Working despite pinpoint pupils caused by accidental exposure to the agent vapor, they analyzed the compound and tested it in laboratory rabbits, watching in stunned fascination as the animals convulsed and died. According to the scientists' laconic report, "The shells were found to contain a markedly potent and hitherto unknown organophosphorus nerve agent." Chemists with the U.S. Army's 45th Chemical Laboratory Company also analyzed Tabun in their mobile laboratory and came to the same conclusion.

The realization that the Germans had secretly developed and produced a new chemical warfare agent of unprecedented power came as a terrible shock to the Allies. Although a few intelligence reports from 1943 and 1944 could be interpreted in retrospect as having referred to a German nerve agent, they had contained no firm evidence or tangible clues about its composition. The U.S. and British Chemical Warfare Services and the U.S. National Defense Research Committee had prepared and investigated approximately 150 organophosphate compounds during the war, but none had approached the lethality of the new German agent. Tabun was at least five times as toxic as DFP, the most effective nerve agent developed by the British.

Had Germany employed Tabun on the battlefield or for strategic attacks against British cities, there was little doubt that the initial effects would have been devastating. According to a 1951 report to the British War Office by Lieutenant Colonel D.J.C. Wiseman, "Although the respirator gives complete protection from [Tabun] vapour to the eyes and lungs, the difficulty of recognition would have been considerable, and the danger from the liquid remained. Only battle experience would have shown the degree of effectiveness of these gases, but from laboratory experiment and extrapolation from animal results they obviously possess great potentialities for the future."

AT THE BEGINNING of May 1945, the U.S. and British governments began to implement a top secret joint operation to arrest and interrogate the elite of German military science and industry. Known as Operation Dustbin, this long-standing plan involved the establishment of a special holding and interrogation center at Kransberg Castle, a medieval fortress in the Taunus region, north of Frankfurt, that had formerly served as Reichsmarshal Hermann Göring's headquarters.

Over the next several weeks, the Americans and British took numerous German military scientists into custody. At the end of June, the first group of scientists arrived at Kransberg Castle, which had been code-named

"Dustbin." The internees lived in relative comfort: they were housed in the former servants' wing, were allowed to move freely about the castle area, consumed adequate amounts of American army rations, and even organized scientific lectures and a weekly cabaret show.

In addition to luminaries such as Albert Speer, Fritz Thyssen, and Wernher von Braun, the list of German scientists at Dustbin included a large number of chemical weapons specialists, among them Wilhelm Kleinhans, Gerhard Ehlers, Heinrich Hörlein, Walter Hirsch, SS Brigadier General Walther Schieber, and most of the chemists and technicians from the Anorgana factory at Dyhernfurth. Numerous American and British intelligence teams visited the internment camp and interrogated individuals of particular interest, often repeating the same questions. Meanwhile, the Allies continued to search intensively for the remaining German chemical warfare specialists who remained at large, including Gerhard Schrader, Otto Ambros, and Albert Palm, the director of Dyhernfurth.

On a sunny Sunday morning in early summer, the British arrested Schrader at his home outside Leverkusen and took him to Dustbin. Always a practical man, he decided that it was now in his interest to cooperate with Germany's new rulers. Under interrogation, he emphasized his scientific motivation for working with the organophosphate compounds, deflected responsibility to Gross and Hörlein for the military applications of Tabun and Sarin, and emphasized his lack of influence over the German Army's decision to mass-produce them. Schrader also shared with CIOS investigators the secret chemical formulas of the nerve agents, described the Tabun production facility at Dyhernfurth, and conveyed the disturbing news that it had fallen into Soviet hands. CIOS published a special report based on the interrogation of Schrader and Gross titled "A New Group of War Gases," which provided a detailed account of the discovery of the nerve agents.

THE FIRST VANGUARD of American troops that rolled into Gendorf, Bavaria, in the spring of 1945 discovered a carefully camouflaged IG Farben chemical plant on the outskirts of the town. On reaching the factory, the lead tank smashed unceremoniously through the plant's entrance, knocking over one of the brick gateposts and trailing yards of wire fence behind it. As the GIs secured and inspected the administration building, they encountered a well-dressed German in his midforties who greeted them in remarkably fluent English. Sporting a neatly groomed mustache, he said that his name was Otto Ambros and that he had no rank or serial number—he was just a "plain chemist" who worked for the IG Farben company. The American officers who met with Ambros found him to be witty, intelligent, and charming.

A few days later, an advance detail of General George Patton's army arrived in Gendorf, and the commanding officer ordered Ambros held for questioning. During the interrogation, the German said that he was the manager of the chemical factory, which produced soap, cleaning powder, paint, and other commercial products. Ambros showed the Americans spectrum cards displaying the many-hued paints manufactured at the plant. Although one part of the factory (the area that had produced mustard) was underground, Ambros denied that it had a military role and noted that if it did, the Allies would have bombed it. He also tried to win over the American troops by handing out bars of soap, cans of cleaning powder, and paint for their vehicles.

It soon became known that Ambros was a person of considerable interest to Allied intelligence, and on May 29, the special Army intelligence unit (G-2) at SHAEF ordered him sent to Dustbin for interrogation. The transfer was delayed, however, while the internment camp was being organized at Kransberg Castle. During this period, Ambros remained under house arrest in Gendorf. Every few weeks, senior U.S. commanders passing through the town had the genial German chemist brought in for questioning. On learning that Ambros was in Gendorf, Lieutenant Colonel Tarr traveled to Bavaria to meet with him.

By this time the dark side of Ambros's personality had become known, including his alleged exploitation of concentration-camp inmates. He had overseen IG Farben's synthetic rubber plant at Auschwitz, where some 30,000 forced laborers had worked until they died or were deemed unfit and sent to the gas chambers. Nevertheless, Tarr's only interest in Ambros was to extract his valuable scientific knowledge, and the two men developed a friendly rapport. The German chemist responded at length to the American officer's technically informed questions, providing a wealth of information about the organization, structure, and capabilities of the German chemical weapons program.

Another American technical intelligence team from the ALSOS Mission interviewed the organic chemist Richard Kuhn, the inventor of Soman. The two investigators, Professor Louis Fieser of Harvard University and Carl Baumann of the University of Wisconsin, were chemists who had worked in Kuhn's laboratory before the war. When they arrived at the Kaiser Wilhelm Institute for Medical Research in Heidelberg, Kuhn greeted them warmly as former colleagues. During the interrogation, the German Nobel laureate claimed that he had not been involved in military research and had spent the war years working on the chemistry of modern drugs. But given Kuhn's close ties with senior Nazi officials and his chairmanship of the German Chemical Society, other ALSOS investigators did not find his denials credible and ordered him kept under surveillance.

DURING WORLD WAR II, the budget of the U.S. Chemical Warfare Service had risen from $2 million in 1940 to $60 million in 1941 and $1 billion in 1942. The CWS had acquired a vast production capacity for mustard, phosgene, and other toxic agents by building thirteen manufacturing facilities, including Pine Bluff Arsenal in Arkansas and Rocky Mountain Arsenal in Colorado. But the fact that poison gas had never been used in battle meant that the chemical troops had been relegated to manning flamethrowers and laying down smoke screens. Now, however, the remarkable properties of the German nerve agents revived the lagging military interest in chemical warfare.

Major General William N. Porter, the chief of the Chemical Warfare Service, ordered the CWS Development Laboratory, located on the campus of the Massachusetts Institute of Technology (MIT) in Cambridge, Massachusetts, to investigate methods for detecting the German nerve agents as well as the effectiveness of standard gas masks in protecting against them. The CWS Development Laboratory had been established in December 1941 under the direction of Captain Jack H. Rothschild and was housed in a new building that had recently been completed for MIT's Department of Chemical Engineering. With the rapid expansion of wartime chemical research, MIT had agreed to construct the laboratory on an accelerated schedule so that the Army could use it for the duration of the war. Staffed with handpicked scientists from the main CWS research and development center at Edgewood Arsenal in Maryland, the MIT facility conducted cutting-edge research on gas mask filters and other areas of chemical defense.

On May 15, 1945, two chemists at the CWS Development Laboratory, Captain Robert D. Coombs III and First Lieutenant Charles W. Sauer, began a study of Tabun, which was completed three and a half months later. The two men purified and analyzed the German nerve agent and found that standard Allied respirators provided complete protection against it. They also assessed the ability of existing U.S., British, and German indicator papers and detector tubes to recognize the new agent. Although mustard gas detectors worked for liquid Tabun, they could not reliably detect the cloud of fine droplets and vapor released by the explosion of a chemical shell. Thus, until an automatic vapor detector could be developed, U.S. troops would have to don gas masks at the beginning of a heavy artillery bombardment and wear them for long periods, impairing their fighting ability.

On May 29, General Porter requested the immediate shipment to the United States "by air, under highest priority," of five German 250-kilogram bombs so that their Tabun filling could be used to charge chemical mortar shells for field testing. The chief of the CWS also asked the Army Air Forces and the chief of ordnance to determine the combat usability of the captured German munitions. After VE-Day, some 530 tons of German chemical munitions, including 3,000 bombs and 5,000 artillery shells, were shipped to the United States to provide adequate stocks of Tabun for large-scale testing.

IN PARALLEL WITH the Americans, the British conducted their own scientific investigations of the German nerve agents. On May 23–24, 1945, a few weeks after the German surrender, the British General Staff sent to Raubkammer a team of fifty chemical weapons experts from the Chemical Defence Experimental Establishment at Porton Down. Known as the "No. 1 Porton Group," this team was to assist the SHAEF authorities in uncovering stocks of German chemical weapons, interrogating key personnel, and confiscating documentary records. Civilian scientists from Porton were granted temporary Army commissions and wore beige uniforms, although they could be distinguished from the regular officers by the absence of badges on their berets.

Some of the German scientists who had worked at the proving ground were willing to assist the British investigators. They revealed the location of a large buried cache of microfilmed research documents, which was recovered and sent back to England for analysis. German chemical engineers also helped to reactivate the Tabun and Sarin pilot plants at Raubkammer and explained the production process, which was documented in manuals. The pilot plants were later disassembled and shipped to Porton Down.

A key task for the Porton scientists was to determine how effective the German nerve agents would have been, had they actually been used on the battlefield. Although a few Tabun-filled bombs were sent back to England for examination, it was more convenient to carry out the field trials at Raubkammer, which had a large testing range that was better suited to the German weapons. Accordingly, the Porton team spent four months there, beginning in June 1945. In addition to static weapons testing, British Typhoon fighter-bombers practiced dropping Tabun-filled bombs and using spray tanks to contaminate large swaths of terrain from the air. Visiting parties from Porton Down arrived every week to tour the experimental station and observe the field trials.

AT THE SAME time that the British and the Americans were carrying out Operation Dustbin, the Soviets also made the hunt for Nazi military secrets a top priority. The Red Army's Sixteenth and Eighteenth Chemical Brigades had captured the nerve agent factory at Dyhernfurth nearly intact, along with substantial quantities of raw materials, and the partially completed Sarin plant at Falkenhagen had also fallen into Soviet hands. Although most key items of production equipment at Falkenhagen had been removed, Soviet military chemists derived valuable information by analyzing the stocks of raw materials and interrogating captured German personnel.

In late May 1945, Professor S. I. Volfkovich, an expert on inorganic phosphorus compounds at the Academy of Chemical Defense in Moscow, traveled to the Soviet occupation zone in eastern Germany. Among the sites he visited was the factory at Piesteritz that produced elemental phosphorus, a key ingredient in the manufacture of nerve agents. Another Soviet chemist from the Academy of Chemical Defense, Professor V. A. Kargin, also visited eastern Germany at the end of the war. In a mine shaft at Rüdersdorf, east of Berlin, he discovered a hidden cache of laboratory notebooks and other documents describing Richard Kuhn's synthesis of Soman, the deadliest of the German nerve agents. Kargin arranged for these files to be sent back to Moscow for analysis.

IN JUNE 1945, the British Chiefs of Staff decided to dispose of all confiscated stocks of German mustard, phosgene, and other standard chemical weapons, with the sole exception of Tabun-filled aerial bombs. Large quantities of chemical bombs and shells were buried or burned in open pits. British sailors also loaded thousands of tons of German chemical munitions onto twenty aging merchant ships, which were towed into the Baltic Sea. Near the coast of Norway, the sailors donned gas masks, wired the ships with explosives, and sent them to the ocean floor. Similarly, from June 1946 to August 1948, the United States conducted Operation Davey Jones' Locker, in which captured German ships and submarines were filled with more than 30,000 tons of chemical weapons and scuttled in the North Sea and the Skaggerak Strait between Norway and the Jutland Peninsula of Denmark. In all, thirty-eight ships containing a total of 168,000 tons of chemical weapons were sunk in the Skaggerak, most containing mustard but also some Tabun.

The U.S. Joint Chiefs of Staff were interested primarily in Sarin, which they considered of greater military value than Tabun because of its superior potency, volatility, and stability. Thus, although the United States retained a small stockpile of Tabun-filled bombs for research purposes and as an operational reserve, the British were allowed to claim the lion's share of the German weapons. The British Chiefs of Staff decided to ship 71,000 Tabun-filled aerial bombs to Britain to serve as a contingency stockpile for possible retaliatory use in the Pacific Theater, where the war still raged. In the view of the British military, the German munitions represented a "considerable technical and productive effort" that "could not be reproduced in this country until after the end of the Japanese war."

The British Ministry of Defence decided to store the Tabun weapons at a Royal Air Force (RAF) base at Llandwrog in northwest Wales. Although this base had been used for training during the war, it was now inactive, and its remote location, far from populated areas, offered both safety and security. RAF Llandwrog was also situated directly on the Welsh coast, where the prevailing winds would blow any toxic fumes from leaking weapons harmlessly out to sea.

The secret transfer of the German Tabun bombs to Britain, code-named

Operation Dismal, began in October 1946. A ship transported the weapons from the port of Hamburg to Newport Docks in South Wales, where they were loaded under tight security onto railway cars, five hundred bombs per trainload. The trains traveled at night, when the rail network was quiet, to the town of Llanberis, where the cars were unloaded and the weapons transferred to trucks for the final run to Llandwrog. The last consignment of German weapons arrived at the RAF base on July 13, 1947. Still in their original wooden crates, the Tabun bombs were stacked on three runways in the open air. To ensure safe storage and reduce the risk of leaks, ordnance experts removed the fuses from the bombs and dipped them in lanolin, a wax-based sealant. A few years later, shelters were built on the runway to shield the weapons from the elements.

ON JULY 13, 1945, SHAEF was disbanded and the technical intelligence work performed by CIOS was taken over by a new organization called the "Field Information Agency Technical," or FIAT, which was based in the headquarters of the Hoechst Company (part of IG Farben), just outside of Frankfurt. After the Potsdam Treaty of August 2, 1945, divided Germany into four Allied occupation zones (American, British, French, and Soviet), FIAT coordinated technical intelligence collection in the British and American zones. Technical exploitation teams collected hundreds of tons of files, documents, photographs, and equipment, which were shipped to London and Washington for analysis. Drawing from meticulous German records, FIAT compiled a comprehensive list of thousands of German scientists, technologists, and industrialists.

Investigators from FIAT's Enemy Personnel Exploitation Section regularly visited the Dustbin internment camp to interrogate German chemists and physiologists who had worked on chemical weapons. With the exceptions of Schrader and Kuhn, the Allied investigators were not overly impressed by the quality of the German research effort. According to a British intelligence report, "One gets the idea that if the IG [Farben] had not been fortunate enough to stumble on Tabun, in the course of other work, the German C.W. research picture would have seemed not to be very advanced, but of course this may not be fair, since the very promise of Tabun may have discouraged the expenditure of effort in other fields."

FIAT investigators also looked into the persistent rumors of human experimentation with nerve agents. Of the German physiologists interned at Dustbin, including Ferdinand Flury, Wirth, Hörlein, and Gross, none admitted to having any involvement with human experimentation. Because these individuals were often evasive during interrogation, however, suspicions remained. As a BIOS team observed, "It does seem to be a matter for serious doubt whether the higher [Nazi] Party organizations would have agreed to the diversion of considerable effort, in difficult circumstances, to the production of a chemical warfare agent which had not been shown unequivocally to be capable of killing men." During an interrogation at Dustbin, Jürgen von Klenck revealed that Karl Brandt had once told him and Ambros that he had witnessed one of Professor Wirth's tests involving Sarin and that the results had been "very impressive." Although Brandt had spoken of "guinea pigs," Klenck said that he had inferred by the manner in which Brandt described the results that the experimental subjects had actually been humans and not laboratory animals.

Finally, in August 1945, FIAT investigators interrogated Fritz ter Meer, a member of the IG Farben board of directors, who eventually admitted that Tabun and Sarin had been tested on concentration-camp inmates in the 100-cubic-meter gas chamber at IG Elberfeld to determine the lethal dosages in man. Such experiments had been conducted initially with monkeys and apes, and later with human beings. According to the interrogation report, "KZ [concentration camp] inmates who had been condemned to death were selected and were allowed to volunteer for the experiments with the provision that in case of survival they would be pardoned." Ter Meer argued with cold-blooded logic that "no harm had been done to the KZ inmates as they would have been killed anyway and were thus offered a chance of survival." He also claimed that the tests had a humanitarian purpose because the goal was to develop an improved antidote for nerve agents, which would have saved countless German lives.

Further investigation revealed that Dr. Karl Wimmer, a physician in the Luftwaffe Health Directorate, and Professor August Hirt, of the University of Strasbourg, had used inmates at the Natzweiler Concentration Camp to test the effects of nerve agents and the effectiveness of various antidotes. After suffering excruciating deaths, the human subjects had been autopsied and their organs examined. During the Nuremberg war-crimes trials, Dr. Wimmer handed over to prosecutors detailed pathology reports on the human experiments.

NUMEROUS INTERROGATIONS OF German scientists suggested that Otto Ambros was the IG Farben official who was most knowledgeable about chemical weapons production, making him a person of keen interest to FIAT. In July 1945, the Control Branch of FIAT issued fresh orders for the transfer of Ambros and several of his associates to Dustbin for interrogation. Lieutenant Colonel Tarr agreed to deliver Ambros to the internment camp, and the two men set off from Gendorf by car. En route to Frankfurt, Tarr suddenly diverted to Heidelberg and obtained permission from FIAT to hold Ambros for another forty-eight hours of questioning. After two days had passed, however, the men did not proceed to Dustbin as planned. Instead, while Ambros remained behind in Heidelberg, Tarr flew to Paris and London, at first requesting and then demanding the release into his custody of Ambros and the other German chemical warfare experts held at the internment camp. When FIAT and the British authorities refused, an angry confrontation ensued.

Tarr sent a telegram from Paris in the name of Colonel J.T.M. Childs of the British Ministry of Supply (the agency responsible for chemical warfare matters) ordering the release into U.S. custody of all German chemical weapons experts held at Dustbin. When informed of this message, Colonel Childs immediately repudiated it and accused Tarr of having perpetrated a forgery. By now it was clear to the British government that Tarr had no intention of turning Ambros over to FIAT. In the confusion, Ambros disappeared, surfacing a few days later at Villa Kohlhof, the IG Farben guesthouse near Heidelberg.

Dr. Wilhelm Hirschkind, a chemist on leave from the Dow Chemical Company, was conducting a study of the German chemical industry for the U.S. Army Chemical Warfare Service. He had spent most of May and June 1945 visiting IG Farben plants in northern Germany and had returned to London to write his report. In July he received orders to travel immediately to Heidelberg and investigate the large IG Farben plant in nearby Ludwigshafen, which was then under American control but would soon become part of the French occupation zone in southwest Germany. When Hirschkind reached Heidelberg, the U.S. Army commander gave him a letter of introduction to Colonel Weiss, the French commander at Ludwigshafen, requesting assistance in the conduct of his mission. Weiss promptly referred Hirschkind to Ambros as the IG Farben official best qualified to provide the desired information. Ambros had lived for several years in Ludwigshafen, forty kilometers from the French border, and had many French friends.

Hirschkind found Ambros and his aide Klenck extremely cooperative. On July 28, 1945, the two IG Farben executives summoned several company scientists and engineers who had been involved in the development and production of nerve agents for meetings at Villa Kohlhof with Hirschkind, Tarr, and other CWS intelligence officers. The meetings extended over a few days, during which the men discussed in detail the chemistry of the nerve agents and their manufacturing processes. Several years later, in a letter to Ambros, Hirschkind warmly recalled these conversations. "In addition to . . . specific Chemical Warfare items, we discussed many chemical developments," he wrote, "and it was only natural for me to tell you in parting that I would look forward after conclusion of the peace treaty in continuing our relations as a representative of Dow."

Meanwhile, FIAT continued to demand the internment of Ambros. On August 16, 1945, a British intelligence officer, Major P. M. Wilson, arrived in Ludwigshafen with an arrest team and discovered to his shock that the French military government had installed Ambros as director of the IG Farben plant and refused to hand him over. The German was allowed to move freely within the French and American occupation zones, and Wilson suspected that Lieutenant Colonel Tarr had "taken steps to assist him to evade arrest." The British officer reported to London his outrage at "the friendly treatment being given to this man who is suspected of war criminality."

Over the next several months, Ambros continued to enjoy the protection of the French government, and FIAT officials made several trips to Heidelberg and Ludwigshafen to arrest him, without success. The IG Farben executive even had the impudence to thumb his nose at his pursuers, writing to FIAT that he was "regretfully unable to meet in Heidelberg because I have to attend important meetings with high-ranking French gentlemen."

IN CONTRAST TO AMBROS, Gerhard Schrader agreed voluntarily to be interned at Dustbin, hoping to work out a deal with the Allied authorities. During an interrogation with FIAT officials on August 30, he said that his research on nerve agents was still incomplete and offered to work for the British. When FIAT asked him to draw up a comprehensive list of nerve agents, Schrader explained that it would take him a while to compile the information because he had prepared sixty compounds in the Tabun group and eighty in the Sarin group. On September 2, 1945, Schrader proposed that the British hire him to develop a new insecticide to combat the Colorado beetle, a major insect pest in Europe. FIAT finally gave Schrader complete access to his files at IG Elberfeld so that he could write a lengthy monograph describing his discovery of the organophosphate insecticides and nerve agents. This report was later published in two versions: an unclassified version, covering only the development of insecticides, and a "Secret" version that included the detailed syntheses of Tabun and Sarin.

Because Schrader had gone out of his way to cooperate with FIAT, he began to resent his prolonged detention at Dustbin. On April 9, 1946, Edmund Tilley wrote in a memo to Lieutenant Colonel Wilson:

> SCHRADER has been one of the most co-operative Germans at Dustbin. He has written many useful reports and has volunteered information on anything that he thought might be of use to us. He had been recommended for transfer to England for work on insecticides and possibly poison gases there, and he probably would have been very useful to us. . . . As a reward for his co-operation SCHRADER feels he has received eight months imprisonment. His resentment may have turned him, or may turn him from a collaborator into an enemy. I hope his speedy transfer to Britain or his immediate release pending such a transfer may be arranged as soon as possible. Many other C.W. experts were released months ago and SCHRADER wonders what crime he has committed to warrant his continued detention.

The charges against Schrader were finally dropped and he was released from the internment camp in mid-1946. Although the British were interested in his research and invited him to England for a month, Schrader turned down a job offer and returned to his laboratory at the Bayer company in Leverkusen. There he worked intensively on a new organophosphate pesticide known as E-605, which was marketed in 1947 under the trade name Parathion. Over the next few years, Schrader sought to develop analogues of the pesticide that were less toxic to humans but more persistent in attacking insects.

Meanwhile, the top executives of IG Farben were put on trial at Nuremberg. After the high-profile tribunal for the major Nazi war criminals, the United States, France, and Britain disagreed over how to treat lower-ranking officials and collaborators. Neither the British nor the French wished to prosecute the German industrialists who had supported the Hitler regime, so the U.S. military government established its own tribunals for this purpose. The IG Farben trial was the second of three trials of leading industrialists; the other two involved the firms of Flick and Krupp. In October 1946, Military Tribunal No. VI was established at Nuremberg to try the principal officers of IG Farben for crimes against peace, humanity, and property rights. The presiding judge was Curtis Shake, a former chief judge of the Supreme Court of Indiana; the chief of counsel for the prosecution was

At Military Tribunal No. VI at the Palace of Justice in Nuremberg, Germany, the twenty-three principal officers of IG Farben were tried in 1947–1948 for crimes against humanity and the planning and waging of aggressive war. In this photograph, eight of the defendants wait for the indictments to be handed out. Seated in the front row (from left to right) are August von Knieriem, Fritz ter Meer, Christian Schneider, and Otto Ambros.

Telford Taylor; and his deputy was Josiah E. DuBois, a lawyer from Camden, New Jersey.

Although the American prosecutors issued a warrant for Otto Ambros's arrest, he was safe as long as he remained in the French occupation zone. A FIAT report noted, "He is wily and will remain there, as he knows the hunt for him is on in the U.S. Zone." Ambros remained at large for several more months, but eventually he grew cocky and dared to travel outside Ludwigshafen. His luck finally ran out on January 17, 1946, when he was arrested by the American occupation forces and handed over to the prosecutors at Nuremberg.

On May 1, 1947, in a sworn deposition in his own defense, Ambros argued that his briefing at Wolf's Lair on May 15, 1943, had aroused doubt in Hitler's mind about whether the Allies had independently discovered nerve agents. "I believe," Ambros said, "through my objective description of the production situation and above all through my objective reference to the possibilities of the enemy side, I significantly contributed to the fact that Germany did not make any use of chemical weapons." This self-serving interpretation conveniently sidestepped the fact that Ambros's negative depiction of Germany's chemical warfare capabilities had been intended to persuade Hitler to *expand* the production capacity for Tabun and Sarin and thereby ensure a qualitative advantage over the Allies. Indeed, after the meeting at Wolf's Lair, Hitler had increased funding for nerve agent production. As historian Peter Hayes later wrote about Ambros and his fellow executives, "Lacking the courage of moral conviction almost as a condition for their professional success, they shut off their consciences, which was tantamount, in this instance, to having no consciences at all."

The IG Farben trial ran from August 27, 1947, to June 11, 1948—152 working days. Although the trial began with twenty-four defendants, it ended with twenty-three because one case was discontinued due to illness. Organized into five main points and 147 individual charges, the prosecution's case was that the company had established an alliance with Hitler and the Nazi Party to become the unchallenged leader of the world chemical industry.

During the trial, the prosecution presented 2,282 documents, 419 sworn depositions, and 87 witnesses, while the defense presented 4,102 documents, 2,394 sworn depositions, and 102 witnesses. Ambros was accused of having personally selected the Auschwitz concentration camp as the site of IG Farben's synthetic rubber plant so that the inmates could be exploited for slave labor. Survivors of Auschwitz testified about the grim conditions at the rubber plant, including beatings, ill treatment, and summary executions.

When the IG Farben tribunal reached its judgment on July 29–30, 1948, the Nuremberg judges found thirteen of the accused directors guilty and the other ten not guilty. Ambros was convicted of the use of forced labor at the Auschwitz plant and sentenced to eight years in prison, minus time already served. All the other charges against him were dropped, however, including the planning, preparation, and execution of offensive war, war crimes, and crimes against humanity. Deputy Prosecutor DuBois considered Ambros's sentence "light enough to please a chicken thief" and later wrote an angry book on the IG Farben trial titled *The Devil's Chemists*.

ALTHOUGH THE ORIGINAL justification for the British to keep a large number of German Tabun-filled bombs at an RAF base in Wales had been

Brigadier General Telford Taylor, chief counsel for the prosecution, delivers the opening statement in an all-American court at Nuremberg, Germany, charging IG Farben executives (including Otto Ambros) with war crimes and crimes against humanity, September 1947.

the need for a chemical retaliatory capability during the war against Japan, the emergence of the Cold War created a new rationale. The continued stationing of large numbers of Soviet troops in the "satellite" nations of Eastern Europe created a military threat to Western Europe and its isolated outpost in West Berlin. Accordingly, the British government came to view the German nerve agent bombs at Llandwrog as a contingency stockpile in case the East-West confrontation ever turned hot. A 1947 report from Porton Down concluded, "It is believed that the Russians know that we possess this gas; therefore, its retention by us in some form of storage as a potential threat or bargaining agent, is recommended, even if present weapons are not very effective."

The feasibility of actually using the German Tabun bombs stored at Llandwrog was doubtful, however. Because the weapons had not been designed for delivery by British bombers, they would first require modifications to their suspending lugs. On July 18, 1950, the British Defence Research Planning Commission asked the Air Ministry to determine the amount of effort and cost that would be required to make the bombs deliverable by British aircraft. The ministry estimated that the modification parts for the 71,000 bombs would cost about 500,000 pounds sterling, and that the new lugs would not be available for nine to twelve months after the order was placed. This estimate did not include the expense of installing the parts, as well as new fuses. Some British military planners argued that because the operational value of the German bombs was limited, the high cost required to modify them was unwarranted; others countered that the bombs represented the only stock of nerve agent in British hands and that scrapping them would be unwise.

In 1952, British officials decided to modify nine thousand of the bombs by March 1954 to improve the country's readiness for chemical warfare. This order was later rescinded, however, because the stocks at Llandwrog were showing signs of corrosion and leakage and the Tabun fill had deteriorated over nearly a decade of storage. Britain was also acquiring nuclear weapons, which were considered to be a far better strategic deterrent than poison gas. Accordingly, in June 1954, the British government decided that the Tabun stockpile had outlived its usefulness and should be disposed of at sea. Under Operation Sandcastle, which lasted from January 1955 to July 1956, the seventy-one thousand German bombs were loaded onto rusting hulks, which were towed out into the North Atlantic and scuttled in deep water.

Meanwhile, the chemical warfare establishments of the Allied powers strove to assess and exploit the German nerve agents. The discovery of Tabun and Sarin had reduced the most effective chemical weapons of World War I—phosgene and mustard—to secondary status, if not obsolescence. Until persistent nerve agents could be developed, mustard would retain some utility for long-term terrain denial, but it had clearly lost its status as "the king of the war gases." The nerve agents' extreme toxicity, rapid action, and difficulty of detection offered new possibilities for offensive use, while posing unprecedented challenges for chemical defense.

Military strategists also debated the role of gas in future conflicts. For most of World War II, the strong personal antipathy of President Roosevelt and many of his top generals toward chemical weapons had limited their integration into U.S. force structure and military doctrine. Strategic considerations also played a role. During the bloody battles for the Pacific islands, where the use of gas against Japanese soldiers holed up in caves and tunnels offered undeniable tactical advantages, Pentagon planners feared that initiating chemical warfare against Japan would give the Germans an excuse to employ gas in the European theater. After Roosevelt's death on April 12, 1945, the inauguration of Harry Truman, and the German surrender in May, U.S. chemical warfare planning against Japan advanced significantly. Pentagon strategists drew up plans to drop mustard and phosgene bombs on Japanese cities in October 1945, prior to an invasion scheduled for early November. But the atomic bombing of Hiroshima and Nagasaki in August 1945 ended the war without the need for a land invasion.

Now, with the discovery of the secret German breakthrough in chemical warfare, several important questions remained to be answered. Although nerve agents had not been used during World War II, would they have a place in future wars? What military missions could these weapons fulfill? And how would the problems of chemical defense be solved with respect to detection, prophylaxis, and therapy?

CHAPTER SIX

RESEARCH AND
DEVELOPMENT

IN 1945, in the immediate aftermath of World War II, the U.S. Army
Chemical Warfare Service decided to focus its research and development
efforts on the German nerve agents, the technological challenges of which
promised to ensure the organization's survival through the period of postwar demobilization and declining military budgets. The CWS closed its
Development Laboratory at MIT and transferred the scientific staff back to
the main research and development center at Edgewood Arsenal, located on
a secluded peninsula jutting into Chesapeake Bay some twenty miles northeast of Baltimore. Beginning in 1940, a major construction effort had
greatly expanded the size of the base, and about ten thousand civilian personnel had been stationed there during the war. Surrounded by high fences
to keep out intruders, the sprawling arsenal included laboratories, engineering shops, a facility for making chlorine (a key ingredient of many chemical
warfare agents), a pilot plant for developing manufacturing processes, and
several large test ranges. Even so, much of the peninsula remained undevel-
oped, with more than three thousand acres of forest, fields, and wetlands
inhabited by a rich array of wildlife, including bald eagles and osprey.

Although the scientists and engineers at Edgewood spent their days
researching poisons of unprecedented power, they lived quiet, middle-class
lives in nearby suburban Maryland communities such as Aberdeen and Bel
Air. Few of them experienced pangs of conscience about their work, which
they justified as creating a credible deterrent against chemical attack. Moreover, many Edgewood scientists considered chemical weapons to be a more

Edgewood Arsenal, near Aberdeen Proving Ground, Maryland, in the late 1950s. The Pilot Plant building is at center-left. A production process for Sarin nerve agent was developed here from 1948 to 1951.

humane means of putting enemy soldiers out of action than blasting them to bits with high explosives or incinerating them with napalm.

In parallel with Porton Down, Edgewood Arsenal conducted a detailed technical assessment of the captured German stocks of Tabun. The CWS research-and-development program for fiscal year 1946, approved on July 5, 1945, included a new activity called "Project A1.13." This effort involved determining the physical constants and characteristics of Tabun, preparing analogues of the compound for toxicity studies, devising new methods of detection and decontamination, and developing an industrial manufacturing process for Tabun "in case field tests indicate its usefulness as a chemical agent." Edgewood also began a program of static testing with several types of Tabun-filled bombs and mortar shells to determine how much of the agent was destroyed during explosive dissemination, the distribution of drop sizes and vapor concentrations, and toxicity under various atmospheric and weather conditions.

A SMALL BUT significant step forward in the study of the German nerve agents was the adoption of a standard nomenclature. Immediately after the war, the American, British, and Canadian armies used different code names for Tabun, Sarin, and related compounds, which frequently gave rise to misunderstandings. Whereas the U.S. Chemical Warfare Service referred to Tabun as "MCE," the British called it "T-2104" or "MCP." Similarly, the U.S. Army code name for Sarin was "MFI"; the British used a different designation. On October 16, 1945, Colonel Jack H. Rothschild, the chief of the CWS Technical Division, proposed creating a uniform system for assigning symbols to the various nerve agents so as to minimize confusion. Drawing on the existing system of letter codes for chemical warfare agents (such as HD for distilled mustard and CG for phosgene), he suggested that the German nerve agents be code-named with the letter "F," followed by another letter of the alphabet. Tabun, Sarin, and Soman would therefore be designated "FA," "FB," and "FC," respectively.

Major James E. McHugh, the head of the CWS Training Division, agreed on the need for a uniform naming system but took issue with Rothschild's proposal; FA, he noted, could be confused with the abbreviation for "Field Artillery," while FD might refer to "Finance Department." McHugh suggested avoiding these misleading code names by referring to Tabun, Sarin, and Soman as "FB," "FC," and "FE," respectively.

Rothschild responded with a memo noting that other abbreviations involving the letter "F," such as "FM," could be problematic. He therefore proposed that the major German nerve agents and their analogues be designated with the letter "G," followed by a second letter not previously associated with it. ("GC" was excluded because it was too easily confused with CG, the existing Army code for phosgene.) According to Rothschild's scheme, Tabun would be designated by the code name "GA," Sarin by "GB," Soman by "GD," Ethylsarin by "GE," Cyclosarin by "GF," and Isopentylsarin by "GH." (Although the Germans had used letter codes to refer to the two formulations of Tabun and the solvent chlorobenzene, with "G" standing for 95 percent Tabun and "GA" for 80 percent, there was apparently no connection between the German codes and the ones devised by Rothschild.)

The Chemical Warfare Service and the British Army signed off on the new naming convention, and the CWS's Chemical Warfare Technical Committee formally approved it on December 20, 1945. This committee also agreed that if the letter codes were used in conjunction with the chemical names or formulas of the nerve agents, they would be classified Secret, but if the code names were unaccompanied by any specific means of identification, they would remain unclassified so that useful information about the nerve agents could be listed in Army field manuals. Rothschild's naming system was widely adopted and is still in use today.

Another shift in nomenclature came in 1946, when U.S. Secretary of War Robert Patterson asked Congress to change the name of the Chemical Warfare Service to the Chemical Corps. This change had been proposed several years earlier, without success. On August 4, 1937, President Roosevelt had vetoed Senate Bill 1284 to establish an Army Chemical Corps because he believed it would institutionalize chemical warfare, which he considered immoral. In 1946, however, Roosevelt's successor, Harry Truman, approved the change in name and status. One reason was that the mission of the Chemical Corps was no longer limited to the military use of chemicals on the battlefield but also encompassed civilian activities such as the development and production of DDT and other pesticides.

Although the Chemical Corps focused its initial research-and-development activities on Tabun, the recognition that Sarin offered superior military characteristics soon led to a change in emphasis. In 1946, the Chemical Corps established Project A1.13-2.1, the development of a method for the mass production of Sarin. Despite the fact that small amounts of the nerve agent could be synthesized fairly easily in the laboratory, scaling the process up to the industrial level posed major technical challenges.

ADDING SPECIAL URGENCY to the U.S. Army's research-and-development program on nerve agents was the fact that the Soviet Union had obtained a significant head start. In September 1946, with the aid of captured German scientists, Soviet engineers and pipe fitters had systematically dismantled the Tabun and Sarin plants at Dyhernfurth, which the Red Army had seized as war booty, and shipped the production equipment to Stalingrad. Much of the German apparatus was custom-made and of high quality, including chemical reaction vessels and pipes lined with silver or nickel to resist corrosion, as well as specialized filters, evaporators, driers, pumps, compressors, distillation columns, and valves of several gauges. The Soviets also confiscated air-handling and containment systems for work with highly toxic chemicals, such as control systems, fans, ducts, and hermetically sealed production compartments made of glass and steel.

The German equipment was shipped to Chemical Works No. 91 near the town of Beketovka, five miles south of Stalingrad on the banks of the Volga River. This factory had begun operation in 1929, under the first Soviet Five-Year Plan, to manufacture both civilian and military chemicals, and during World War II it had mass-produced mustard agent for the Red Army. Until the chemical apparatus from Dyhernfurth could be reassembled, it was stored in a guarded warehouse in a secure part of the factory grounds, behind high walls and barbed wire.

To assist in rebuilding the nerve agent production lines, in 1948 the Soviets brought to Stalingrad about a dozen German chemists and process engineers from Dyhernfurth who had been detained by the Red Army. The most senior of the German scientists was Dr. Bernd von Bock, the former production manager. Soon after his arrival, he was ordered to write a detailed technical report on Tabun production and was questioned at length about the metal corrosion problems associated with the manufacture of Sarin. Although Sarin was known to be militarily more effective, the Soviets decided to produce Tabun first because of the relative simplicity of the process.

Dr. Mikhailov, the chief engineer at Chemical Works No. 91, died shortly before the Tabun plant went into operation. He was replaced by Professor Leonid Zaharovich Soborovsky from Scientific Research Institute No. 42 (NII-42) in Moscow, the Soviet Union's leading center for chemical weapons research and development. The reconstruction of the Tabun production line went smoothly, and by 1949, the Soviet *Military Chemical Textbook* listed Tabun as one of the chemical weapons stockpiled by the Red Army.

As East-West tensions ratcheted upward, the implications of the Soviet lead in nerve agent technology triggered alarm bells in Washington. For American military planners, the Soviet Union's production of Tabun, combined with its potential acquisition of intercontinental-range bombers, posed an emerging strategic threat to the U.S. homeland. In an article published in March 1946 in *The Saturday Evening Post,* Major General Alden H. Waitt wrote, "Today, with the development of the long-range bomber, no area, however remote, is immune from gas attack."

Given these perceptions, a chemical arms race between the superpowers was almost inevitable. In 1947, President Truman withdrew the 1925 Geneva Protocol banning the use of chemical weapons in war from the docket of the Senate Foreign Relations Committee, where it had languished for two decades, indicating that the United States did not intend to ratify it.

THROUGHOUT the postwar years, Britain, Canada, and the United States intensified their collaboration on chemical weapons research, which had begun during World War II. In 1940, the British government had sought an open-air proving ground somewhere in the Commonwealth. This search had led in 1941 to the founding of the British-Canadian experimental station at Suffield in southeastern Alberta, about 30 miles northwest of Medicine Hat and 150 miles southeast of Calgary. By displacing about a hundred families, the Canadian government set aside a thousand square miles of semiarid short-grass prairie for the open-air testing of chemical weapons. Named the Suffield Experimental Station (SES), the Canadian base was remote enough to be secure from enemy attack but still reasonably accessible and close to support facilities. Initially, the research staff at SES consisted of about forty Canadian scientists, twenty British scientists, and a hundred technicians. Additional scientific teams visited from Britain and, after the Japanese attack on Pearl Harbor in December 1941, from the United States as well. By the end of World War II, the SES had a staff of nearly 600 scientists and technicians who were conducting research on chemical warfare agents, smoke, flame weapons, biological warfare, and ballistics.

In 1946, Britain, Canada, and the United States decided to formalize their collaboration by holding annual research conferences on offensive and defensive chemical warfare. The following year, the three governments signed a Tripartite Agreement giving each country access to the findings of the others and mandating a rational division of labor to avoid redundancy. They agreed, for example, that British scientists would develop a manufacturing process for Sarin (GB), while their American counterparts worked on a similar process for Ethylsarin (GE).

Another priority for the Tripartite program was to develop improved defenses against nerve agents, including detectors, protective suits, respirators, and chemical antidotes. A Royal Air Force report dated September 6, 1947, noted that the standard British gas mask gave good protection against Tabun and Sarin "provided that it is put on in time and does not leak." Unfortunately, a significant percentage of gas masks were defective and did not provide a high level of protection. Since the toxicity of nerve agents left no margin for error, it was essential to develop an improved mask with a leakproof seal and a more efficient activated-charcoal filter. Designing this new mask involved complex technical problems that were only gradually overcome. Porton Down also developed and tested a concertina-type bellows system to give artificial respiration to soldiers whose breathing muscles had been paralyzed by exposure to nerve agent on the battlefield. Called the "Porton Resuscitator," the device went through several stages of development before it was adopted for service use.

In addition to the extensive testing of nerve agents on experimental animals, Porton and Edgewood conducted low-dose trials involving human subjects. Edgewood drew roughly thirty volunteers a month from the 1,500 enlisted men stationed at the base. Although these individuals participated willingly in the trials, they were not fully informed of the potential risks and side effects. In 1948, for example, Edgewood scientist L. Wilson Greene reported that several soldiers who had been exposed to low concentrations of Tabun vapor had been "partially disabled from one to three weeks with fatigue, lassitude, complete loss of initiative and interest, and apathy."

BASIC RESEARCH AT Edgewood Arsenal and Porton Down also sought to elucidate the mechanism of action of the nerve agents at the molecular level. These studies confirmed Richard Kuhn's finding that the diverse symptoms of nerve agent poisoning result from the inhibition of cholinesterase in the peripheral and central nervous systems. Knocking out this key enzyme causes acetylcholine to accumulate in excessive amounts, disrupting the neural regulation of various target organs and wreaking havoc on the body as a whole.

Further scientific investigation revealed that many organs and tissues in the body have protein "receptor" sites to which acetylcholine binds specifically, in lock-and-key fashion, to trigger various physiological effects. Two basic types of acetylcholine receptors were identified: muscarinic receptors, which are activated selectively by the drug muscarine, and nicotinic receptors, which are activated selectively by the drug nicotine. Muscarinic receptors for acetylcholine are present on the smooth muscles that surround the airways of the lung and the gastrointestinal tract, the ciliary muscles of the eye (which control the size of the iris), and the salivary and sweat glands. Nicotinic receptors, in contrast, are present on the skeletal muscles and on certain nerve cells in the spinal cord. Both muscarinic and nicotinic acetylcholine receptors exist in the brain.

The discovery of two broad classes of acetylcholine receptors in the human body shed new light on the action of nerve agents. By inhibiting cholinesterase, nerve agents result in a surfeit of acetylcholine, which in turn overstimulates the muscarinic or nicotinic receptors on the cells of diverse target organs. The smooth muscles surrounding the bronchial tubes tighten, reducing the flow of air and causing an asthmalike shortness of breath; the muscles of the gastrointestinal tract go into spasm, resulting in abdominal cramps, nausea, vomiting, and diarrhea; and the ciliary muscles in the eye constrict, reducing the iris to pinpoint dimensions with a concomitant dimming and blurring of vision. Too much acetylcholine also causes several glands to become overactive and secrete excessive amounts of nasal mucus, saliva, and sweat.

The nicotinic effects of nerve agents chiefly involve the skeletal muscles. At first, localized contractions called "fasciculations" appear, resembling ripples or worms under the skin, after which the large muscle groups begin to twitch in an uncoordinated manner. At higher doses of nerve agent, the skeletal muscles contract violently, causing convulsions, and then become fatigued, leading to flaccid paralysis. In the brain, nerve agents induce a mixture of nicotinic and muscarinic effects, which are manifested by seizures, loss of consciousness, generalized depression, and suppression of the breathing center. Even low-dose exposures may cause neurological and psychological disturbances that persist for days, such as an inability to think clearly, insomnia, poor concentration, and emotional swings.

It was also discovered that repeated low-level exposures to nerve agents, spaced over a period of days or weeks, have cumulative effects by progressively depleting the body's supply of cholinesterase faster than it is replenished. Eventually a threshold of enzyme depletion is reached at which the individual begins to suffer acute symptoms of nerve agent poisoning, usually when the body's level of the enzyme has been reduced by about 50 percent. Complete physiological recovery takes place only when the nervous system has restored its reservoir of cholinesterase, a process that may take several months.

Armed with this knowledge, British and American researchers sought to develop a more rapid and efficient means of treating nerve agent casualties on the battlefield. The standard antidote for organophosphate poisoning is the drug atropine, which blocks the muscarinic receptors for acetylcholine on the smooth muscles and glands but has less of an effect on the nicotinic receptors of the skeletal muscles. In routine medical practice, small doses of atropine are used to dilate pupils and treat heart-rhythm abnormalities, but larger doses (2 to 6 milligrams) injected intramuscularly will reduce glandular secretions and relax the smooth muscles of the bronchioles and the gastrointestinal tract.

In devising a system to administer atropine to soldiers exposed to nerve agents on the battlefield, the main challenge was speed of delivery. To provide maximum benefit, atropine must be injected into the thigh muscle within a few minutes of exposure. Because nerve-gassed soldiers would have no time to wait for a medic to arrive, they would have to self-administer the drug. Beginning in 1950, U.S. troops were issued syrettes, small collapsible metal tubes filled with a solution of atropine, with a hypodermic needle at one end. If exposed to nerve agent, a soldier had to jab the needle into his thigh muscle and inject the atropine by squeezing the tube, repeating the process at five-minute intervals until the symptoms diminished. Needless to say, many individuals were uncomfortable with this approach.

CHEMICAL ENGINEERS at Edgewood Arsenal also worked on an industrial manufacturing process for Sarin. After failing to develop a simplified four-step method, they decided to adopt the German five-step approach known as the DMHP (dimethyl hydrogen phosphite) process. But this technique entailed the use of highly toxic and corrosive ingredients, such as hydrogen fluoride gas, resulting in numerous leaks and other damage to the apparatus. In addition, a small fire in the pilot plant destroyed some of the control instruments and required rebuilding the ventilation system, causing further delays. The numerous technical problems plaguing the development of the Sarin production process at Edgewood ultimately led the U.S. Army to seek the help of German scientists. In this effort, Colonel Charles E. Loucks, a career officer in the Chemical Corps, played a prominent role.

During World War II, Loucks had managed the construction of a mustard agent production plant at Rocky Mountain Arsenal near Denver. Immediately after the war, he had served as the chief chemical officer in Tokyo during the U.S. occupation of Japan and had then returned to Washington for a tour of duty at the Pentagon. In June 1948, he began a new assignment as chief of the Chemical Division of the U.S. European Command (EUCOM), based in Heidelberg, where he was responsible for collecting intelligence on the chemical warfare programs of the major countries of Western Europe. Soon after his arrival in Heidelberg, Loucks was ordered to dispose of 6,000 tons of German chemical shells stored at the Sankt Georgen depot in southeastern Bavaria. He organized the transport by train of 350 to 500 tons of munitions per day to the northern German port of Bremerhaven, where the weapons were loaded onto U.S. Navy ships and dumped into the North Sea. This operation, carried out by German workers under the supervision of American officers, was completed without serious injuries.

A few months later, Colonel Loucks received a telephone call from a lieutenant in U.S. Army intelligence. "A man here says he can be helpful," the officer said. "His name is Walther Schieber. Is there any information that you would like to get from him?"

Loucks had heard of Schieber, a former brigadier general in the SS who had worked for Speer's Armaments Ministry. He had been detained at Dustbin and transferred in autumn 1946 to the Nuremberg Military Tribunal, where he had been held as a prosecution witness. But he had obtained early release by writing reports on the German chemical warfare program for U.S. Army intelligence. With no further charges pending against him, Schieber had recently returned to his hometown of Bopfingen, eighty miles southeast of Heidelberg.

Loucks expressed interest in meeting the former SS general, and on October 14, 1948, he attended a meeting at EUCOM headquarters in Heidelberg at which Schieber was present. The German had a stout build and shook Loucks's hand firmly. He seemed eager to prove that he was on the side of the United States and its allies in the emerging confrontation with the Soviet Union. "I want you to know that if there is anything I can do to help the West, I shall do it," he said. "Anything I can do or any information I have, I will help."

That night, Loucks scribbled in his desk diary, "Attended conf. with Lt. Col. Taylor, Lt. Moller, and Dr. Walter [sic] Schieber—classified matters. No particular info but hope for more later, possibly when better acquainted. I'll try to see him next time he reports in to Div. of Intell. He directed production of war gases on a rather high echelon so doesn't have the detailed knowledge that I want, but possibly I can get the names of useful people from him. Took him to the house for a drink."

Loucks had heard that the chemical engineers at Edgewood Arsenal were having trouble developing an industrial production process for Sarin, even with the help of experts from leading U.S. universities. He concluded that Edgewood would benefit by receiving technical advice from the German chemists and engineers who had developed the DMHP process. During his next meeting with Schieber at EUCOM, Loucks took the general aside for a private chat. "Perhaps you can help us with a problem," he said. "Could you describe the German process to make Sarin and put it down on paper with drawings, specifications, tables, and safety measures?"

Schieber nodded. "Yes, I could arrange that."

"What do you know about Sarin production?" Loucks asked.

Schieber explained that a full-scale Sarin manufacturing plant at Falkenhagen near Berlin had been about 80 percent finished when the war ended. To escape the advancing Red Army, most of the German chemists and engineers at Falkenhagen had fled west into the American- and British-controlled zones. "I know these people," Schieber added. "They worked with me during the war. I could get in touch with them."

Loucks was intrigued by this suggestion. "Would you be willing to do that? We wouldn't expect you to do it for free. We could put you on retainer and also cover the scientists' expenses and pay them something for their work."

Schieber smiled and nodded.

"Well, think it over," Loucks said. "I'd like to meet with you again after you've had a chance to talk to your friends."

Two weeks later, Loucks and his wife hosted a dinner party at their home

Dr. Walther Schieber, a chemist, brigadier general in the SS, and senior official in the Speer Ministry. Colonel Charles Loucks of the U.S. European Command (EUCOM) hired him as a consultant in 1948. Schieber in turn recruited several German chemical engineers who had worked at Falkenhagen to prepare a detailed report for the U.S. Army on the Sarin manufacturing process.

in Heidelberg and invited Schieber. That evening, Loucks jotted down his impressions of the German in his desk diary:

> Schieber is interesting—an independent-thinking, intelligent and very competent man. He related much of his experiences with the Russians. A prisoner after 1st World War for a year. He was an honorary (?) Brigade Fuehrer of SS this last war. In confinement at Nuremberg for seven months. Quartered next to Goering until the latter killed himself. Was an admirer of Todt, later worked for Speer, was directed to report to Hitler frequently. He has many anecdotes and is a loyal German. Is willing to do anything for the future of the world and Germany.

The next morning, Friday, October 29, 1948, Loucks and a colleague drafted a cable to the chief of the Army Chemical Corps proposing that EUCOM hire Schieber and the Falkenhagen scientists to prepare a detailed report on the Sarin production process. Later that day, Loucks wrote in his desk diary, "Hope the chief will support us. If he does we'll be able to get all of the German CW technical ability on our side and promptly. They know on what side they belong. All we need to do is treat them as human beings. They recognize the military defeat and the political and ideological defeat as well and accept it."

On November 3, Loucks traveled to England for consultations at Porton Down, where he met with the director and other British chemical warfare experts. They discussed the types of technical information on Sarin production the German scientists might provide. Loucks noted in his diary, "It appears much is needed and that the Brits feel that engineering data has not been obtained."

A week later, the chief of the Chemical Corps approved the proposal to hire Schieber as a consultant to EUCOM, at a generous salary of 1,000 marks per month. After Loucks had drawn up the contract, Schieber recruited six chemists and process engineers who had worked at Falkenhagen, and on December 11, Loucks hosted the first meeting of the group at his home in Heidelberg. Although the Army approved the project on January 17, 1949, the initial funding arrangements were ad hoc. Loucks complained in his diary, "I'm to get my marks to pay for the work . . . being done by Schieber by signing for 100 cartons of cigarettes. Intelligence will sell the cigs on the black market! What a way to do business. In the meantime, Washington bigshots not responsible for getting info will debate learnedly and do little or nothing. Someone should shake up the Pentagon." The next day, however, a cable arrived from Washington stating that additional money would be forthcoming for the Schieber project.

Over the next three months, Schieber and the six German chemical engineers came to Loucks's home in Heidelberg every other Saturday for all-day work sessions. Loucks met periodically with one of the Germans who spoke excellent English to keep track of how the work was progressing. The scientists prepared a detailed report on the Sarin production process at Falkenhagen that was illustrated with numerous drawings and charts, including a manning table and a complete list of equipment and materials. Loucks sent the finished report to Washington for review by Chemical

Corps officials. Although the engineers at Edgewood benefited from the information, they insisted on making extensive modifications to the German manufacturing process at a cost of about $1 million. For example, instead of using silver-lined reactors and pipes, they built the production apparatus out of other corrosion-resistant materials that were available in the United States but not in Germany.

In addition to Schieber, EUCOM hired about thirty German chemical warfare experts, some of whom were later transferred to Edgewood Arsenal to continue their work on American soil. This recruitment effort, initially code-named Operation Overcast, was part of the postwar competition among the victorious Allies for the cream of Nazi Germany's scientific and industrial brainpower. In September 1946, President Harry Truman authorized the expanded recruitment of German scientists and engineers in areas deemed vital to U.S. national security. The program was dubbed Project Paperclip because the files of German scientists who were of interest to the U.S. government were marked with paper clips. The War Department's Joint Intelligence Objectives Agency (JIOA) conducted background investigations on the chosen scientists. In February 1947, JIOA director Bosquet Wev submitted the first set of personnel files to the Departments of State and Justice for review. When several of the recruited German scientists turned out to be former Nazi Party members, in violation of federal immigration policy, JIOA simply cleansed their files of Nazi references. The U.S. government provided each Paperclip scientist brought to the United States with a house, a car, and a generous salary.

Parallel to and partially in coordination with Project Paperclip, the British government ran Operation Matchbox, under which German scientists and technicians were identified and recruited for defense work to enhance Britain's military potential at Germany's expense. Three German chemists who had played key roles in the development and production of nerve agents—Eric Traub, Max Gruber, and Friedrich "Fritz" Hoffmann—were offered research positions at Porton Down. Gruber was known to have been an ardent Nazi, and Hoffmann had synthesized poison gases for the Chemical Warfare Laboratories at the University of Würzburg and the Luftwaffe's Technical Research Institute near Berlin. In some cases, British intelligence offered the German scientists immunity from war-crimes prosecution in exchange for their knowledge. On September 18, 1945, Britain invited Australia to participate in Operation Matchbox, and between 1946 and 1951, the Australian government recruited at least 127 German scientists and technicians, including thirty-one known Nazi Party members.

Several of the German chemical weapons specialists recruited by Porton Down later moved to the United States, where they were granted U.S. citizenship and prominent scientific positions. At Edgewood, Paperclip scientists conducted research on nerve agents, including tests on laboratory animals and human volunteers, and developed new gas masks, protective clothing, and antidotes. Fritz Hoffmann went to Edgewood in 1947 and was initially assigned to work on organophosphate insecticides, but the next year the Chemical Corps allowed him to start a classified research program on nerve agents. According to a former colleague, Hoffmann was a large, gentle man who spoke softly in a thick German accent and had an encyclopedic knowledge of organophosphorus chemistry. Another German chemist and former Nazi Party member, Theodor Wagner-Jauregg, arrived at Edgewood in 1948. For his part, Walther Schieber continued to work for the Chemical Division of EUCOM in Heidelberg. Despite growing suspicions that he was involved with arms smugglers, fugitive Nazi war criminals, and Soviet spies, he remained on the U.S. government payroll until 1956.

DURING HIS TOUR in Germany, Colonel Loucks also established close ties with his counterparts in the French Army. Although France had ratified the Geneva Protocol in 1926, it had reserved the right to use chemical weapons in retaliation and therefore maintained an offensive development program at the Poudrerie Nationale du Bouchet, near the town of Vert-le-Petit outside of Paris. On April 1, 1945, the arsenal was attached to the French Army's Chemical Warfare Service (Service de l'Arme Chimique) and was renamed the Research Center du Bouchet (Centre d'Études du Bouchet, or CEB). After the end of World War II, a team of chemists at CEB analyzed Tabun obtained from seized German weapons. In 1948, French scientists succeeded in synthesizing Tabun, ensuring an ample supply of the agent for testing purposes.

In addition to conducting laboratory research on the German nerve agents, the chemical armaments division of the Technical Service of the French Army performed field trials with captured German munitions at an open-air testing site in Algeria, which was then a French colony. The Algerian site had been established in 1935 for large-scale experiments with chemical warfare agents that could not be performed for safety reasons in metro-politan France. Shortly before World War II, the French Foreign Legion had expanded the testing facilities in Algeria. In 1945, immediately after the war, the French Army began to conduct trials there of confiscated German munitions containing Tabun.

Known as the Seasonal Experiment Station (Centre d'Expérimentations Semi-Permanent), the French chemical weapons testing complex in Algeria conducted open-air trials during the winter and early spring, in annual cam-paigns. It consisted of a support base at Beni Ounif code-named "B1," and a proving ground code-named "B2-Namous," located near the valley of Oued Namous about 100 kilometers east of the Moroccan border. With the exception of a few nomadic herders, this part of the Algerian desert was uninhabited and largely denuded of vegetation, making it ideal for the test-ing of chemical weapons. Despite the area's remoteness, Beni Ounif was eas-ily accessible by military aircraft and was on the rail line from Oran to Colomb Béchar.

In early 1949, the French Army invited Colonel Loucks to visit the chemical weapons testing site in Algeria, and he made arrangements through the U.S. military attaché in Paris. On February 23, having received his travel orders only that morning, Loucks traveled to the French air base at Wiesbaden and boarded a C-47 with a pilot and a crew of six, which took off at about 1:00 p.m. The plane made a refueling stop at Istres in southern France and then flew on to Maison Blanche Airport in Algiers, where Loucks spent the night. The next morning, after breakfast in the French officers' mess, he made a courtesy call on the base commandant.

Around 11:00 a.m., a three-propeller Junker arrived from Beni Ounif. Loucks was introduced to the pilot, Captaine Holl, a French chemist who spoke fairly good English. As soon as the plane had been refueled and loaded with boxes of technical supplies, fresh vegetables, and other cargo, the two men took off. After flying for four and a half hours, they landed on the air strip at Beni Ounif, an oasis about 100 kilometers north of Colomb Béchar. When the Junker had rolled to a stop on the landing strip, several French officers came out to greet them, including the base commander, Lieutenant Colonel Bonnard. Because the desert air was quite cold, the French officers wore bernouses made of heavy red or white flannel over their uniforms, and the commander arranged for Loucks to receive one as well. They then enjoyed a cup of hot tea, made a brief inspection tour of the experiment station, and sat down for an excellent dinner with a total of nineteen people—seventeen officers and two civilians.

Lieutenant Colonel Bonnard explained to Loucks that the elevation of the testing site, at 800 meters above sea level, gave the area a temperate climate remarkably similar to that of the potential battlefields of central Europe. Although the desert heat was intense in the summer months, the temperature during the first third of the year was cool during the day and quite cold in early morning—the optimal time for open-air testing because of the stability of the atmosphere. During the annual testing campaigns, a large number of French personnel traveled to Beni Ounif and lived under fairly rustic conditions. The staff included chemists, physicians, nurses, toxicologists, mathematicians, meteorologists, mechanics, construction crews, laboratory assistants, and maintenance workers. The base also had a garrison of French troops, who secured the testing site before each open-air trial to prevent nomadic Bedouins from wandering into the exclusion zone, which covered more than 6,000 square kilometers. In one unfortunate incident, a herd of camels had accidentally been killed by a toxic cloud.

The next morning, February 25, 1949, everyone awoke at five, when the sky was still dark. Loucks shivered in the cold air, and an Arab servant brought him a basin of hot water to wash with. After breakfast, they climbed into jeeps and drove 70 kilometers on a heavily potholed dirt road out to the proving ground at B2-Namous. The vast, desolate plateau had a flat desert floor pockmarked with millions of smooth stones. Loucks was shown an artillery battery from which the chemical shells were fired. The impact zone had been demarcated with a circular grid pattern drawn over 1,800 square kilometers. The grid lines farthest out from the center were spaced one kilometer apart, but as one approached the target zone where the chemical cloud was most concentrated, the distance between the lines shrank to a hundred meters and then to ten meters.

During the early trials of Tabun after the war, the French technicians had limited open-air testing to small-scale releases because they were uncertain how the German nerve agent would behave in the atmosphere. By 1949, however, trials of Tabun-filled munitions had become routine. The French tested German 150 mm shells, American 105 mm phosphorus shells that had been emptied and recharged with German nerve agents, and artillery rockets, a type of delivery system that was particularly well suited to chemical attacks.

Loucks was taken to the observation post to observe the firing of Tabun-filled shells. Watching from a safe distance, he saw the shells burst on the target grid, releasing white clouds that faded into invisibility as they exposed a series of sheep and pigs tethered at various distances downwind. Loucks was disappointed by the test results, which were not as dramatic as he had expected. Because a large proportion of the nerve agent was destroyed by the explosion, only the animals positioned within a few hundred feet of the explosion received a lethal dose.

The next morning, February 26, Loucks inspected the laboratory at Beni Ounif. Then, after a hearty lunch of couscous and michoui washed down with red wine, Loucks, Bonnard, and five other French officers took off in the Junker for the short flight to Colomb Béchar, a frontier-style town in the desert. After spending the night, Loucks returned to Germany.

DESPITE THE CLOSE collaboration between Porton and Edgewood, the British government did not want to become overly dependent on the United States in an area it considered vital to its national security. As an "insurance policy" against the emerging Soviet chemical threat, London decided to acquire an independent Sarin production capability. The first step was to build a pilot plant to test manufacturing techniques, produce enough Sarin for offensive and defensive research, and preserve the option to acquire a deterrent stockpile in the future.

Although Porton Down was responsible for laboratory R&D, a second British government–owned facility known as the Research Establishment at Sutton Oak specialized in process development for the manufacture of chemical warfare agents and the filling of munitions. Founded in the early 1920s, Sutton Oak was located near the city of Saint Helens in Lancashire, a heavily industrialized part of northern England. At the request of the British Ministry of War, Sutton Oak began to develop an improved production process for Sarin. Because the Research Establishment was located in a densely populated area, any production of nerve agent beyond the laboratory scale raised obvious safety concerns. The pilot plant would therefore have to be built in an isolated area where any accidental release of the lethal agent would not endanger the local population.

The British Ministry of Supply, which was responsible for the development, testing, and production of chemical weapons, began in September

1947 to look for a safer location. This search was based on several criteria, including remoteness, the prevailing winds, and the availability of electricity, water, and a local labor force. Gradually, fourteen candidate sites were narrowed down to two. The best option appeared to be a Royal Air Force (RAF) base on Anglesey, an island off the northern coast of Wales that had served during the war as a filling station for mustard bombs. Not only was Anglesey close to the peninsula where Sutton Oak and most of the British chemical industry were located, but an offshore site was desirable for safety reasons. The other promising location was a disused RAF base called Portreath, near the village of Nancekuke on the north coast of Cornwall. The 800-acre base was at the top of a coastal cliff, so that any toxic gases released accidentally would be blown out to sea. Nancekuke had one major drawback, however: it was located near the southwestern tip of England, whereas Sutton Oak and the chemical industry were in the north.

The deciding factor turned out to have nothing to do with logistics. Nancekuke was fifteen miles from Saint Ives, a famous artists' colony, and the wife of the director-designate of the Sarin pilot plant was a painter who had her heart set on living in Saint Ives. Her preference apparently won out, and in February 1949, the British Ministry of Supply approved the founding of the Chemical Defence Establishment at Nancekuke. Ironically, most of the staff were transferred from Sutton Oak and spoke with North country (Lancashire) accents.

Construction of the Sarin pilot plant at Nancekuke began in August 1951 and was completed two years later. The design provided for continuous rather than batch manufacturing and included innovations in automatic process control that enabled the plant to operate largely unmanned. Production capacity was about one ton of Sarin per week, or two and a half tons when operated on a three-shift basis. This output provided enough of the agent for experimental purposes and to build up a small stockpile. Nancekuke also had research laboratories, service buildings, engineering workshops, stores, and welfare facilities.

Meanwhile, the Cold War between the United States and the Soviet Union was intensifying, bringing with it competition in all categories of armament, including the German nerve agents.

CHAPTER SEVEN

BUILDING THE STOCKPILE

As the Cold War deepened, the United States relied heavily on its nuclear monopoly to deter a Soviet invasion of Western Europe or Japan. Because U.S. policy makers viewed nuclear weapons as a panacea for America's security problems, the Army Chemical Corps went into decline. Its budget was slashed by nearly two thirds in 1947, and there was even a short-lived proposal to downsize Edgewood Arsenal and move it to Camp Siebert, Alabama.

Nevertheless, development of the nerve agents continued. Researchers at Edgewood Arsenal evaluated several members of the so-called G series—Sarin (GB), Soman (GD), Ethylsarin (GE), Cyclosarin (GF), Isopentylsarin (GH), and other structural analogues. The Edgewood scientists sought to identify the agent with the best combination of militarily useful characteristics, including high toxicity, stability, nonflammability when explosively dispersed, ease of decontamination, availability of antidotes or protective drugs, and feasibility and economy of production.

A debate ensued over whether Sarin or Soman should be chosen as the standard U.S. nerve agent. An Edgewood chemical engineer, Benjamin L. Harris, did a study of Soman and determined that it was superior to Sarin in both toxicity and persistence. The synthesis of Soman required only a minor alteration in the manufacturing process for Sarin, namely the replacement of one alcohol with another. Still, whereas the production of Sarin used ordinary isopropyl (rubbing) alcohol, which was cheap and widely available, Soman required pinacolyl alcohol, which was difficult and costly to produce. Since no large-scale manufacturing facility for pinacolyl alcohol existed, a dedicated plant would have to be designed and built at considerable expense. Despite this hurdle, Harris argued to Edgewood technical director Seymour Silver that the military advantages of Soman outweighed the drawbacks.

Saul Hormats, a weapons developer, disagreed. He noted that Sarin was not only easier to manufacture than Soman but was superior for attacking enemy troop concentrations because it evaporated more readily to form a lethal vapor. Another drawback of Soman was the lack of a reliable antidote. The agent inactivated ("aged") cholinesterase irreversibly within two minutes of exposure, making it hard to treat friendly troops who might be exposed accidentally. Unless and until an improved antidote was developed, Hormats argued, Soman would be too dangerous to produce, transport, and handle. In view of these considerations, in May 1948 the Chemical Corps Technical Committee endorsed the adoption of Sarin as the standard U.S. nerve agent.

Meanwhile, tensions in Europe were rising. On June 24, 1948, after the three Western powers had introduced a new currency, the deutsche mark, in their occupation zones, the Soviet Union cut off all land and rail traffic to West Berlin in an effort to starve the western enclave into submission. Over the next year, a massive U.S. and British airlift kept the city's population supplied with food and other vital goods, ultimately leading Stalin to lift the blockade on May 12, 1949. At the same time, however, Moscow tightened its grip over the rest of Eastern Europe. U.S. intelligence agencies estimated that the Soviet Union had a large numerical edge in conventional forces in Europe and was modernizing its chemical arsenal, even as it worked feverishly to develop a nuclear weapon. According to a top secret assessment in January 1949 by the Joint Chiefs of Staff, "The Soviet Union possibly possesses limited stockpiles of German nerve gases, and has the ability to produce them, but probably could not engage in large-scale nerve gas warfare before mid-1950."

In August 1949, the Soviet Union successfully tested an atomic bomb on the remote steppes of Kazakhstan, achieving a nuclear weapons capability years earlier than Western intelligence services had predicted and radically shifting the global balance of power. Given the transformed strategic situation, U.S. policy makers were forced to reevaluate their heavy reliance on nuclear deterrence. If it was no longer possible for Washington to threaten the use of atomic weapons without inviting mutual destruction, what other weapons could help win a future global conflict?

At this juncture, the Chemical Corps saw an opportunity to regain some of its lost influence. Stressing the need to reinforce nuclear deterrence by other means, the Corps called for an end to the "retaliation-only" chemical warfare doctrine that had been in effect since 1943, when President Roosevelt had stated that the United States would "under no circumstances resort to the use of such weapons unless they are first used by our enemies."

On the afternoon of February 1, 1950, President Truman met in the Cabinet Room of the White House with the members of his National Security Council to discuss U.S. chemical warfare policy. Secretary of Defense Louis Johnson gave an oral presentation and distributed a memorandum from General Omar N. Bradley, the chairman of the Joint Chiefs of Staff, arguing against a change in the "retaliation-only" posture. "While the United States must be at all times prepared for the initiation of gas warfare by our enemies—the time, place and purpose of initiation to be chosen by them—it is doubtful if the United States Government should adopt a policy of unrestricted gas warfare excepting in retaliation," General Bradley wrote. Such a policy shift, he explained, would be unacceptable to the European NATO allies because of their "vulnerable and comparatively defenseless position" and would also be opposed by the American people, who remembered the horrors of gas warfare in World War I.

Swayed by General Bradley's arguments, President Truman reaffirmed the "retaliation-only" chemical doctrine in policy memorandum NSC-62 of February 17, 1950. This decision was a setback to the ambitions of the Chemical Corps, which had sought to remove any constraints on the acquisition and use of chemical weapons by the armed forces. Nevertheless, the U.S. Army moved forward with plans to acquire a stockpile of Sarin nerve agent. The Chemical Corps Technical Committee had decided in June 1949 that Tabun (GA) was inferior to Sarin and should not be mass-produced, although the Tabun-containing bombs and shells confiscated from Germany were still in good condition and would be retained as an emergency war reserve.

The next step was to procure a stockpile of Sarin. Major General Anthony C. McAuliffe, the chief of the Chemical Corps, had become famous during World War II as the commander of U.S. forces at Bastogne who had said "Nuts" when the Germans told him to surrender. In a secret memorandum dated April 14, 1950, he estimated the U.S. stockpile requirement for Sarin at 48,000 tons. If the Army built a production facility with a capacity of 25 tons per day, it would have to operate continuously for five and a half years to produce that quantity of agent and could then be placed on standby status. In wartime, the estimated military requirement for Sarin would be 2,000 tons per month, requiring two additional production plants.

In late April 1950, General McAuliffe gave a speech at a meeting in Detroit of the American Chemical Society in which he revealed publicly for the first time that the U.S. Army was working on a new generation of chemical warfare agents that attacked the nervous system. "Our use of them would be purely retaliatory," he explained. "It is a well-known fact that many German scientific experts on toxic chemical warfare are being exploited by Soviet Russia. It must be assumed, therefore, that we are not the sole possessors of the offensive and defensive secrets of the new nerve gases."

After General McAuliffe's disclosure, the nature and composition of the nerve agents became a topic of intense speculation among journalists and armchair military strategists. Because of the dense veil of secrecy surrounding the new weapons, misconceptions were widespread, such as the belief that nerve agents caused temporary incapacitation rather than death. According to a May 1950 article in *Time* magazine, "Presumably [nerve gas] would be sprayed over enemy cities by planes in the same way that whole areas are sprayed with mosquito-killing DDT, paralyzing the whole population. Then the attacking army, equipped with protective masks, would march in and take over."

ALTHOUGH THE U.S. nerve agent program was proceeding at a deliberate pace, world events suddenly transformed the situation overnight. At approximately 4:00 a.m. on June 25, 1950, the North Korean Army began to fire artillery and mortar shells at South Korean military positions south of the 38th Parallel, which marked the border between the two countries. Soon massive columns of North Korean tanks and infantry poured across the demarcation line at multiple points, and at 11:00 a.m., Pyongyang issued a formal declaration of war. Responding decisively to the Communist surprise attack, President Truman quickly organized a military intervention in Korea under United Nations auspices.

On June 30, 1950, only five days after the North Korean invasion, a group of civilian advisers to the Pentagon called the Ad Hoc Committee on Chemical, Biological and Radiological Warfare submitted its report. The panel was chaired by Earl P. Stevenson, president of the consulting firm Arthur D. Little, Inc. Rejecting the notion that poison gas was uniquely immoral or inhumane, the Stevenson committee argued that chemical arms might be "exceedingly important as a supplement to weapons now in general use for holding back the advance of enemy ground forces." Although the timing of the report was fortuitous—Secretary of Defense Louis A. Johnson had appointed the blue-ribbon panel back in December 1949—the sudden outbreak of war in Korea greatly enhanced its political impact.

The Stevenson committee concluded that the United States lagged far behind the Soviet Union in chemical warfare capabilities. Whereas intelligence reports indicated that the Soviets had captured entire German factories for the manufacture of nerve agents, the United States possessed only limited stocks of mustard and phosgene and had not yet begun to produce nerve agents or suitable delivery systems. The committee blamed Roosevelt's "retaliation-only" policy for putting the United States in a position of dangerous inferiority vis-à-vis the Soviet Union. "Such a policy has resulted in the assignment of low priorities to the research, development, and production of chemical weapons," Stevenson wrote in his cover letter. "The security of the United States demands that the policy of 'use in retaliation only' be abandoned."

The Stevenson committee's recommendations sparked an intense debate within the Joint Chiefs of Staff. Some senior military officials favored switching to a policy of chemical first use, both to deter the Soviet Union and to counter human-wave infantry attacks in Korea. They warned that the Chinese Communists led by Mao Zedong, who had seized power on the mainland in 1949, represented a new "Yellow peril" that might intervene on the North Korean side. Nerve agents, they argued, would be highly effective against the Chinese People's Army, a technically backward force that lacked modern protective gear and whose chief asset was large reserves of manpower.

Although both the U.S. Army and Air Force favored scrapping the retaliation-only policy, the Navy opposed an expanded role for chemical weapons and strongly defended the status quo. Aware that the Army wanted the other services to share the burden of delivering chemical weapons, Navy officials were concerned about the problem of storing nerve agent munitions on aircraft carriers and other warships. Because the cramped quarters of a ship at sea provided nowhere to run, a single leaking chemical bomb would be disastrous for the crew. For this reason, the Navy resisted a change in chemical warfare posture and ultimately prevailed in the internal Pentagon debate.

On September 7, 1950, the Joint Chiefs informed the Office of the Secretary of Defense that they accepted all of the Stevenson committee's recommendations except for the proposed change in U.S. chemical warfare doctrine. Not only did the United States lack the stockpiles and delivery systems needed to employ nerve agents on a large scale, but Great Britain supported the "retaliation-only" policy and a unilateral change in the U.S. posture would risk alienating America's closest ally. General Bradley concluded that improved preparedness to conduct chemical warfare "can and must be achieved under a policy of retaliation-only" and suggested that any consideration of a change in doctrine be "deferred pending further developments." On October 27, 1950, the new Secretary of Defense, General George C. Marshall, signed off on the JCS position.

Despite this decision, the advocates of chemical warfare achieved most of their objectives. Without challenging the "retaliation-only" policy directly, Secretary of Defense Marshall directed the Pentagon to implement all the other recommendations in the Stevenson report, including actions needed "to make the United States capable of effectively employing toxic agents at the outset of a war." During the fifteen months following the outbreak of the Korean conflict, the research-and-development budget of the Chemical Corps (which also included smoke and incendiary munitions) tripled in size, and the number of researchers grew from 2,100 to 3,700. Several private companies and universities also received government contracts to perform related R&D. To support this expanded effort, the Pentagon authorized the open-air testing of advanced chemical weapons and delivery systems, which required a vast amount of open space far from populated areas. During the summer of 1950, the Army reactivated and expanded Dugway Proving Ground, a chemical and biological testing site in the Utah desert that had been established in 1942 and placed on standby status after World War II.

THE PRIMARY TASK facing Edgewood Arsenal was the design and construction of a full-scale Sarin manufacturing plant based on the German DMHP process and drawing on the technical details provided by Schieber and his team of Falkenhagen scientists. On October 31, 1950, Secretary of Defense Marshall, using contingency funds available to him, secretly authorized $50 million for the initial design, engineering, and construction of the Sarin plant under the code name "Gibbett." The Chemical Corps established a task force to oversee all aspects of the project, including procurement, funding, security, and administration. Kellex Corporation (which later changed its name to Vitro Corporation) was selected as the prime contractor, and in November 1950 the design process began under the management of the Army Corps of Engineers.

Earlier, the Chemical Corps had contracted with Monsanto Chemical Company to build and operate a small pilot plant to test the DMHP process, but corrosion had damaged the apparatus so badly that it could not be salvaged. It was therefore clear that the design, construction, and operation of the full-scale Sarin plant would demand a high level of engineering expertise. According to a memorandum by Major Stanley Levy, chairman of the Chemical Corps's Industrial Mobilization Review Committee, "The chemistry of the G Agents embraces an entire new field and much time must be given to the instruction of any contractor selected for this work."

At the same time the Chemical Corps began designing the full-scale Sarin production plant, it was developing a specialized delivery system called the M34 cluster bomb. A metal cylinder with seventy-six Sarin-filled bomblets neatly packed inside, the bomb weighed a total of 1,000 pounds. Development of this weapon proceeded slowly until the outbreak of the Korean War, when the Pentagon authorized a "crash" acquisition program. In August 1950, even though the engineering and testing of the M34 were not yet complete, the Chemical Corps froze the design in order to move it rapidly into production. This telescoping of development and procurement gave rise to numerous technical problems later on.

INITIALLY, the Army assumed that the entire Sarin manufacturing process would be carried out at a single location. As planning progressed, however, it was considered prudent to reduce the vulnerability of the production complex to attack or sabotage by dividing the five manufacturing steps between two separate facilities, designated Site A and Site B. Site A would perform the initial three-step process in which elemental phosphorus was converted into a chemical intermediate called methylphosphonic dichloride [$CH_3P(O)Cl_2$], known by the short name "dichlor." The dichlor manufactured at Site A would then be shipped by rail in special tanker cars to Site B, where the final two-step conversion into Sarin would take place.

The Chemical Corps decided to build Site A on forty-five acres of land purchased from the Tennessee Valley Authority near the town of Muscle Shoals, in the northwest corner of Alabama. This site, on the TVA's Wilson Dam Reservation, was chosen because the land was government-owned and had an existing plant for converting phosphate ore into elemental phosphorus, ample electrical power and water, and a pool of trained operating personnel. The location selected for Site B was on ninety acres in the north-central portion of Rocky Mountain Arsenal near Denver, Colorado. Founded in 1942 to produce mustard agent and incendiary weapons for World War II, the arsenal sprawled over twenty-seven square miles of flat scrubland and cottonwoods along the foothills of the Rockies.

The U.S. program to mass-produce Sarin took on new urgency in the light of an ominous assessment of the Soviet chemical warfare threat. On December 15, 1950, the Central Intelligence Agency issued a top secret National Intelligence Estimate (NIE) on the possibility of a Soviet chemical or biological attack against the United States. This report concluded that if the Soviets decided to launch a chemical attack, they would almost certainly use nerve agents. By 1952, the CIA predicted, the Soviets would have at their disposal "sufficient nerve gas for sustained extensive employment" on the battlefield, and by 1954 they would possess "new agents in sufficient quantity for limited mass lethal attacks on selected military or industrial targets in the US."

The NIE also discussed the possibility that Soviet covert operatives might smuggle nerve agents into the United States for sabotage attacks against key military installations. "Since the agents are odorless, colorless liquids, they can be transported in glass or suitably lined containers," the report noted. "Hence, the agent could be shipped in any desired quantity disguised as innocuous liquids, such as champagne or perfume." Because the unique characteristics of nerve agents would make it possible to identify the Soviet Union as the source of an attack, the CIA did not consider it likely that the Kremlin would resort to such weapons prior to the outbreak of general war. Nevertheless, the CIA's alarmist assessment of the Soviet threat gave the Chemical Corps a powerful rationale to accelerate the production of nerve agents in order to deter a Soviet chemical attack.

IN THE SPRING of 1951, Vitro Corporation began to build the Sarin production facilities at Site A (Muscle Shoals) and Site B (Rocky Mountain Arsenal). Given the urgent military demand for Sarin-filled weapons, every effort was made to expedite construction. To save time, the various production facilities were designed and built concurrently, without the usual exhaustive development and testing at the bench-scale and pilot-plant levels. Because much of the information needed to scale up dichlor production was lacking, Vitro engineers were forced to make numerous judgment calls, turning Site A into what was effectively a "pilot plant" of gigantic proportions. Although the Army had planned for Site A to be operational in November 1951 and Site B a month later, at a total cost of $30 million, these projections proved to be wildly optimistic.

The Muscle Shoals complex began limited operation in June 1952. It consisted of a series of chemical plants, one for each step in the conversion of phosphorus to dichlor, plus a dedicated facility for the production of chlorine. Except for high fences and other security measures, Site A resembled an ordinary chemical factory. Mounted on steel superstructures in the open air were numerous corrosion-resistant reactor vessels interconnected with stainless-steel pipes, pumps, and valves. Local residents referred to the mysterious industrial facility at the TVA reservation as "The Thing" and speculated that the Atomic Energy Commission was using it to process uranium ore. To conceal its real purpose, the Army gave Site A the innocuous name of "Phosphate Development Works."

Because the Muscle Shoals facility had been a crash program based on incomplete development, Vitro Corporation and the Chemical Corps faced numerous technical problems in getting it fully operational. The highly corrosive chemicals used to make dichlor caused leaks at expansion joints and numerous failures in valves, lines, and other components, necessitating equipment changes, process modifications, and lengthy downtimes. In 1953,

The Phosphate Development Works at Muscle Shoals in northwestern Alabama. This facility manufactured dichlor, the main precursor chemical used to produce Sarin, from 1953 to 1957.

A production plant for phosphorus trichloride—the main starting material for dichlor production—in the Phosphate Development Works at Muscle Shoals.

A plant for the Step 1 process in dichlor production in the Phosphate Development Works at Muscle Shoals.

a runaway reaction in the phosphorus trichloride plant (Building 101) caused an explosion in which five workers died.

Another problem arose during the third step of dichlor production, which generated an unwanted by-product called phosphorus oxychloride in quantities too large to be disposed of by sale on the commercial market. Even when the Muscle Shoals facility was running at only half capacity, it produced about 55 tons of phosphorus oxychloride per day. It was therefore necessary to build a separate reprocessing plant to convert the by-product back into phosphorus trichloride, which could then be reused as a chlorinating agent in steps 1 and 3 of the production process. The reprocessing plant cost about $9 million to build, was costly, difficult, and dangerous to operate, and never kept pace either with the volume of by-product or with the demand for raw material.

IN PARALLEL WITH the construction of the troubled dichlor plant at Muscle Shoals, Vitro Corporation, supported by hundreds of subcontractors and suppliers, built the North Plants complex at Rocky Mountain Arsenal to perform the final two steps in the production of Sarin. The manufacturing process was complex and extremely dangerous. First, dichlor was reacted with hydrogen fluoride (HF) gas to yield a roughly 50:50 mixture of dichlor and difluor [$CH_3P(O)F_2$]. This so-called di-di mixture was then combined with isopropyl alcohol to produce Sarin, giving off hydrochloric acid (HCl) as a gaseous by-product. Sarin was purified by passing it through a distillation column in which the temperature was precisely controlled; it was then stabilized with tributylamine or triethanolamine and loaded into munitions.

Because the Sarin manufacturing process involved two highly corrosive chemicals that could erode stainless steel (hydrogen fluoride as a reactant and hydrochloric acid as a by-product), the design, construction, and operation of the Rocky Mountain plant posed unique engineering challenges. To resist corrosion and prevent the gradual destruction of the process equipment, Vitro used parts made of a high-nickel-steel alloy called Hastalloy. The company also developed new fabrication and welding techniques to make optimal use of this special material. Construction of the Sarin production facility ultimately consumed 150 tons of nickel, fifteen tons of Hastalloy, thirty-five tons of carbon steel, thirty-five tons of copper, and five tons of aluminum.

The Sarin production equipment was installed in three bays inside Building 1501, a windowless, five-story blockhouse that was sealed to contain the lethal fumes. One of the largest poured-concrete structures in the United States at the time, the blockhouse was designed to withstand a major earthquake and hurricane-force winds of 100 miles per hour. The Rocky Mountain complex also included a munitions filling line, an administration building, a hospital, quality control laboratories, utilities, and a waste treatment plant. In view of the extreme hazards posed by nerve agents—breathing air containing only one part per million of Sarin vapor for ten minutes could be fatal—several U.S. government agencies and industrial safety experts analyzed every conceivable risk from the production facility to the surrounding communities. Vitro engineers developed methods to prevent leaks and ensure the safe operation of the plant that went far beyond all previous safety requirements.

By late 1952, Site B at Rocky Mountain Arsenal, which the Army gave the deceptive cover name of "Incendiary Oil Plant," was ready to begin operation. Production was delayed, however, by the ongoing technical and management problems at Site A, which had become a major thorn in the side of the Chemical Corps. Although various parts of the Muscle Shoals complex were operational, the facility as a whole could not achieve the planned sustained rate of dichlor production. In addition to persistent technical problems, inadequate supervision and poor personnel management caused further delays.

The Air Force was becoming increasingly impatient over the lack of Sarin to fill the M34 cluster bombs. Although the weapon had been scheduled to enter production in May 1951, the start date had been delayed five times. In order to work around the chronic problems at Muscle Shoals, the Chemical Corps hired the Shell Chemical Company to manufacture dichlor at Rocky Mountain Arsenal, using an alternative to the DMHP method called the aluminum phosphorus complex (APC) process. Dichlor produced by the APC process was ultimately used to produce about a third of the Sarin in the U.S. stockpile. Although the APC method worked fairly well, it had several liabilities: the process was complex, created an explosive hazard, and generated a large volume of hazardous waste for each pound of product.

In 1952, Otto Ambros was released from prison after serving only two years of his eight-year sentence at Nuremberg. The sentence had been reduced at the request of the West German government and John J. McCloy, the U.S. High Commissioner for Germany. After Ambros's release, EUCOM offered him a job advising the Army Chemical Corps. Dr. Wilhelm Hirschkind, the Dow Chemical scientist who had interviewed Ambros in July 1945, also renewed contact and arranged several meetings for him with Dow executives. As a result, Ambros soon became a successful consultant to both the German and American chemical industries.

Meanwhile, the Korean War was still raging. In March 1953, the U.S. Army Chief of Staff recommended shipping a stockpile of chemical weapons to Okinawa, an island off the coast of Japan that had been controlled by the U.S. military since the end of World War II. Deploying chemical munitions on Okinawa would enable the Army's Far East Command to retaliate if the North Koreans or their Red Chinese allies resorted to chemical attacks. According to a memo to the Joint Chiefs from the

Army Chief of Staff, the deployment of chemical arms to Okinawa required "the utmost secrecy . . . in order to forestall disclosure as long as possible. However, it should be noted that disclosure becomes an increasing possibility after the shipments of munitions are set in motion from United States depots."

The Korean War ended on July 27, 1953, without any use of chemical weapons. Several years later, however, Brigadier General Jack Rothschild revealed that U.S. Army field commanders in Korea had requested permission to use poison gas to break the military deadlock but had been turned down.

Shortly after the end of the war, the Chemical Corps launched a public relations campaign to educate ordinary Americans about chemical warfare and civil defense. Army officials invited well-known journalists and authors to visit the Muscle Shoals and Rocky Mountain production plants and receive briefings on the Soviet chemical warfare threat. One of the more famous writers to accept this invitation was Cornelius Ryan, a popular military historian and the author of the best-selling World War II epic *A Bridge Too Far.* Ryan's article, titled "G-Gas: A New Weapon of Chilling Terror," was the cover story in the November 1953 issue of *Collier's Magazine.*

Citing Chemical Corps experts, Ryan reported that a single Soviet TU-4 bomber could drop seven tons of Sarin-filled bombs on a major American city, potentially inflicting a death toll in the millions. A map accompanying the article compared the effects of a nerve agent attack on Washington, D.C., with the detonation of a small atomic bomb. Whereas the A-bomb would kill everyone within a three-mile radius, the Sarin cloud would drift up to fifty miles downwind, blanketing an area of a hundred square miles with lethal vapor. After painting this grim picture, Ryan quoted Major General E. F. Bullene, the chief of the Chemical Corps, who argued that the best way to deter a Soviet chemical attack was for the United States to possess the means to retaliate. "At this time," General Bullene warned, "the only safe course is to be prepared to defend ourselves and ready to use gas in over-powering quantities."

However improbable the scenario that Moscow would launch a massive chemical first strike against U.S. cities, the idea took root in the fertile soil of Cold War paranoia. The federal Civil Defense Administration prepared a thirty-minute color film showing how enemy aircraft might spray clouds of lethal nerve gas to kill and demoralize the U.S. civilian population, and how ordinary citizens could protect themselves. Because nerve agent vapors tended to hug the ground, the narrator intoned, the best way to survive a chemical attack on one's home was to close the windows on the lower floors and shelter in an upstairs room.

By December 1953, Site A was two years behind its original projected date for full-scale operation and millions of dollars over budget. It was not until 1954 that the Muscle Shoals facility finally began producing dichlor at full capacity. Because the chemical was a solid at room temperature and highly corrosive, it had to be transported to Rocky Mountain Arsenal in special railroad tank cars lined with nickel. When the trains arrived at their destination, heating coils inside the tank cars melted the dichlor into liquid form, which was then pumped into glass-lined tanks for storage.

Site B operated twenty-four hours a day, seven days a week, converting dichlor into thousands of tons of Sarin per year. The interior of the block-house was divided into three operating bays that contained the chemical-processing units; in an emergency, each bay could be automatically sealed off from the other parts of the building. In addition to these physical barriers, powerful ventilation fans kept the interior of the blockhouse at a nega-

An aerial view of the North Plants complex at Rocky Mountain Arsenal, which produced Sarin from 1953 to 1957.

Rocky Mountain Arsenal blockhouse, which contained the final steps in the Sarin production process. The building was hermetically sealed to prevent the escape of deadly gases.

Soldiers guard Sarin-filled ton containers at Rocky Mountain Arsenal.

tive atmospheric pressure so that in case of a breach in containment, the deadly fumes would be retained inside.

Because of the extremely hazardous nature of the Sarin production process, technicians rarely entered the manufacturing areas while the plant was in operation. Instead, they manipulated valves with long-handled levers from outside the sealed enclosures, while monitoring gauges that provided a continuous readout of the reaction temperatures and pressures. Between runs, however, occasional visits to the manufacturing bays were necessary to perform adjustments, repairs, and maintenance. Whenever workers entered the "hot zone," they wore full-body rubber suits and masks, carried syrettes filled with atropine, and worked in pairs so that they could assist one another in an emergency. Over the years, the plant workers developed a healthy respect for Sarin's insidious power. They had their cholinesterase levels checked once a week at the base dispensary; if the value fell below a specified threshold, they stopped working in the hot zone until the test returned to normal.

Despite these elaborate safety precautions, numerous mishaps occurred. In 1954, more than seventy technicians at Site B received low-level exposures to Sarin and had to be treated with atropine at the base hospital. A few individuals were hospitalized for several days with "small eye" (pinpoint pupils and blurred vision), cramps, chest pain, shortness of breath, and nausea. Some of them experienced wild dreams, extreme anxiety, and an inability to make decisions. These mental symptoms suggested that exposure to even extremely low doses of nerve agents could cause psychological disturbances and distort the judgment of commanders and troops in combat.

At the end of the production process, the distilled Sarin was pumped from the blockhouse into the munitions-loading plant (Building 1601), a narrow, windowless, bunkerlike structure about 600 feet long. Airtight, this building contained filling lines for the various types of munition, including artillery shells (155 mm, 8-inch, 105 mm), aerial bombs, and submunitions for cluster weapons. Each filling line had four or five stations enclosed inside sealed metal cabinets to prevent the escape of toxic fumes. At the first station, the machine loaded a projectile or bomb with a preset amount of liquid Sarin pumped from an underground storage tank. A conveyor belt carried the munition to the next station, where an overlay of helium gas was injected into the space remaining inside. The filling aperture was then capped, welded with a double seal, and vacuum-tested for leaks using a helium detector. Because helium is an extremely small molecule that can penetrate the slightest leak, the presence of the inert gas in the air was a reliable indicator of defective welds.

Next the filled shells or bombs moved along the conveyor belt to another automated station that decontaminated the outside surfaces. Finally, the finished munitions were transported to an open packing area, where workers weighed, painted, stenciled, assembled, and crated them for storage. Because of a shortage of empty bomb casings and projectiles, thousands of gallons of bulk Sarin were stored temporarily in steel ton containers, two feet wide by eight feet long. Hundreds of these containers, painted silver, were lined up in rows in the storage yard.

Elaborate safety measures were designed to protect the workers at the Rocky Mountain Arsenal facility. Thirty M5 Automatic G-Agent Fixed Installation Alarms, each seven feet high and weighing 725 pounds, were installed throughout the blockhouse and the filling plant to monitor the concentration of Sarin vapor in the air. The detectors contained a solution that reacted with Sarin to yield a fluorescent compound; a photometer measured the fluorescence and triggered an alarm in about ten seconds. Human operators then shut down production and sealed off the affected unit. To back up the electronic alarms, cages containing canaries and white rabbits were placed at strategic points around the facility. Because these animals were more sensitive than humans to nerve agent exposure, they would provide a few minutes' warning of an accidental leak; the rabbits' large pink eyes made it easy to see when their pupils were constricted. The Sarin plant also incorporated an advanced pollution abatement system in which contaminated effluent air from the blockhouse and the filling plant passed through a series of four caustic Venturi scrubbers before being exhausted out a 200-foot stack.

THE AMOUNT OF TIME, money, and effort invested in the Sarin production program turned out to be vastly greater than anticipated. Whereas the original cost estimate for construction of the two facilities had been $30 million, the actual total was well over $100 million. Nevertheless, there was no outcry from Congress or the public over the huge cost overruns because the entire project was shrouded in secrecy. It was not until July 1954 that Major General William M. Creasy, Bullene's successor as chief of the Chemical Corps, disclosed to the public that the real purpose of the Phosphate

Development Works at Muscle Shoals was to produce an intermediate chemical used in the manufacture of nerve gas.

Even as the mass production of Sarin was under way, chemists at Edgewood Arsenal continued to assess a variety of novel compounds as potential nerve agents. In 1953, the Advisory Committee on New Agents requested the toxicity screening of 157 candidate chemicals, selected from a list of about 400. Edgewood scientists also studied agent cocktails such as Sarin and mustard, and experimented with thickeners and other additives that could modify the droplet size and physical properties of Sarin, increasing its persistence or ability to penetrate clothing or skin.

The Chemical Corps issued numerous research contracts for studies of Sarin production, stabilization, detection, and decontamination to outside entities such as the National Bureau of Standards, the Standard Oil Development Company, the University of Kansas, and Louisiana State University. The Air Force also commissioned Project Big Ben, a study group of statisticians, mathematicians, and engineers at the University of Pennsylvania, to analyze optimal ways of dispersing Sarin over area targets such as a military formation.

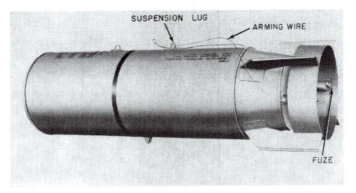

The M34 aircraft-delivered cluster bomb was filled with 76 Sarin-filled bomblets that were dispersed over the target area.

In 1954, the Air Force and the Army Chemical Corps conducted a series of ten field tests of the M34 cluster bomb at Dugway Proving Ground. During these trials, which were performed at night, two B-47 bombers flew from Eglin Air Force Base in Florida to Dugway and each dropped two M34 bombs from an altitude of 35,000 feet. Fire pots laid out on the desert floor delineated the target area: a 6,000-foot square superimposed on a circle 8,000 feet in diameter and crosshatched with grid lines. As each cylindrical bomb fell to earth, it deployed a parachute to slow its descent. The metal casing then burst open and scattered its cargo of seventy-six bomblets, which detonated on impact with the ground and discharged their content of Sarin (2.6 pounds). Distributed over the test grid were sampling devices designed to collect Sarin vapor and droplets, and cages containing test pigeons. The target area also included two family-style model homes and a slit trench to measure the ability of the toxic cloud to penetrate these structures. In addition to the M34 cluster bomb, the Chemical Corps standardized two artillery projectiles in 1954: a light 105 mm shell that held 1.6 pounds of Sarin, and a heavy 155 mm shell that held 6.5 pounds.

Sarin-filled munitions were stockpiled at several Army depots on U.S. soil, and some were secretly deployed overseas. Although the first chemical weapons transferred to Okinawa in 1953 had been mustard-filled, in 1954 the stockpile was augmented with munitions containing Sarin. Outside the Chemical Corps, however, Army field commanders viewed chemical weapons as more trouble than they were worth, contaminating the battlefield and forcing troops to wear clumsy protective gear that degraded their fighting efficiency. This antipathy led to a strong resistance to integrating chemical arms into the Army's force structure and war-fighting doctrine. As a result, apart from the small stockpile deployed on Okinawa, the vast majority of chemical munitions were stored within the continental United States, mostly at or near the original production locations and far from coastal ports where they could be readily deployed to Europe or the Pacific. Moreover, a large fraction of the

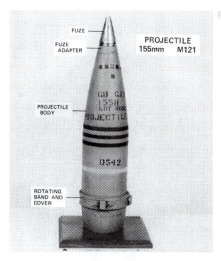

This 155 mm artillery shell was loaded with Sarin (GB) and marked with three rings, the symbol for chemical nerve-agent munitions.

nerve agent stockpile was not loaded into munitions at all, but remained in bulk storage tanks.

Throughout the 1950s, the Pentagon continued to justify the mass production of nerve agents with the specter of a large-scale Soviet chemical attack. The Joint Strategic Plans Committee of the JCS predicted, "The Soviets have been producing at least one of the 'G' agents since 1949 and hence, by 1956, will probably be capable of extensive employment of nerve gases." Kremlin leaders, for their part, saw the United States moving forward aggressively with Sarin production and were determined not to be left behind in the chemical arms race between the superpowers.

CHAPTER EIGHT

CHEMICAL ARMS RACE

ONCE THE SOVIET UNION had successfully produced Tabun at Chemical Works No. 91 in Stalingrad, it moved on to the more challenging task of manufacturing Sarin on an industrial scale. One of the chemical engineers who was deeply involved in developing the Sarin production process was Boris Libman. Although only twenty-seven years old when he joined the development team at Stalingrad, he had already overcome a lifetime's worth of adversity.

Born in 1922 to affluent Jewish parents in the Latvian capital of Riga, Boris had grown up during the brief period between the wars when Latvia and the other Baltic republics were independent and relatively prosperous. In 1940 the Red Army invaded Latvia; it was subsequently incorporated into the Soviet Union and the Soviet authorities confiscated the Libman family's land and assets. Conscripted into the Red Army, Boris was wounded in a battle with the Germans near the Latvian border. After a year of rehabilitation, he was sent back to the front. During a major military operation south of Leningrad, he was severely injured and left for dead, but once again he miraculously survived his wounds.

After his recovery, Libman received an honorable discharge as a disabled veteran and sought to resume his studies. He sent a letter to the Leningrad authorities requesting permission to study at the Moscow Institute of Chemistry. A few weeks later he received an official reply denying his request on the grounds that "Boris Libman" had been killed in action; the authorities apparently believed that he was trying to impersonate a fallen soldier. After much effort, Libman managed to persuade the officials of his true identity. He then began his studies in the military division of the Moscow Institute, where he obtained a candidate's degree (equivalent to a master's) in chemical engineering. A firm believer in the Communist Party, he wanted to do his part to contribute to the nation's defense.

In 1949, Libman was hired by Professor Soborovsky at the Stalingrad branch of Scientific Research Institute No. 42 (known by its Russian acronym NII-42) to assist with the development of an industrial-scale manufacturing process for Sarin. The Soviets had decided to adapt the German DMHP process, which involved the production of dichlor as an intermediate. Soborovsky and Libman began with laboratory studies and progressed to a pilot-scale facility, drawing on the knowledge of Dr. von Bock and the other captive German scientists from Dyhernfurth. Because the Germans had no love for their Soviet hosts, they provided information only grudgingly and often deliberately tried to mislead. Although Libman spoke fluent German, he pretended not to understand the language so that he could eavesdrop on the private conversations of the Dyhernfurth scientists and glean useful tidbits. Once the Soviet engineers had extracted as much information from the Germans as they could, Bock and his colleagues were sent back to West Germany in 1954.

In May 1952, the Soviet Council of Ministers and the Central Committee of the Communist Party passed a secret resolution authorizing the construction of a full-scale Sarin production facility at Chemical Works No. 91, with a planned capacity of 2,000 metric tons of agent per month. To implement this resolution, the engineers at Stalingrad faced a daunting series of bureaucratic hurdles. Importing foreign-made production equipment required the approval of the Council of Ministers, which was granted only if the requester could prove that an item could not be manufactured in the Soviet Union. Although much of the production apparatus for the Sarin plant had been confiscated from Dyhernfurth, some items had to be built from scratch, a difficult and time-consuming process. The first challenge was to obtain corrosion-resistant titanium and high-nickel steel, which were generally reserved for more favored industries, such as nuclear power, aircraft, and submarines. Because specialized materials were in short supply, Soborovsky had to persuade a committee reporting to the Council of Ministers that no substitutes were possible, a task that took several months.

Once the materials had been secured, it was necessary to negotiate with the equipment manufacturers. The state-owned factories often insisted that a certain piece of apparatus could not be made with the available machine tools and demanded modifications in the design. Although Soborovsky tried to resist such demands, in the end he always had to compromise. Another anomaly of the Soviet centrally planned economy was that the price of machinery was determined strictly by weight, giving manufacturers no incentive to produce small lots of high-quality items. Instead they preferred to make large orders of simple, heavy equipment that could be produced easily and with high labor productivity. For this reason, Soborovsky had great difficulty obtaining specialized pieces of chemical production apparatus. Even when he got his way, he had to order larger quantities than he really needed.

The Sarin plant at Stalingrad started out as a copy of the one at Dyhernfurth, but for reasons of national pride Soborovsky and Libman tried to improve on the German production process. In so doing, they ran into serious technical problems that resulted in lengthy delays. To get the development effort back on track, NII-42 in Moscow sent to Stalingrad a brilliant chemical engineer named Simion Levovich Varshavsky, but it took him several years to work out all the complexities of the Sarin manufacturing process. Although the development had begun in 1948, it was not until 1959—more than a decade later—that Chemical Works No. 91 was churning out large amounts of Sarin with a satisfactory level of purity and stability.

The Sarin production plant consisted of a main technical building and three annexes, including a filling line where the nerve agent was loaded into artillery and mortar shells, aerial bombs, missile warheads, and spray tanks. To manage the plant operations, the Soviet government recruited young male engineers over eighteen years of age from the institutes of technology in Moscow, Leningrad, and Stalingrad, which had special programs to train specialists in chemical weapons production. After World War II, women could not work in chemical weapons plants and were restricted to auxiliary operations or laboratory research.

Chemical Works No. 91 was a dual-purpose industrial complex in which chemical weapons accounted for about 35 percent of production and commercial chemicals for about 65 percent. The Soviet authorities had decided to integrate military and civilian activities at the sprawling site along the Volga in order to generate needed income and create a legitimate cover story tht would mislead foreign intelligence services. Moreover, many of the raw materials and intermediate chemicals used to make organophosphate pesticides and fire retardants could also serve as precursors in the manufacture of nerve agents.

Field trials of Sarin-filled munitions took place at the Central Military Chemical Testing Site ("Polygon") of the Red Army near the town of Shikhany on the Volga, some twenty kilometers northwest of the city of Volsk. Formerly known as Tomka, it was where the Soviet Union and Germany had secretly collaborated on chemical weapons development and testing from 1928 to 1933. In 1937–38, the Soviets had enlarged the polygon to an area of 600 square kilometers, an effort requiring the evacuation of four large towns, and in 1941–42 it had been expanded further, to 1,000 square kilometers.

By the early 1950s, the Shikhany proving ground comprised a large laboratory complex, workshops, garages, administration buildings, pilot production facilities, storage bunkers, a test range, housing for the commandant and the station personnel, barracks for visiting experimental teams, stalls for experimental animals, a chemical school, a military hospital, and an airfield with hangars. The permanent staff numbered about 100 military officers, 850 noncommissioned officers and enlisted men, and 250 civilian chemists, physicians, biologists, and engineers. Visiting teams of five officers and 200 soldiers often visited Shikhany to conduct open-air trials of chemical weapons.

MEANWHILE, an accidental discovery in the course of industrial pesticide development opened the way to a new class of chemical nerve agents. In 1951, during the Korean War, U.S. Army personnel delousing North Korean refugees and prisoners of war found that the lice had become resistant to DDT, which had first been marketed in 1942 by the Swiss firm Geigy. On learning that the world's best-selling insecticide was losing its effectiveness, the major chemical companies saw a lucrative market opportunity and launched an intensive effort to develop a substitute. One of these firms was Imperial Chemical Industries (ICI), which had been formed in 1926 by the merger of four of Britain's largest chemical concerns.

In 1952, Dr. Ranajit Ghosh, a chemist of East Indian ancestry working in ICI's Plant Protection Laboratory, and his colleague J. F. Newman synthesized a new organophosphate compound containing sulfur and nitrogen that was later marketed under the trade name Amiton. Although it proved to be a potent insecticide, particularly against the red spider mites that infest fruit trees, Amiton had the drawback of being a potent cholinesterase inhibitor that was quite toxic to humans. In early 1953, a Porton scientist presented toxicology data on Amiton and related compounds to a tripartite meeting of American, Canadian, and British military scientists.

As often happens in science, chemists in other countries independently developed compounds that were structurally related to Amiton. One of those scientists was Gerhard Schrader, then working at Bayer in Leverkusen. He and his colleagues Ernst Schegk and Hanshelmut Schlör developed a new family of insecticides that had a basic molecular structure similar to that of Amiton and were effective against flies, mites, and leaf lice.

Also in 1952, Lars-Erik Tammelin, a chemist at the Swedish Institute of Chemical Defense, discovered a class of sulfur-containing molecules with potent anticholinesterase activity that came to be known as the "Tammelin esters." Although considerably more toxic than Tabun or Sarin, these compounds appeared to be too unstable for military use. Accordingly, in 1957 the Swedish government allowed Tammelin to publish his findings in the journal *Acta Chemica Scandinavica.*

EVEN AS THE new generation of nerve agents was emerging from the laboratories, the British government intensified its research on the physiological effects of Sarin and the other G agents. This effort included extensive human experimentation under the Porton Down Volunteers Program, which dated back to 1916. During the late 1940s and early 1950s, British service members were encouraged to volunteer for nerve agent trials by the promise of extra pay and leave, yet they were not informed about the toxic effects of the chemicals being tested on them nor warned of the risk of lasting harm.

The first nerve agent experiments on these "human guinea pigs" sought to determine the minimum dose of Sarin required to trigger miosis (pinpoint pupils) and the cumulative effects of low-level exposures. In 1950, Porton scientists began to test higher doses of Sarin on human subjects to measure the severity of initial symptoms, such as runny nose, headache, vomiting, miosis, and eye pain. A 1952 study sought to determine the effects of nerve agent on mental performance by exposing twenty airmen to Sarin vapor and then measuring how they performed on intelligence and aptitude tests. This experiment showed that low doses of nerve agents worsened visual coordination but had no effect on reasoning and intellectual ability.

In May 1953, Porton scientists conducted a large trial on 396 men in order to estimate the dosage of Sarin and two other G-series nerve agents that, when applied to the skin, would cause incapacitation or death. The scientists planned to expose groups of volunteers to various sublethal amounts of the agents, measure the degree to which these exposures reduced the level of cholinesterase in the subjects' blood, and then extrapolate from these data points to estimate the lethal dosage in man. Unfortunately, these experiments were based on the flawed assumption that a linear relationship existed between the depression of blood cholinesterase levels and the severity of clinical symptoms.

One experiment in the series, conducted at Porton Down on May 6, 1953, aimed to determine the extent to which Sarin evaporated before it could penetrate the skin, including the effects of clothing on the rate of absorption. To maintain secrecy, the volunteers were misled into believing that they were helping to find a cure for the common cold. Ronald Maddison, a twenty-year-old Royal Signal Corps engineer from the town of Consett in County Durham, had decided to volunteer after seeing an advertisement stating that the study participants would not be harmed and would receive a payment of 15 shillings. Ronald planned to use the money to buy an engagement ring for his girlfriend, Mary Pyle.

In the research laboratory at Porton Down, Maddison and five other subjects were fitted with respirators and placed in a sealed gas chamber. The room was hot and

Ronald Maddison, a twenty-year-old Royal Air Force engineer, died during a human trial of Sarin at the Porton Down chemical warfare establishment in England on May 6, 1953.

airless, and the gas mask smelled strongly of rubber and created an unpleasant sense of claustrophobia. Beginning at 10:17 a.m., a technician deposited twenty drops, one at a time, on a swatch of material from an Army uniform that had been wrapped around each subject's arm. Since a drop was equivalent to 10 milligrams of Sarin, twenty drops provided a cumulative dose of 200 milligrams. The subjects were then placed under observation for half an hour.

Twenty minutes after receiving the drops of liquid on his arm, Maddison complained of feeling ill. Three minutes later, he slumped over and began to gasp audibly for breath. Concerned by this development, the scientists took the young man out of the chamber and removed his gas mask, but Maddison's condition continued to worsen. He went deaf, started to wheeze as if suffering from a bad asthma attack, and suddenly fell to the floor and began to convulse. Finally recognizing the seriousness of the situation, the scientists injected Maddison with atropine and called for an ambulance.

Alfred Thornhill, a nineteen-year-old national service trainee, had arrived at Porton Down a few days earlier for a month-long posting as an orderly with the base ambulance service. On the morning of May 6, 1953, he answered an emergency call to the research laboratory and witnessed a scene that would haunt him for the rest of his life. A young man about his age, in blue RAF trousers and a boiler suit, lay unconscious on the floor outside the gas chamber, his body thrashing with violent spasms while several scientists in white coats looked on helplessly. Although Thornhill had seen epileptic seizures before, the young man's convulsions were much more violent, as if he were being electrocuted. The muscles under his skin were vibrating visibly and a thick foam, resembling frog spawn or tapioca, poured from his open mouth. Almost as terrifying was the look of panic in the scientists' eyes.

Thornhill and his fellow orderlies loaded the young man's still-shaking body into the ambulance and sped to the medical unit at Porton, arriving shortly before 11:00 a.m. As Thornhill wheeled the gurney into the clinic, he could smell the sour odor of sweat and fear. The emergency unit had been cleared of other patients, and several doctors and scientists in white coats were waiting around the bed, looking pale and distraught. After the orderlies placed the inert body on the bed, a doctor injected more antidote into Maddison's thigh and administered oxygen, but the young man's heart had stopped beating and he had no pulse.

Watching in horrified fascination, Thornhill saw a bluish tint appear at the young man's exposed ankle and spread slowly up his white leg, as if a blue liquid were being poured into a glass. The Sarin overdose had constricted Maddison's bronchial tubes and blocked the flow of air to his lungs, starving his brain and tissues of oxygen. As Thornhill stood gaping, a doctor filled a large syringe with adrenaline and plunged the needle into the young man's chest in a desperate attempt to restart his heart. At that point, a nurse screamed at the orderlies to leave.

The next morning, Thornhill was deeply saddened to learn that the unnamed patient had died. He was ordered to pick up the body at the Porton medical unit. When he arrived at the clinic, which reeked of disinfectant, a sullen doctor called him into an office, ordered him to sign a security form, and warned that if he ever spoke as much as a word about what he had seen, he would be put away for years. The doctor then ordered him to drive the body to the mortuary at Salisbury Hospital, taking a circuitous route along the back roads. Although troubled by the secrecy and intrigue surrounding the young man's death, Thornhill was deeply cowed and decided to keep his mouth shut. Still, having recently become engaged, he could not help wondering if the poor fellow had a wife or girlfriend, and if she would ever learn what had happened to him.

The British Home Office ordered an inquest into Ronald Maddison's death, which was kept secret on grounds of national security. Ten days after the fatal experiment, the inquest concluded that the subject's death had been accidental ("misadventure") and that the Ministry of Defence could not have predicted the fatal outcome. In fact, there was evidence to the contrary: a few days before Maddison's death, another volunteer named John Kelly had nearly died during a similar experiment.

Only one member of the Maddison family—the deceased's father, John—was allowed to attend the official government inquest. The authorities swore him to secrecy and allowed him to say only that Ronald's death had been "an unfortunate accident while on duty." Maddison's mother, Jane, and his four brothers and sisters received neither an explanation nor an apology for his death. For two years after the tragic incident, Porton Down suspended all testing of nerve agents on human subjects while an expert committee chaired by Professor E. D. Adrian, a physiologist at the University of Cambridge, developed a set of guidelines for human experimentation with nerve agents. From then on, trials were allowed to resume at lower doses.

WHILE THE SARIN pilot plant at Nancekuke was under construction in the early 1950s, the British armed forces prepared a military requirement for a nerve agent stockpile that would help to offset the Soviet Union's marked superiority in conventional armed forces in Europe. According to a memorandum by the British Chiefs of Staff, nerve agents were "likely to be particularly effective against armoured vehicles, since a nerve gas shell can produce death or severe disability to the crew of a tank without penetration of the armour. In addition, nerve gas weapons can fill an outstanding need for a land weapon for tactical use against the Russian Hordes."

In December 1952, British military planners estimated that because of the diluting effects of wind, weather, and terrain, it would be necessary to disperse about one metric ton of Sarin per square kilometer to inflict 25 percent casualties on unmasked enemy troops. Calculated in terms of munitions, delivering a ton of Sarin would require about 300 heavy (155 mm) artillery shells or about seven 500-pound bombs. Accordingly, an adequate war reserve of Sarin-filled weapons would consist of 7,600 cluster bombs and 10,000 aerial bombs. Producing a stockpile of that size would require a full-scale plant with an output of fifty metric tons of Sarin per week. At that rate, meeting the British military requirement would take about a year of sustained production, after which the plant could be placed in "mothballs" until the outbreak of war required its remobilization.

In 1953, however, technical problems and financial pressures caused the British government to rethink the planned construction of a full-scale Sarin plant at Nancekuke. With the end of the Korean War and the easing of international tensions, chemical weapons were no longer considered an urgent defense priority, and the plans were put on hold. Apart from the high cost of the plant, the delays in development meant that the Royal Air Force would not be able to procure a stockpile of Sarin-filled bombs until 1955 or later.

At this juncture, the United States offered some assistance. Washington had a clear interest in helping its closest ally to acquire a more modern chemical arsenal, bolstering deterrence of the Soviet Union. The U.S. government proposed to sell Britain a stockpile of bombs containing 2,500 tons of Sarin—equivalent to a one-month war reserve—at a cost of $10 million, on the condition that the British would replace the stocks when they began manufacturing their own. By purchasing the American weapons, the RAF would be able to acquire an operational nerve agent capability years earlier than would otherwise be possible. In April 1953, the British Ministry of Defence expressed interest in the American offer and inquired whether the U.S. Air Force could provide the bulk of the weapons in the form of cluster bombs standardized to British requirements.

By September 1953, however, the U.S. government had begun to distance itself from the proposed deal. One reason was that the Chemical Corps was having trouble getting the dichlor production facility at Muscle Shoals to work properly and had therefore fallen far behind its own Sarin production schedule. According to a memorandum by K. N. Crawford, the chairman of the British Chemical Warfare Sub-Committee,

> So far little progress seems to have been made in our approaches to the Americans, and I personally feel that there is small chance of a favourable solution to the problem on these lines. The American development of a satisfactory production process has met a very sticky passage and I have good reason to believe that they are looking to us to prove our own superior production process. In addition, the transport of this highly dangerous material from America presents major difficulties and will undoubtedly be an expensive business. . . .
>
> It therefore seems to me that a firm decision is needed in principle whether we must have these chemical weapons in our armoury or not. Doubts have been expressed whether we should spend large sums of money on weapons which we are pledged by international convention not to use in war. I cannot, however, conceive that in a major war where atomic weapons are used, such a weapon as nerve gas would remain excluded for long. It seems to me that the important thing to decide is whether nerve gas weapons are likely to prove a sufficiently powerful adjunct to the equipment of the Army and the Tactical Air Force to be worth the expenditure involved.

On March 1, 1954, Lieutenant Colonel C. A. Morgan, Jr., an official at Edgewood Arsenal (temporarily renamed the U.S. Army Chemical Center), wrote a memo pointing out the drawbacks of supplying 2,500 tons of Sarin to Britain immediately. At the current rate of production, fulfilling the British order would delay meeting the U.S. stockpile requirement for another six months. Morgan also estimated that the cost of 2,500 tons of Sarin would be $20 million, rather than the $10 million that Washington had quoted. The higher figure was in line with the price paid by the U.S. Navy of $4.00 per pound. In April 1954, the British Cabinet's Defence Committee finally dropped the idea of purchasing a war reserve of Sarin-filled bombs from the United States and instead authorized the construction of a full-scale production facility at Nancekuke. Until that factory went on line, the Sarin pilot plant, which had begun operation in January, would produce enough agent for field trials.

FRANCE ALSO MOVED forward with nerve agent production. Chemical engineers at the Centre d'Études du Bouchet (CEB) built a pilot plant that, from 1950 to 1952, produced twenty metric tons of Tabun for testing purposes. Nerve agent trials continued at B2-Namous in Algeria and at three testing sites in metropolitan France: the Camp de Mourmelon near Reims, Cazaux in the Landes region south of Bourdeaux, and Bourges in central France.

In November 1951, the General Staff of the French Army created a new Special Weapons Command [Commandement des Armes Spéciales], which was responsible for all matters relating to the acquisition and use of chemical, biological, and nuclear arms. Its first commander was Lieutenant Colonel Charles Ailleret, a hero of the French resistance who had survived a period of imprisonment in Buchenwald concentration camp. To coordinate all chemical weapons research-and-development efforts, in August 1952 the French Army established the Commission for Chemical and Biological Studies and Experiments (Commission des Études et Expérimentations Chimiques et Bactériologiques) under the direction of the Special Weapons Command.

In the early 1950s, chemists at the CEB succeeded in synthesizing Sarin. France also collaborated on nerve agent research and development with Belgium and the Netherlands. In 1956, however, the defense minister, Maurice Bourgès-Maunoury, launched the French nuclear weapons program, which soon siphoned resources away from chemical and biological weapons development. Members of the Commission for Chemical and Biological Studies and Experiments criticized the budget cuts, especially in view of the rapid advances being made in the chemical weapons field by both the Soviet Union and the United States.

ALTHOUGH ICI MARKETED Amiton commercially in 1954, the new insecticide soon proved too dangerous for routine agricultural use. Not only was it a potent cholinesterase inhibitor that caused pinpoint pupils, shortness of breath, and other symptoms of organophosphate poisoning, but it was highly persistent in the environment and could readily enter the user's bloodstream through the skin. In view of these "toxicological disadvantages," ICI was forced to withdraw Amiton from the market. The same properties that deprived the insecticide of commercial value made it attractive for military use, however, and Porton Down put the compound and its structural relatives through a battery of tests. Because the members of the Amiton family were highly toxic and readily penetrated the skin—characteristics similar to those of snake venom—these compounds were named V agents, for "venomous." From then on, all compounds of the Amiton class were identified with a two-letter code beginning with V, using the same convention developed earlier for the G-series agents. (Amiton itself, for example, was designated VG.)

Preparations also continued in Britain for the production of Sarin at Nancekuke. In 1955, anticipating the construction of a full-scale manufacturing facility, the pilot plant was shut down after having produced some 20 metric tons of agent over two years of operation. Technicians decontaminated and mothballed the equipment to preserve the option of restarting it in the future. In 1956, however, the Cabinet Defence Committee abandoned its plan to mass-produce Sarin and, in a major shift in policy, renounced the possession of an active chemical weapons stockpile. (A decade earlier, London had decided to retain the existing stocks of mustard and phosgene weapons until nerve agents became available.) The considerations that led to this landmark decision were largely financial. As a party to the 1925 Geneva Protocol, Britain could not justify the expense of developing a means of warfare that it was legally prohibited from using except for retaliation. Having tested its first atomic bomb in 1952, the British government now chose to rely on nuclear weapons to deter any Soviet use of chemical arms. As a consequence of this policy shift, a large fraction of the research staff at Nancekuke was laid off in 1957 and 1958.

Even after Britain renounced the possession of an active chemical weapons stockpile in 1956, scientists at Porton Down remained involved in nerve agent research and development and continued to collaborate with their American and Canadian colleagues under the Tripartite Agreement and the U.S.-U.K. Mutual Weapons Development Program. The rationale was that the British government sought to retain the know-how to manufacture G and V agents, both to test defensive equipment and as an insurance policy should the nation's defense ever require the acquisition of a nerve agent stockpile. Washington also remained an eager customer for British innovations in military chemical technology.

On March 15, 1956, President Eisenhower approved an amended decision memorandum on "Basic National Security Policy" that had been prepared by the staff of the National Security Council. Designated NSC-5602/1, this top secret document moved for the first time beyond the existing retaliation-only policy for chemical warfare. The operative paragraph 12 read: "To the extent that the military effectiveness of the armed forces will be enhanced by their use, the United States will be prepared to use chemical and bacteriological weapons in general war. The decision as to their use will be made by the President." If time and circumstance allowed, the United States would consult with its allies before resorting to this option.

The 1956 memorandum gave future U.S. presidents the flexibility either to employ chemical weapons strictly for retaliation or to initiate their use during a conventional war. Moreover, whereas President Roosevelt had publicly declared the retaliation-only policy, the new posture set out in NSC-5602/1 was classified. For Eisenhower, who remained personally committed to "no first use," the purpose of the policy shift was to encourage research and development on chemical and biological weapons so as to ensure an adequate retaliatory capability. Nevertheless, by implementing the recommendations in the Stevenson report, the Army and the Air Force acquired the capability to employ chemical weapons early in an armed conflict.

In the fall of 1956, a high-level civilian advisory panel headed by Otto N. Miller, a vice president of Standard Oil of California, recommended to Secretary of the Army Wilber M. Brucker that chemical weapons had a proper place in military planning and should be developed for "actual use" if necessary. Not long after the submission of the Miller report, the Department of Defense issued a top secret directive on October 6, 1956, stating that the military services should maintain a comprehensive research-and-development effort on chemical weapons, including field testing, with an emphasis on exploitation of the V-series agents and the "development of munitions that will achieve optimal large area dispersion and dissemination of nerve gases by aircraft and missiles scheduled to be available in 1960 and beyond."

The Army's interest in the possible first use of chemical weapons was also reflected in a new edition of the *Field Manual on the Law of Land Warfare,* published in 1956 but not made public until 1959. The section on chemical warfare included the following categorical statement: "The United States is not a party to any treaty, now in force, that prohibits the use in warfare of toxic or nontoxic gases. The Geneva Protocol for the prohibition of the use in war of asphyxiating, poisonous, or other gases, and of bacteriological means of warfare, is not binding on this country."

In 1957, after three years of full-scale manufacturing activity, the U.S. military's stockpile requirements for Sarin had finally been met. The production lines at Muscle Shoals and Rocky Mountain Arsenal were shut down in July and August, respectively, and both plants were placed in mothballs so that they could be restarted in the future should additional quantities of Sarin be required. In the meantime, the Army leased out part of Rocky Mountain Arsenal to the Shell Chemical Company for the production of the insecticides Aldrin and Dieldrin.

Even after the large-scale manufacture of Sarin came to an end, the toxic legacy of the program remained. Because of the intense pressure at the height of the Cold War to turn out thousands of tons of Sarin "product," little effort had been made to control pollution. Environmental laws were weak during the 1950s, and the Army and Shell Chemical exploited the secrecy surrounding the nerve agent program to cut corners even further. As a result, the toxic by-products of Sarin and pesticide production were simply buried in unlined pits on the grounds of Rocky Mountain Arsenal or pumped into "Basin F," a ninety-three-acre artificial pond lined with asphalt and sand. It was assumed that evaporation would gradually decontaminate the foul-smelling, coffee-colored brew of toxic wastes.

In fact, the environmental problems at the arsenal only worsened over time. Migrating waterfowl that had the misfortune to land on the waste-water pond were poisoned and died, and hazardous chemicals that had been buried in landfills seeped into the ground, contaminating water supplies outside the arsenal fence. Wells in south Adams County became tainted with methylphosphonic acid, a Sarin breakdown product. In one incident, a farmer sprayed fifty acres of sugar beets with toxic well water, killing his entire crop. By the late 1950s, the several hundred acres at the heart of Rocky Mountain Arsenal had acquired the dubious distinction of being the most polluted piece of real estate in America.

In 1961, the Army drilled a 12,000-foot well down to the hard rock underlying the sediment of the Denver basin and on March 8, 1962, began injecting liquid wastes. A month later, after four million gallons had been pumped underground, Denver experienced its first earthquake in eighty years. More toxic waste was forced down the hole in May and the next month a series of minor quakes occurred, some reaching 4 on the Richter scale. Although the pumping declined toward the end of 1962, it increased in March 1963—and again, the tremors followed with a one-month delay. Critics warned that the pumping had destabilized the geological strata under Denver, but it was not until February 1966 that the Army finally acknowledged the problem and halted the operation. Over the four-year period, the Army had injected a total of 163 million gallons underground and the Denver area had experienced some 1,500 tremors. Because some geologists feared that removing the wastes from the deep well might worsen the earthquake situation, they were simply left in place.

Even as the Army struggled to deal with the toxic legacy of Sarin production, it moved forward with the development and production of the next generation of nerve agents: the V series. A new phase of the superpower chemical arms race was about to begin.

CHAPTER NINE

AGENT VENOMOUS

DURING THE MID-1950S, the U.S. Army Chemical Center (formerly Edgewood Arsenal) synthesized approximately fifty V-series nerve agents—including those code-named "VE," "VG," "VM," "VP," "VR," "VS," and "VX"—and screened them for the best combination of militarily desirable characteristics, such as toxicity, stability in storage, persistence on the battlefield, and ease of manufacture. In February 1957, the Army Research and Development Command selected VX as the V agent on which to concentrate further work, including pilot-plant development and dissemination studies.

Pure VX was an odorless liquid with a density slightly greater than that of water, a viscosity similar to that of 30-weight motor oil, and a color that varied from clear to amber depending on purity. In tests with experimental animals, VX proved to have a toxicity three times that of Sarin when inhaled and a thousand-fold greater when absorbed through the skin. Extrapolating these results to humans, it was estimated that less than ten milligrams of VX—a small drop of fluid on the skin—could kill a grown man in fifteen minutes. A liter of VX contained enough individual lethal doses, theoretically, to kill one million people.

VX could be disseminated either as a fine airborne mist or a coarse spray of viscous droplets that clung to whatever they touched, contaminating equipment, buildings, vegetation, terrain, and unsheltered troops. As a result, both a gas mask and a full-body suit were needed to protect troops from VX contamination. The agent was also highly persistent: whereas a cloud of Sarin vapor would dissipate in fifteen minutes to an hour, depending on weather conditions, liquid VX sprayed on the ground would remain lethal for up to three weeks. It could therefore serve to contaminate large areas of the battlefield and channel enemy forces into "killing zones." Airborne clouds of VX, consisting of droplets large enough to impinge on the skin but too small to settle out rapidly, could also penetrate buildings and field fortifications. These attributes suggested that VX would eventually replace mustard as the standard persistent agent in the U.S. chemical inventory.

Scientists at the U.S. Army Chemical Center developed an industrial production process for VX. Led by Sigmund R. Eckhaus, Bernard Zeffert, and Jefferson C. Davis, Jr., a team of about thirty chemical engineers and Army draftees worked in shifts in Building 2345, a four-story structure that contained five large engineering bays. For safety reasons, the building incorporated a negative-pressure ventilation system and an air lock with decontamination showers that separated the "clean" areas from the "hot" areas where lethal chemicals were in use. Inside the engineering bays, the engineers worked under giant fume hoods fifteen feet high, equipped with powerful fans to remove the toxic gases.

The Edgewood team developed a method for VX production called the "transester process." It involved reacting phosphorus trichloride (PCl_3) with methane gas (CH_4) at high temperature to form the intermediate CH_3PCl_2, referred to by the code name "SW." Because this compound reacted violently with water, caught fire in the presence of moist air, and was highly corrosive, it had to be synthesized inside a coil of high-nickel steel from which the oxygen had been purged and replaced with inert nitrogen gas. SW was then combined with ethanol in an inert gas blanket to form a "diester," which underwent a third reaction to yield a liquid phosphorus intermediate known as "transester," or "QL." Finally, QL was mixed with powdered sulfur, reacting spontaneously to produce VX and a great deal of heat.

The transester process had the advantage of generating a highly pure product that did not need to be distilled. To test the new process on an industrial scale, the Edgewood engineers built two pilot plants: one capable of making twenty pounds of VX per eight-hour shift, and a larger version that produced 250 pounds. Based on experience with the two pilot plants, plans were drawn up for a full-scale VX production facility with an output of ten tons of agent per day.

In April 1957, Dugway Proving Ground organized a "V-Agent Team" to handle testing and evaluation of new delivery systems for VX, including self-dispersing submunitions dropped from aircraft or guided missiles. Engineers sought to design bomblets that would generate VX droplets of optimal size for military purposes, taking into account the downwind transport of the agent cloud, absorption of the droplets through clothing and skin, and other factors. From the standpoint of chemical defense, the V agents posed difficult technical challenges because of the need for complete protection of the skin and decontamination of vehicles and equipment.

WHILE ONE TEAM of chemical engineers at Edgewood was working on a manufacturing process for VX, another team began to develop "binary" chemical weapons, in which two relatively nontoxic ingredients were combined inside a bomb or shell to yield a lethal nerve agent. The basic principle of binary weapons dated back to 1885, when military chemists had attempted to produce nitroglycerin, a dangerously unstable explosive, by combining its two chemical constituents (nitric acid and glycerin) inside an artillery shell. In the late 1940s, scientists at Edgewood Arsenal had studied binary munitions as a possible way to reduce the hazards associated with the production, storage, and handling of unitary chemical weapons. Leading this research effort was a hulking German chemist named Fritz Hoffmann. He had worked in the Nazi nerve agent program and, after the war, had been recruited by the U.S. government under Project Paperclip. After first experimenting with a design for a binary mustard bomb, Hoffmann switched to nerve agents in the early 1950s. Because the last step in the synthesis of Sarin, Soman, and VX involved the reaction of two precursors, he realized that it would be possible to store these chemicals in separate compartments inside a bomb or shell and cause them to mix and react to produce the lethal agent while the munition was still in flight to the target. (For technical reasons, Tabun could not be produced in a binary system.)

In 1954, building on Hoffmann's work, three engineers in the Weapons Research Division at Edgewood—Ted Tarnove, Gene Bowman, and Marty Sichel—developed a concept for a binary VX bomb in which powdered sulfur would be injected into a liquid solution of QL. The technology looked promising, but Colonel Jim Hebbeler, the head of the Chemical Warfare Laboratories, and Dr. Ben Witten, the director of weapons research and development, were skeptical about its military utility. Although it was true that the storage and transport of binary chemical munitions would entail fewer risks, that difference meant little to Chemical Corps officers, who prided themselves on their ability to handle unitary weapons safely. Furthermore, because the reaction of two chemical precursors inside a binary bomb or shell would inevitably be incomplete, such a munition would actually deliver less nerve agent to the target than a unitary weapon, reducing its military effectiveness. Given the lack of institutional support for binary weapons, Sigmund Eckhaus, the director of the VX pilot plant at Edgewood, had to "smuggle" samples of QL to the binary development team. Ultimately, the lack of interest from senior Army officials led to the cancellation of the binary R&D program after a few years of work.

In parallel with the development of V agents at Edgewood, scientists in U.S. government laboratories and academic institutions worked on improved medical defenses against nerve agents. This area became a high priority when it was discovered that atropine, which could counter the effects of Tabun and Sarin, was far less effective at treating exposures to Soman or VX. In the mid-1950s, the Chemical Corps awarded a secret contract to Dr. David Nachmansohn, a professor of biochemistry at Columbia University's School of Physicians and Surgeons in New York, to investigate the toxicology of the nerve agents and develop new antidotes.

Nachmansohn chose to work with electric eels from the Amazon basin, which generate high voltages to stun their prey. These animals provide an excellent experimental model of the human nervous system because the electrochemical phenomena are greatly amplified and therefore much easier to study. Moreover, because the nervous system of the electric eel is highly enriched in cholinesterase, it is possible to purify the enzyme in relatively large quantities. To obtain enough cholinesterase for his experiments, Nachmansohn asked the Army to procure a hundred electric eels. The Army contracted with the New York Aquarium, which in turn gave the assignment to J. Auguste Rabaut, a middle-aged Frenchman and expert fisherman who lived in the upper reaches of the Amazon basin. Rabaut trapped more than a hundred electric eels and kept them alive in water-filled tanks, which were shipped by plane to New York City.

At Columbia University, Nachmansohn and his coworkers purified cholinesterase from the nervous tissue of the eels. Because samples of Sarin or VX were not available for experimental use, the scientists used the less potent nerve agent DFP, which was easier to synthesize and safer to work with. In test-tube experiments, the investigators found that DFP completely inactivated eel cholinesterase but did not destroy it. This observation suggested that it might be possible to develop an effective treatment for nerve agent poisoning by restoring the activity of the enzyme.

Subsequent experiments by Nachmansohn and his colleague Irwin Wilson determined that the nerve agent molecule attaches to the "active site" of cholinesterase, a groove in the surface of the enzyme where catalysis occurs, blocking the further breakdown of acetylcholine. Once the Columbia researchers had elucidated this mechanism, they sought to develop a drug that would displace the nerve agent molecule from the active site of cholinesterase. In this way, it would be possible to restore the normal activity of the enzyme and counteract the effects of the poison. Through a long and tedious process of trial and error, Nachmansohn and his team synthesized dozens of novel compounds and screened them for the ability to reactivate cholinesterase. Finally, a compound called PAM (pyridine aldoxime methiodide), a member of the class of drugs known as oximes, proved to be highly effective at displacing DFP from the active site of cholinesterase and restoring the normal function of the enzyme, at least in the test tube.

Excited by this finding, Nachmansohn devised an experiment to test the activity of PAM in living animals. He divided forty white mice into two groups of twenty and injected all of them with DFP. The first group of mice was then given a prompt injection of PAM, while the twenty "control" mice received only a saline solution. Within five minutes, all of the control mice were dead, but the treated mice continued to scamper about unharmed. Subsequent studies on laboratory animals with more potent nerve agents, such as Sarin and VX, showed that PAM was of limited benefit when given alone but highly effective when administered together with atropine. Whereas treatment with PAM raised the lethal dose of VX two- or threefold, PAM plus atropine increased the level of protection tenfold. The explanation for this difference was that the two antidotes worked synergistically through different but complementary mechanisms. By blocking the receptors for acetylcholine and thereby counteracting its physiological effects, atropine served to tide the animals over until PAM restored the normal activity of their cholinesterase.

Based on the knowledge that the basic operation of the nervous system is similar in all mammals, the Columbia scientists expected that PAM would also be effective in humans. To test their prediction, they administered the antidote to human volunteers exposed to low concentrations of Sarin vapor. These experiments demonstrated that PAM itself was relatively nontoxic and that it was highly effective at counteracting the symptoms of nerve agent exposure, especially when given together with atropine.

The next challenge was to develop an effective means of administering the two antidotes into the thigh muscle of a soldier within minutes of exposure. Although metal syrettes containing atropine had been issued to U.S. troops in the 1950s, they posed numerous problems. Not only did soldiers wearing gas masks and thick rubber gloves have difficulty manipulating the metal tubes, but many individuals were understandably reluctant to plunge a needle into their own body. To solve these problems, British scientists at Porton Down developed an automatic syringe called the "Autoject." In the United States, Stanley J. Sarnoff, a professor of physiology at Harvard University, invented a similar device called the "Ace autoinjector," which the Army standardized in 1959. Both systems consisted of a cigar-shaped syringe containing a premeasured dose of atropine or PAM and a recessed, spring-loaded needle. A soldier who had been exposed to nerve agent simply released the safety catch and pressed the end of the device against his thigh. The pressure triggered the spring mechanism,

A U.S. soldier self-administers nerve-agent antidote with an autoinjector during a chemical warfare exercise.

driving the needle through layers of protective clothing and injecting the correct dose of antidote.

Another top development priority for chemical defense was a portable battlefield detection system for nerve agents. Since Sarin was effectively odorless, an automatic detector and alarm system were needed to give troops advance warning of an approaching agent cloud so that they could don their gas masks in time. Early field detectors were slow and unreliable, and until the early 1960s, the British Army continued to use caged canaries—which are more sensitive to nerve agents than humans are—as their primary detection system. In the United States, the Chemical Corps developed the M6 automatic G agent field alarm, which was standardized by the Navy in 1958. When any G agent came in contact with a reagent in the device, it produced a color change in a paper tape that was detected by a photo cell, triggering a buzzer alarm. The M6 weighed twenty-five pounds, fit into a portable aluminum case, and could operate unattended for twelve hours, at which time it required fresh solutions and new tape. Nevertheless, the detector had major drawbacks: it ran off a battery, did not operate below freezing, and required frequent maintenance. Thus, although the Navy procured over five hundred of the units for its dockyards and ten for shipboard use, the Army rejected it for standardization.

UNDER THE TRIPARTITE AGREEMENT, British, American, and Canadian scientists shared technology and coordinated research at biannual conferences on toxicological warfare. At the Eleventh Tripartite Conference in May 1957, for example, participants from the three countries agreed to intensify their work on the V-series agents. Research topics judged to be of highest priority were studies on the deposition and toxicity of agent droplets of various sizes on skin, clothing, and respiratory passages; the persistence of and residual hazard from agents in the field as influenced by climatic conditions; and the development of new agents or additives that could more readily penetrate the skin. At the Twelfth Tripartite Conference in the fall of 1957, the participating scientists agreed on a division of labor, with the Americans continuing to develop land and air munitions for the dispersal of V agents, the British evaluating the military potential of these agents with model systems, and the Canadians taking the lead in determining the secondary hazard from contaminated terrain.

Under a second agreement called the U.S.-U.K. Mutual Weapons Development Program, British process engineers collaborated with their colleagues at Edgewood Arsenal to develop industrial production techniques for VX and other V agents. At Nancekuke, the British built a small pilot plant that produced kilogram quantities of VX with a method known as the "water process," which they considered superior to the U.S. transester process. The bilateral collaboration was highly productive, but British scientists complained about what they perceived as a "one-way street": they felt that they told the Americans everything and received relatively little information in return.

Along with the expanded research and development on the V agents, the U.S. Army Chemical Corps began to plan for the large-scale production of VX. In 1957, the corps appointed a V-Agent Committee to study the problems of VX manufacture and to select a suitable site for the facility. For safety reasons, the plant could not be located near a densely populated area. The Chemical Corps also did not want to produce VX at an existing site, such as Muscle Shoals or Rocky Mountain Arsenal, on the principle of "not putting all of your eggs in one basket."

The V-Agent Committee finally decided to locate the VX plant at the Wabash River Ordnance Works near the town of Newport, Indiana, a small farming community in the western part of the state, about thirty miles from Terre Haute. The Ordnance Works had originally been built in 1941 to produce plastic explosive for World War II. It was also the site of an inactivated government facility, the Dana Heavy Water Plant, which the Atomic Energy Commission had built in 1952 to produce heavy water (deuterium oxide) for the U.S. nuclear weapons program. The abandoned AEC plant included sixty giant extraction columns that had served to refine heavy water from huge volumes of natural water drawn from the Wabash River. Although most of the columns were eventually torn down, the few left standing were incorporated into the design of the VX plant as structural support elements for pumps, pipes, and reactors.

Major General William M. Creasy, the chief of the Chemical Corps, decided to contract with private industry for the design, construction, and operation of the VX factory and the associated munitions filling line. On May 9, 1958, the Chemical Corps asked the Army Corps of Engineers to solicit proposals from qualified industrial firms. A year later, on June 23, 1959, the Department of Defense signed contracts with the Lummus Company for the design of the plant and with the Food Machinery and Chemical (FMC) Corporation for its construction and operation. After two years of work and an expenditure of $8 million, the Newport Army Chemical Plant was completed in April 1961. Because of a number of technical problems, full-scale production and loading of munitions did not get under way until 1962, with a planned output of ten tons of VX per day.

Meanwhile, several new delivery systems for nerve agents were under development. As its "workhorse" weapon for delivering VX or Sarin on the battlefield, the Army developed the M55 rocket. Made of rolled and welded sheet aluminum, the six-foot-long, fin-stabilized rocket had a range of six miles. It was an integrated package that included a solid-fuel motor, a warhead filled with five quarts of nerve agent, an impact fuse, and an explosive burster charge that would disperse the agent as a spray or vapor. In 1960, the Army Chemical Corps let a production contract for the M55 to the Norris Thermador Company of Los Angeles, which ultimately manufactured

A M55 rocket with a nerve-agent warhead is test-fired from a multiple-tube launcher system. Hundreds of thousands of these rockets were produced and filled with Sarin or VX in the early 1960s. Many of the Sarin-filled rounds soon began to leak, creating a huge disposal problem for the U.S. Army.

about 478,000 rounds. Each filled rocket weighed fifty-five pounds and was packed in a fiberglass shipping tube that also served as a launching tube. A single mobile launcher held up to forty-five rockets that could be fired in salvos, theoretically drenching a target with nerve agent. From the outset, however, the M55's performance in testing was highly erratic and unreliable.

In addition to the M55 rocket, new weapons in the U.S. chemical arsenal included a 750-pound aerial bomb containing 220 pounds of Sarin, a land mine that dispersed 11.5 pounds of VX, and VX-filled artillery shells. The Air Force, the Army, and the defense industry also developed self-dispersing cluster warheads for three mid-range tactical rockets: the Little John, the Honest John (range: 16 miles), and the Sergeant (range: 75 miles). Standardized in 1960, the Honest John warhead contained 356 spherical bomblets made of ribbed aluminum, each 4.5 inches in diameter and containing about a pound of Sarin. The warhead was designed to break open in flight over the target area and release the bomblets, which armed themselves by spinning as they fell to earth. Detonating on impact with the ground, the

An Honest John rocket designed to deliver a nerve-agent warhead is shown being test-fired around 1964.

The Honest John rocket warhead contained 356 spherical aluminum bomblets, each of which contained about a pound of liquid Sarin.

Cutaway of an M139 Sarin bomblet shows internal cavities that were filled with liquid Sarin. At its center are a fuse and an explosive burster that detonated on impact, generating a fine mist of nerve agent.

bomblets released clouds of Sarin vapor that merged to blanket a large area. Field tests indicated that a single Honest John warhead could generate a lethal concentration of Sarin over a radius of 500 meters, not including downwind spread.

A full-scale model of the 4.5-inch-diameter M139 Sarin bomblet, similar to the ones used in cluster warheads.

ALTHOUGH FRANCE did not participate in the Tripartite Agreement, it launched an independent effort to develop and produce V-series nerve agents. In January 1958, Defense Minister Jacques Chaban-Delmas transformed the Special Weapons Command, which had been subordinated to the General Staff of the French Army, into the Joint Special Weapons Command (Commandement Interarmées des Armes Spéciales), reporting to the Joint Chiefs of Staff. At the same time, the organization for chemical and biological weapons R&D was restructured so that it served all of the French armed services.

Throughout the Algerian war of independence (1954–62), the French continued to test chemical weapons at B2-Namous. Despite the bloody colonial war taking place nearby, the testing campaigns in Algeria focused exclusively on the Warsaw Pact chemical threat to Western Europe. Indeed, the nerve agent trials were conducted by technical elements of the French Army that had no involvement in local combat operations.

In 1959, chemists at the Centre d'Études du Bouchet synthesized VX, which they gave the code name "A4." Three years later, in 1962, U.S. Army scientists transferred data on the V agents to their French colleagues after first obtaining permission from the British government. Over the next few years, French process engineers developed industrial production methods for Sarin and VX. During the mid-1960s, France manufactured several dozen tons of Sarin and 400 kilograms of VX for testing purposes at a pilot-scale ("semigrand") production facility in Braqueville, near Toulouse in southern France.

When Algeria finally gained its independence in 1962, it was expected that France would abandon B2-Namous. However, the Évian Accords that Paris concluded with the new Algerian government allowed France to retain its nuclear weapons testing sites and support bases in the country for another five years, until 1967. Under this agreement, B2-Namous was considered as an annex of the French nuclear testing complex.

SHORTLY AFTER John F. Kennedy was sworn in as president of the United States in January 1961, Secretary of Defense Robert McNamara established a senior interagency task group to review how the Pentagon was organized and the armed forces were structured and equipped. This policy review was broad in scope, involving approximately 150 sequentially numbered studies prepared by the national security bureaucracy. Project 112 examined the utility of chemical and biological weapons for U.S. strategic and limited warfare in view of the military threat posed by the Soviet Union and Red China.

The Project 112 study concluded that with Moscow's acquisition of a potent nuclear arsenal, the United States could no longer rely exclusively on the threat of massive nuclear retaliation to deter a Soviet invasion of Western Europe or Japan. Instead, it would be necessary to invest more heavily in chemical and biological weapons to bolster U.S. conventional forces and provide an alternative to all-out nuclear war. Chemical weapons offered a number of military advantages: they were relatively cheap to produce, the scale of their use could be tailored to a limited conflict, and they could be designed to incapacitate rather than kill. Ironically, the fact that chemical arms were *less* destructive than nuclear or biological weapons made them more attractive for limited warfare. According to a 1962 study by the Joint Planning Staff, the use of nerve agents offered a promising means of delaying the enemy's advance without resorting to tactical nuclear weapons and risking escalation to an all-out thermonuclear exchange. This strategic concept, known as "graduated deterrence," would buttress the credibility of the nuclear balance of terror and, if deterrence failed, help keep the battle under control and gain time for negotiations. In response to this assessment, the annual budget of the Army Chemical Corps nearly tripled between 1961 and 1964, including funds for VX production.

The Project 112 Task Group issued six recommendations to the Joint Chiefs of Staff, including the need to "immediately evaluate and modify operational plans, as necessary, to provide for the specific employment of chemical and biological weapons." Furthermore, the group advised that "stockpiles of modern munitions should be strategically positioned to support these operational plans as soon as possible." Responding to this directive, the Pentagon negotiated storage rights for U.S. chemical weapons in West Germany, France, and Italy at munitions depots controlled by the European Command.

On December 17, 1962, General Maxwell Taylor, the chairman of the Joint Chiefs of Staff, sent a memorandum to Defense Secretary McNamara noting that a token retaliatory stockpile of chemical weapons had been based since 1959 in the Federal Republic of Germany (FRG) under the control of the Seventh Army. This stockpile, consisting of artillery shells filled with mustard and Sarin, and bulk containers containing 12 tons of Sarin, had initially been stored at the Rhein Ordnance Depot in Kirchheimbo-laden and later transferred to a depot at Gerbach in southwestern Germany. Although the U.S. government had possibly notified Chancellor Konrad Adenauer at the time of the original deployment, the existence of U.S. chemical weapons in West Germany remained a closely guarded secret. General Taylor noted, "The storage of these munitions has never been the subject of formal US/FRG negotiations. It is considered desirable, therefore, that storage rights for these munitions be more firmly established by their inclusion in the proposed US/FRG negotiations."

The Project 112 study also found that the U.S. armed forces lacked adequate knowledge of the military effects of nerve agents under realistic field conditions, including the operational challenges of decontaminating vehicles and personnel in various climates and terrain. Accordingly, the study recommended an extensive program of field trials to assess the United States' vulnerability to chemical attack and to test defensive equipment, procedures, and tactics. This recommendation led to the establishment in June 1962 of the Deseret Test Center to coordinate the field-testing program. Jointly staffed and funded by the four armed services, the center was based at Fort Douglas, Utah, and drew on the facilities and personnel of Dugway Proving Ground, about sixty miles away.

In addition to Deseret and Dugway, the individual services each established their own chemical weapons testing sites. The Army conducted nerve agent trials in the hot desert at Fort Huachuca in Arizona, the tropical jungle at Fort Clayton in the Panama Canal Zone, and the frozen arctic at Fort Greely in Alaska. The Air Force's main testing site for chemical munitions was Eglin Air Force Base in Florida, and the Navy's main testing facility was the Naval Weapons Test Station at China Lake, California.

ALTHOUGH RUMORS circulated for years among the residents of Newport, Indiana, about the mysterious Army installation at the Wabash River site, its purpose remained secret until May 1962, when Mearlin Sims, the FMC plant manager, disclosed at a monthly meeting of the Clinton Chamber of Commerce that the Newport facility was producing a powerful chemical warfare agent that was lethal in tiny amounts. Sims's statement, the first official acknowledgment of VX production, was reported in a short news item in the *Terre Haute Star.*

The VX production complex was located three miles south of Newport on State Highway 63, an area of rolling hills, woods, and farms. Protecting the installation from intruders were security checkpoints, heavily armed guards, and a high fence topped with barbed wire that partially concealed rows of weathered yellow barracks. Inside the fence, the sprawling facility consisted of a series of chemical plants containing some forty miles of pipes, furnaces, pumps, mixing tanks, and reaction vessels. The last step in the production process, in which QL was combined with powdered sulfur to yield VX, took place inside a three-story, windowless concrete blockhouse that was hermetically sealed to prevent leaks. Because the QL-sulfur reaction required cooling but no final distillation step, the plant used a continuous rather than batch process to improve productivity and enhance the purity of the final product. On the ground floor of the blockhouse, technicians wearing white jackets, pants, and gloves sat at an eight-foot control panel covered with gauges, recording graphs, and colored warning lights.

The workforce at the Newport plant numbered about 400 people, reaching as high as 550 during periods of peak production. Despite the extreme hazards involved, FMC Corporation had no trouble recruiting workers because of the lack of job opportunities in the economically depressed region. Most of the technicians and laborers came from Vermillion County, where the town of Newport was based, and the adjacent Parke County. Successful applicants had to be in excellent health and pass an FBI security background investigation and a battery of psychological tests.

Although the VX plant was fully automated, every ninety minutes inspectors donned gas masks and full-body rubberized suits to check the

The Newport Army Ammunition Plant in Newport, Indiana, produced the nerve agent VX from 1961 to 1968.

Two extraction columns from the former Dana Heavy Water Plant were incorporated into the VX production facility at Newport.

equipment inside the blockhouse and take samples from the production line. After each inspection, the technicians were required to take a hot shower—a total of three per day—to remove any toxic residues, and they underwent frequent blood tests to monitor their cholinesterase levels. Ten percent of the operating budget of the Newport plant went to safety measures, including a fully equipped hospital with a nine-person staff that was ready at all hours to treat an accidental exposure. The environmental controls at Newport were also superior to those at Rocky Mountain Arsenal: the butane solvent from the reaction sequence was flared, and the liquid wastes were pumped into a 5,500-foot-deep well.

After VX emerged from the manufacturing process, the oily, supertoxic liquid was pumped to an elongated one-story building, where an automatic filling line loaded the agent into bombs, artillery shells, mines, and rockets. As a conveyor belt carried the munitions through an airtight metal cabinet, a series of machines loaded them with VX, added an overlay of helium gas, welded the filling port shut, and checked for leaks. If helium was detected leaking from a shell, an alarm bell rang, lights flashed, and the defective round was splashed with purple paint, pulled off the assembly line, and eventually destroyed. Finally, the shells were X-rayed to measure their contents, and their outer surfaces were decontaminated and dried. Technicians standing outside the closed filling lines monitored the automatic loading and sealing operations. They did not wear gas masks but kept them close at hand in case of emergency.

In the packing area at the end of the production line, workers painted and stenciled the finished munitions, weighed them for firing calculations, and stacked them on wooden pallets for shipment to Army storage depots. The weapons loaded with VX included two types of artillery shells (155 mm and 8-inch), M55 artillery rockets, land mines, and spray tanks holding 1,300 pounds of agent. Over the lifetime of the plant, nearly 1,000 TMU-28B spray tanks were filled with VX and shipped by air for storage at Tooele Army Depot in Utah. The Navy and the Marine Corps also developed a spray tank called the Aero 14B for the A-4 aircraft that was designed to be loaded on the flight line with a field filling machine.

The Newport VX plant operated on a round-the-clock manufacturing schedule for three years, but its output fell sharply after 1963 and continued at a low rate until 1968, when the facility was shut down. By that time, it had produced a total of 5,000 tons of VX. During the Vietnam War, the

This elongated building at the Newport Army Ammunition Plant contained filling lines for loading VX into a variety of munitions.

VX filling lines for 8-inch and 155 mm shells.

Filling line for loading spray tanks with VX nerve agent.

In the "pack-out" room at the Newport Army Ammunition Plant, VX-filled artillery shells were loaded onto pallets for shipment to U.S. chemical weapon depots.

Aboveground storage tanks for bulk VX at the Newport Army Ammunition Plant. A short-age of munitions during the Vietnam War caused large amounts of bulk nerve agent to be stored on the facility grounds.

increased demand for high-explosive artillery shells led to a shortage of munitions. As a result, a large fraction of the VX produced at Newport was never filled into shells but was stored in bulk on the plant grounds in 10,000-gallon storage tanks and one-ton carbonized steel containers. Because of the extremely low volatility of VX, this type of aboveground storage was not considered a safety hazard at the time. Eventually, the "temporary" storage of bulk VX became permanent.

DURING THE EARLY 1960S, Edgewood Arsenal took a second look at binary chemical weapons technology. Although the Weapons Research team had completed the initial proof-of-concept studies on a binary VX bomb in the mid-1950s, this work stayed on the shelf until a young engineer named William C. Dee joined the technical staff. Because his father had been a pipe fitter at Edgewood, Dee had heard about the arsenal throughout his childhood. As a student at Johns Hopkins University in Baltimore, he studied chemical engineering and worked at Edgewood during the summers in the Air Munitions development group, gaining valuable hands-on experience. The only drawback of the job was that it required a security clearance, which meant waiting for a few tedious weeks after the end of classes for his background investigation to be completed. When Dee finished his engineering degree in 1959, the chemical industry was in the depths of a recession. Although he received job offers from a few companies, none was particularly attractive. He therefore accepted a position at Edgewood that was low-paying but promised interesting work, planning to stay for a year or two until a good opportunity opened up in private industry. As it happened, he ended up spending his entire career at the arsenal because of a series of promotions, the technical challenge of the work, and the generous federal pension program.

Soon after Dee arrived at Edgewood, Charles Walker, the chief of the air munitions branch in the Munitions Division, called him into his office and handed him a stack of old research reports from the binary program. "I always thought this was a good idea," he said. "See if you can do something with it." Dee studied the reports and concluded that the binary concept was technically promising, but because the Army continued to deny the need for such weapons, no development funds were available.

In 1960, however, the Pentagon ordered the four armed services to accelerate their development of chemical arms. The Navy was worried about the hazards associated with storing and handling chemical weapons on board aircraft carriers, where even a small leak from a nerve agent bomb stored in the ammunition hold would pose a grave threat to the crew. Using carrier-based aircraft to deliver unitary chemical munitions would therefore require installing an ammunition magazine on each carrier that was hermetically sealed or equipped with special air filters, at an additional cost of about $1 million per ship. As an alternative, the Navy expressed interest in an inherently safe chemical weapon design. Seizing this opportunity, Dee and his colleagues met at the Pentagon with Navy officials, who agreed to provide funds for the development of a 500-pound VX bomb using binary technology.

Once the project had been approved, Dee and an assistant tried to replicate the earlier experiments by Tarnove, Bowman, and Sichel, who had sought to determine the optimal mixing rate of the two VX precursors, powdered sulfur and liquid transester (QL). Dee followed the standard chemical engineering practice of building a small model system that could later be scaled up to a full-size prototype. To that end, he built a two-liter horizontal reactor that was shaped like a miniature bomb. He then designed a research plan calling for twenty experiments, which he believed would be sufficient to define all the relevant parameters.

At first, however, Dee was unable to get the binary system developed by the Weapons Research team to work at all. The earlier group had used a propellant to fire a few grams of particulate sulfur directly into the QL solution, triggering the spontaneous chemical reaction that yielded VX. In Dee's hands, however, the two precursors simply failed to react. He suspected that the source of the problem was the propellant gases being injected into the QL along with the powdered sulfur, since it was known that water vapor could impede the reaction at high temperatures. To solve this problem, Dee contracted with a firm called Aircraft Armaments to develop a telecartridge, an injection device that deflected the hot propellant gases so they did not enter the reaction chamber. Although the telecartridge worked as intended, the binary components still failed to react. Dee racked his brain to understand why and finally realized that he was using a different physical form of sulfur from that employed in the original experiments. When Dee made this change, the reaction worked perfectly, going to completion in about five seconds and generating VX and a great deal of heat.

The initial research plan of twenty experiments proved to be a gross underestimate, however. Over the period of a year, Dee and his small development team did a series of trials with the 2-liter reactor and then scaled up to a 10-gallon reactor and a huge 50-gallon reactor (to simulate an aircraft spray tank) before settling on a 20-gallon reactor, equivalent in liquid capacity to a 500-pound bomb. These experiments were conducted in a reinforced test chamber that had been designed for the explosion of chemical bombs and could be easily drained and decontaminated. Wearing a protective rubber suit and an M9 gas mask, Dee and a technician took turns testing the various reaction parameters. The laboratory was not air-conditioned, and on humid summer days the protective rubber suit became unbearably hot; when Dee removed his boots, sweat poured out. They performed one experiment per day, three days a week, and had to wait a few days for the analytical results because the instruments available at the time were fairly primitive.

Dee found that to scale up the binary reaction to the 20-gallon reactor, it was necessary to install propellers inside the vessel that agitated the liquid QL at the same time that a set of large telecartridges in the rear fired powdered sulfur into the mix. Although mechanically complex, this system generated VX with a purity of 90 percent or more. By 1963, the Edgewood team had developed a binary VX bomb that was dubbed the "Bigeye." Dee built six full-size prototypes, which were shipped from Edgewood to Dugway Proving Ground in Utah for testing in a series of open-air trials. At Dugway's Granite Mountain test site, the prototype binary bombs were suspended on cables from a metal frame and detonated over a sampling grid. Each set of trials lasted one or two weeks, during which the scientists stayed in barracks near Granite Mountain because the dirt roads were too poor for them to commute daily from the main post.

In the initial design for the Bigeye, the VX produced by the binary reaction was dispersed with an explosive burster charge. During the trials at Dugway, however, about one in every three prototypes "flashed" after dissemination, meaning that the cloud of VX droplets ignited into a fireball that consumed most of the lethal agent. To solve this problem, the engineering team switched to a system in which the bomb expelled a pressurized spray of VX as it glided to earth, spreading a rain of lethal droplets over a third of a square mile.

One of the questions studied during the outdoor trials at Dugway was what would happen if a Navy carrier pilot could not release an activated Bigeye from the wing of his aircraft during a bombing run. In that case, the bomb might explode on the wing, contaminating the aircraft with VX. Even if the Bigeye did not go off, it would be too dangerous to land on the carrier deck with the live bomb still attached. The pilot would either have to jettison the bomb rack or, as a last resort, eject from the aircraft and let it crash into the sea. Dee finally decided to avoid these contingencies by designing the Bigeye so that the fuse became primed only after the bomb had been released from the aircraft.

COLD WAR anxieties drove Moscow as well as Washington to upgrade its chemical arsenal. In a speech to the Twentieth Party Congress in 1956, Soviet Deputy Minister of Defense Marshal Georgii Zhukov warned that the imperialists were likely to employ chemical weapons in an attack, creating the need for an effective retaliatory capability. The Soviets were also increasingly concerned about the Communist behemoth to the east. After an ideological split developed between the Soviet Union and Red China in 1960, Moscow moved to modernize its chemical warfare capability in order to offset the numerical superiority of the Chinese People's Army in the event of a Chinese invasion of Siberia or Central Asia.

At NII-42 in Moscow, Simion Varshavsky directed the research-and-development department, which employed more than 250 scientists. Under his leadership, teams of military chemists developed a series of nerve agents with improved toxicity, stability, and persistence, and synthesized small amounts for testing and evaluation. Nevertheless, the generals of the Soviet Chemical Troops lacked confidence in their own scientists and preferred to acquire the same agents and munitions as the United States, to which they attributed superior technical capabilities. During the late 1950s, for example, an espionage operation by Soviet military intelligence obtained the secret chemical formula of VX. The Soviet leadership ordered NII-42 to put its own research on hold and focus its efforts on reproducing the American nerve agent.

Over the next few years, Sergei Ivin, Leonid Soborovsky, and a female chemist named Ia Danilovna Shilakova jointly developed a synthetic method for an analogue of VX, which they termed R-33. The three scientists completed their work in 1963 and were later awarded a Lenin Prize for their achievement. Ivin, Soborovsky, and Shilakova initially claimed to have replicated VX, but their version of the molecule differed from the original in several important respects. Although R-33 had the same number of atoms of carbon, hydrogen, oxygen, sulfur, and nitrogen, they were in a different three-dimensional arrangement. For example, whereas VX had a two-carbon (ethyl) group linked to the central phosphorus through an oxygen atom, the Soviet version had a four-carbon (isobutyl) group in that position. These structural differences gave rise to a distinct set of chemical, physical, and toxicological properties.

The usual explanation for the discrepancies between VX and R-33 is that Soviet military intelligence had obtained the chemical formula of the American nerve agent but not a diagram of its molecular structure, leading Soviet chemists to guess wrongly about its three-dimensional configuration. A more likely hypothesis is that the Soviets knew the correct structure of VX but were incapable of manufacturing it with their available chemical technology. Because the U.S. transester process was very demanding technically, the Soviets chose instead to synthesize a structural variant of VX using a different method: reacting phosphorus trichloride with two chemicals (aminomercaptan and chloroester) in a chloroform solvent. Although the

Soviet production method was less complex, it had the serious drawback that R-33 had to be purified from a large volume of contaminated solvent, an extremely hazardous process. Ironically, toxicological studies later showed that R-33 "aged" cholinesterase—inhibited the enzyme irreversibly—much faster than VX, making the Soviet V agent more lethal and less treatable than the American one. A third hypothesis is that the Soviets deliberately developed a novel analogue of VX in the belief that U.S. field detector-alarms for V agents would not recognize the chemical spectrum of R-33.

At the same time that the Soviets were developing R-33, they began work on a manufacturing process for Soman. In 1960, Soviet military intelligence concluded incorrectly that the United States intended to mass-produce Soman, leading the Kremlin to follow suit. Boris Libman, who in 1958 had been promoted to chief engineer at Chemical Works No. 91, oversaw the development of the Soman manufacturing process. The most challenging step was the synthesis of pinacolyl alcohol, a key ingredient. Working under Libman, an electrochemical engineer named Andrei Petrovich Tomilov devised a production method for pinacolyl alcohol involving five stages of electrolysis, for which he and his coworkers later won a Lenin Prize. In May 1964, the Central Committee approved the construction of a Soman production facility at Volgograd, but numerous technical and organizational hurdles delayed the start of production. (The name "Stalingrad" had been changed to "Volgograd" by the Twenty-second Party Congress in November 1961. Subsequently, Chemical Works No. 91 was renamed the S. M. Kirov Chemical Works, or "Khimprom.")

While building the electrolysis unit for the production of pinacolyl alcohol, Libman encountered numerous problems with shoddy manufacturing. He had ordered nickel cathodes for the electrolyzers from a Soviet company called Uralchimmash. When the cathodes arrived, he found that they had not been made of high-grade nickel, as specified in the contract, but from a low-quality alloy that was much harder to weld, causing many of the cathodes to leak. Libman sent the defective items back to the factory and insisted on receiving new ones, resulting in a six-month delay in production. Problems also arose with the procurement of corrosion-resistant production equipment. Although Plant No. 5 in the city of Sverdlovsk made silver-coated pipes and fittings, it was not capable of producing silver-clad reactors, heat exchangers, columns, or stills. Instead, these items were ordered from the Degussa company in West Germany. Later, the Komsomolets Machine Works in Tambov, Russia, mastered the production of silver-lined equipment, but Libman found that it was inferior in quality to the German apparatus that had been confiscated from Dyhernfurth in 1946. Because of the lengthy delays, Konstantin Alexeievich Guskov, the first deputy director for engineering at NII-42 in Moscow, was dispatched to Volgograd to fix the technical problems with the Soman plant. Under his effective oversight, large-scale production of Soman finally began in 1967 at Unit No. 30 at Khimprom, and the agent was loaded into munitions at Unit No. 60.

The Soviet Union also developed a variety of weapon systems to deliver nerve agents, including Scud ballistic missiles, FROG unguided tactical rockets, artillery shells, multiple rocket launchers, aerial bombs, and chemical mines. Filled chemical weapons were stockpiled at several depots in Russia. In the event of war in central Europe, the Soviets would have used trains to transport empty chemical munitions and tank cars containing bulk nerve agent to army and division levels in its Warsaw Pact allies East Germany, Poland, Czechoslovakia, and Hungary. There the munitions would have been filled with agent and kept under the strict command and control of the Red Army. Some evidence also suggests that the Soviets forward-deployed a small stockpile of filled chemical munitions in East Germany to counterbalance the U.S. chemical stocks in West Germany.

Beginning in 1963, Warsaw Pact battle plans for war against NATO called for the surprise, massive use of chemical weapons on the battlefield to inflict large-scale casualties and demoralize the enemy. Because nerve agents harmed only living beings, such munitions would have been employed instead of tactical nuclear weapons in areas where material damage to buildings and infrastructure was to be avoided. The Soviet military subjected nerve agents to the same strict political controls as those for nuclear arms, so that a decision to authorize a chemical attack would have been made at the highest levels. All information related to Soviet offensive chemical warfare plans was tightly held, using special code words that were changed every six months.

DURING THE 1960s, the U.S. intelligence community conducted a series of National Intelligence Estimates (NIEs) of Soviet chemical warfare capabilities. These highly classified studies were prepared by the CW/BW Intelligence Committee (CBIC) of the U.S. Intelligence Board. Chaired by the Director of Central Intelligence, CBIC included representatives from the CIA, the Defense Intelligence Agency (DIA), the National Security Agency (NSA), and the State Department's Bureau of Intelligence and Research (INR). According to its critics, CBIC tended to exaggerate the Soviet chemical threat because it was staffed by "true believers," most of them reserve officers in the Chemical Corps who had grown up professionally in the U.S. chemical warfare program.

To prepare estimates of the Soviet chemical weapons stockpile, CBIC analysts relied heavily on the "Hirsch report," a detailed overview of Soviet chemical warfare activities and facilities that had been prepared shortly after World War II by Dr. Walter Hirsch, a former chief of the chemical warfare section of the German Army Ordnance Office. CBIC analysts also used overhead reconnaissance photographs of suspected Soviet chemical weapons storage bunkers taken by high-flying U-2 aircraft and orbiting satellites. Such bunkers were generally identified by means of telltale "signatures," such as roof ventilation systems or the presence of decontamination trucks.

To prepare an estimate of the Soviet stockpile, CBIC analysts calculated the interior volume of known and suspected chemical weapons bunkers, making assumptions about whether the agent was in bulk or weaponized form (which could result in a tenfold difference in weight) and how densely the munitions were stacked. Based on these assumptions, they estimated the tonnage of agent contained in each bunker. Summing the contents of all known and suspected bunkers, plus a "fudge factor," led to a total of more than 150,000 agent tons of blister and nerve agents. Because of the uncertainties inherent in this methodology, some senior U.S. government officials believed that CBIC's estimates of the Soviet chemical inventory were grossly inflated and viewed them with considerable skepticism. Nevertheless, advocates of chemical warfare, such as the Army Chemical Corps, frequently cited the CBIC figures when lobbying Congress for higher budgets and stockpile requirements.

In early 1963, John Kerlin, an analyst with the CIA's Office of Scientific Intelligence, was assigned a rotation in the Office of National Estimates, where he was given the task of updating the figures for the Soviet chemical arsenal. Because Kerlin was rigorous about defining analytical assumptions, his estimate of the Soviet chemical weapons stockpile was significantly lower than in previous years. The Chemical Corps and other interested parties were displeased by the lowball figure and harshly criticized Sherman Kent, the director of the Office of National Estimates. A few months after Kerlin issued his report, he died in his thirties of a heart attack, leaving a wife and children. Because Kerlin was no longer around to defend his methodology, the next CIA estimate of the Soviet chemical stockpile returned to the old, inflated figures.

One Cold War scenario that particularly worried U.S. strategists was the possible Soviet use of nerve agents for covert operations against American and Canadian military personnel working for the North American Air Defense Command (NORAD), which was responsible for the early detection and warning of a nuclear attack by Soviet strategic bombers (and later intercontinental ballistic missiles) flying over the North Pole. U.S. military planners feared that Soviet Spetsnatz special forces units might use nerve agents to kill the crews of the early-warning radar stations in Alaska and Greenland, thereby "blinding" NORAD and opening the way for a disarming nuclear first strike against U.S. Strategic Air Command (SAC) bases.

On August 22, 1960, the commander in chief of NORAD wrote a memorandum to the Army Chief of Staff in which he laid out a requirement for "a system to detect and report enemy employment of biological and/or chemical agents which might affect air defense personnel and equipment from carrying out their assigned mission. This system must be capable of providing positive and timely detection of the agent or agents employed, and the instantaneous reporting of such employment to the NORAD Combat Operations Center for assessment and dissemination of appropriate information to air defense agencies in time to place into effect timely protective and defensive actions." A few years later, in a book titled *Tomorrow's Weapons,* Brigadier General Jack Rothschild, the former chief of research and development for the Chemical Corps, warned that the Soviets might use nerve agents for a "sabotage attack against our missile sites and SAC bases preceding a nuclear strike." In response to these concerns, the Pentagon took steps to reduce the vulnerability of its nuclear weapons installations to covert chemical attack.

In May 1963, the British government again changed its chemical weapons policy. Reversing the 1956 decision to renounce an active chemical arsenal, Prime Minister Harold Macmillan decided that Britain should acquire a modest stockpile of nerve agents in order to have a retaliatory option in the event of a limited war in Europe between NATO and the Warsaw Pact. To this end, the Cabinet Defence Committee authorized Porton Down to launch a five-year exploratory program of offensive R&D and limited production. Porton scientists studied eight candidate V-series agents, including the Soviet R-33. At the end of this process, the British Ministry of Defence issued a military requirement calling for 22,000 Sarin-filled 105 mm artillery shells for the British Army and 320 VX-filled spray tanks for the Royal Navy and Air Force.

After the Conservative Party was defeated in October 1964 and Labour Prime Minister Harold Wilson took power, he slashed funding for British chemical rearmament. In July 1965, the British Chiefs of Staff conducted a policy review and concluded that the use of chemical weapons in a general war in Europe would not significantly influence the outcome of the battle or delay the use of tactical nuclear weapons. Even so, the generals continued to recommend the acquisition of a chemical weapons stockpile to deter the Warsaw Pact from employing such weapons in a limited war. In November 1965, the British government debated whether to manufacture Sarin and VX domestically or procure them from the United States. By 1968, however, the plan to acquire a nerve agent stockpile had been shelved indefinitely.

Throughout the 1960s, Britain, Canada, and the United States continued to collaborate on V-agent research and development under the Tripartite Agreement, which was renamed the Tripartite Technical Cooperation Program (TTCP). In 1964, Australia began to participate in TTCP meetings under an army standardization agreement, and it became a formal member in July 1965. Thereafter, the name of the group was changed to "The Technical Cooperation Program," while retaining the same acronym. New Zealand also joined the TTCP in October 1970.

IN 1965, after seven years as chief engineer in Volgograd, Boris Libman suffered a major professional setback. For some time, the Khimprom plant had been discharging toxic wastes from Sarin and Soman production into a holding pond on the factory grounds known as the "White Sea," where the chemicals were neutralized with sodium hydroxide. This method of neutralization turned out to be slow and ineffective, resulting in concentrations of toxic phosphonates—breakdown products of nerve agents—a hundred times higher than permitted. In early February 1965, flooding caused by melting snow caused a levee bordering the wastewater pond to collapse, allowing a large volume of toxic wastes to drain into the Volga River.

The spill had no immediate environmental consequences, and the levee was repaired in a day and a half. Four months later, however, on June 15, tens of thousands of sturgeon in the Volga suddenly died and floated belly up, turning the river white for fifty miles downstream. A possible explanation for the delay between the toxic spill and the fish die-off is that it took four months for the chemicals to build up to lethal levels in the fishes' tissues. Several years later, the Institute of the Soviet Fishing Industry was still finding traces of toxic phosphonates in sturgeon caught in the Caspian Sea.

Responding to public outrage over the environmental disaster, Soviet Prime Minister Alexei Kosygin insisted that the managers of the Volgograd plant had to be punished as an example. Six officials were fined, but the harshest punishment was reserved for Chief Engineer Libman, then forty-three. On March 9, 1966, he was convicted of negligence, stripped of his Lenin Prize, fined 10,000 new rubles (the equivalent of two years' salary), and sentenced to two years in a labor camp in the nearby city of Volsk. During his imprisonment, Libman worked during the day as the foreman of a construction crew that built houses and later at a chemical plant, returning to the prison at night. After serving one year of his two-year sentence, however, he was released because no one else was capable of overseeing Soman production at Volgograd.

ON MAY 27, 1967, the governments of France and Algeria signed a secret framework agreement prolonging France's use of the chemical weapons testing site at B2-Namous for another five years, until 1972. Although the French Army had sent commissions of inquiry to several of France's overseas territories to find a suitable replacement, these efforts had been unsuccessful. No substitute location had been found that offered climatic conditions similar to those of Central Europe. Additional problems were the exorbitant cost of building a new testing ground and the excessive distance of the candidate sites from metropolitan France. When it became clear that there was no alternative to retaining B2-Namous, the French government was prepared to pay almost any price the Algerians demanded.

In exchange for the continued use of the chemical weapons testing site,

France sold Algeria its nuclear testing installations and military equipment for 21 million francs—less than half the estimated value of 50 million francs. Moreover, the French government allowed Algerian chemical weapons specialists to attend the Military School for Special Weapons (École Militaire des Armes Spéciales) in Grenoble and to observe the open-air trials at B2-Namous, although the Algerians were given limited access to the resulting data. The Algerian government also demanded that all French personnel and matériel be transported directly by plane to the airfield at B2-Namous and that no uniformed French soldiers be present at the site. Accordingly, the French Ministry of Defense hired Thomson, a private defense contractor, to operate the base and liaise with the Algerian Army.

During the late 1960s, the French Army became increasingly concerned about the expansion of the Warsaw Pact's chemical warfare capabilities. Accordingly, President Charles de Gaulle decided to improve France's chemical defense posture by testing new military equipment with "live" chemical warfare agents at B2-Namous. These experiments were conducted with "maquettes" (mock-ups) that realistically simulated weapon effects. Although only a few open-air trials with nerve agents were conducted each year, they enabled the French military to assess its vulnerabilities to chemical attack and develop improved defenses. Also during the 1960s, French and American military officials cooperated on chemical weapons research and development under a Mutual Weapons Development Data Exchange Agreement.

Although the vast stockpiles of nerve agents accumulated by the United States and the Soviet Union were sufficient to kill millions of people in the event of World War III, mutual deterrence prevailed in the East–West balance of terror. Nevertheless, each side sought to achieve a marginal advantage in the chemical arms race, at times resorting to deception to do so. According to a history by David Wise, from early 1966 until July 1969 the U.S. Department of Defense ran an intelligence operation called Operation Shocker, which involved providing the Soviets with disinformation on chemical weapons. Sergeant Joseph Cassidy, who worked at Edgewood Arsenal, was recruited by Soviet military intelligence (the GRU) and supplied his handlers with some 4,500 documents describing a purportedly successful effort by the U.S. Army to develop a nerve agent called GJ that was many times more potent than Sarin or Soman and existed in binary form.

In fact, Cassidy was a double agent engaged in a deception operation. Edgewood had attempted to develop agent GJ and failed, and the technical reports that Cassidy handed to the GRU had been partially falsified. The aim of Operation Shocker was to lead Soviet military chemists down a technological cul-de-sac, causing them to waste time and money on an illusory objective. Ironically, the operation later backfired when the Soviets, spurred on by the supposed U.S. breakthrough, successfully developed a new generation of supertoxic nerve agents in the 1970s. In the meantime, the technology and know-how to produce standard nerve agents was spreading to less stable regions of the world, where their actual use became more likely.

CHAPTER TEN

YEMEN AND AFTER

DURING THE EARLY 1960S, Egyptian President Gamal Abdel Nasser, an ardent nationalist who aspired to leadership of the Arab world, decided to pursue an indigenous chemical weapons program under the code name "Izlis." At first Egypt relied heavily on foreign technical assistance. High-ranking Egyptian Army officers traveled to Moscow for training in offensive chemical warfare tactics at the Red Army's Academy of Chemical Defense. The Egyptian government also recruited a group of West German scientists and engineers, many of whom had developed weapons for Hitler during World War II. Although Nasser claimed that the Germans were working on peaceful projects such as a commercial jet engine, they were actually developing ballistic missiles and chemical arms.

In the fall of 1962, Egypt intervened in Yemen, a small country on the Arabian Peninsula, after the death of the country's ruler, Imam Ahmed, and the succession to the throne of his thirty-five-year-old son, Imam Mohammed el-Badr. On September 26, a group of Yemeni military officers, with Egyptian support and encouragement, staged a coup d'état to replace the monarchy with a republican government modeled on the Nasser regime. Yemeni Army tanks bombarded the royal palace in the capital, Sana'a, forcing the young imam to flee to the mountains in the north of the country. There he mobilized loyal tribesmen to fight for the restoration of the monarchy.

Two days after the coup, an advance guard of Egyptian troops was airlifted to Yemen to bolster the republican forces. This initial deployment was followed by a larger contingent of 28,000 Egyptian soldiers, along with military aircraft. Meanwhile, the deposed imam solicited assistance from other Arab monarchies in the region. In October 1962, the government of Jordan flew 150 tons of light arms to Saudi Arabia for distribution to the royalist forces in Yemen. The Saudi royal family also became a major supplier of arms to the imam, despite a long history of conflict with the Yemeni monarchy.

The royalist insurgents in Yemen operated out of caves in the northern mountains that were largely invulnerable to attack with conventional bombs. In 1963, the Egyptian intervention forces, seeking to root the guerrillas out of their mountain redoubt, began to experiment with the use of chemical weapons delivered by Soviet-made Ilyushin-28 bombers. Although the initial chemical attacks involved tear gas, they soon escalated to bombs containing phosgene and mustard. These weapons were either old British munitions that had been abandoned in Egypt during World War II or new ones supplied by the Soviet Union, Egypt's superpower patron. In June 1963, Egyptian aircraft dropped mustard-filled bombs on the royalist village of Al-Kawma, producing numerous casualties who suffered from nausea and skin burns.

On July 9, Saudi Arabia filed a formal complaint with U.N. Secretary-General U Thant that Egypt was using chemical weapons in Yemen. The U.S. and British governments began to investigate the allegation, and Britain asked the U.N. observer group in Yemen to conduct an inquiry. If the Saudi claim was true, the Egyptian chemical attacks were a clear violation of the Geneva Protocol, which Cairo had signed in 1925 and ratified in 1928. President Nasser, for his part, categorically denied the use of poison gas in Yemen.

Because of Saudi and Jordanian assistance to the royalist guerrillas, the Yemen civil war dragged on inconclusively for several years. Beginning in March 1966, Egypt launched a major offensive in northern Yemen. In December, Egyptian bombers dropped fifteen chemical bombs on the royalist village of Halbal, about thirty miles north of Sana'a. The journalist Marquis Childs of the St. Louis *Post-Dispatch* reported another chemical attack on January 4 and 5, 1967, against the villages of Hadda and Kitaf and nearby caves, killing more than a hundred people. At that time, British Prime Minister Harold Wilson told the House of Commons that he believed chemical weapons were being used in Yemen.

On January 29, 1967, a classified cable from the U.S. Embassy in Beirut to the State Department in Washington relayed an eyewitness account of the January 5 chemical attack on the village of Kitaf. The Associated Press correspondent David Lancashire reported that immediately after exposure to the toxic cloud, several victims had begun to vomit, collapse, and die. Such a rapid onset of symptoms suggested the presence of a nerve agent, perhaps combined with other chemicals to mask its identity. If Lancashire's reporting was accurate, the Egyptian attacks may have involved the first combat use of nerve agents in history. The U.S. Embassy cable also stated that CIA agents had collected "bomb fragments and soil samples" from the vicinity of the chemical attack for analysis, although the results were not described.

Beginning in April 1967, the government of Saudi Arabia submitted a series of medical reports to the United Nations alleging that Egypt had employed nerve agents—described as "anticholinesterase gases of the organophosphorus family"—in its air strikes on northern Yemeni villages. The Saudis claimed that between January 4 and May 16, Egyptian aircraft had dropped bombs containing nerve agents at least four times, killing more than 400 persons. Some accounts suggested that the substance used in the attacks had been Sarin, but Marquis Childs wrote that samples of contaminated sand had been smuggled out of Yemen and analyzed at an independent U.S. laboratory, which had found traces of a V-series nerve agent. Reportedly, the agent was so new that it matched none of the thousands of chemical spectrographs on file with the Army Chemical Corps. British operatives also took soil samples but were unable to confirm the presence of a nerve agent.

Egyptian chemical attacks on royalist villages in north Yemen continued throughout the spring of 1967. On May 28, the town of Sirwah was bombed with a lethal gas and high explosives, causing at least seventy-two deaths and many wounded. About thirty of the injured were evacuated to Jiddah, Saudi Arabia, for medical treatment. Physicians who examined the victims confirmed the use of nerve agents and sent reports to the International Committee of the Red Cross (ICRC), a nongovernmental organization based in Geneva that seeks to enforce the humanitarian laws of war. An ICRC team conducted its own investigation in north Yemen and found that a royalist village had been attacked with chemical weapons in early June and that some of the bomb fragments bore markings in Cyrillic letters. Subsequently, an Egyptian military aircraft attacked an ICRC convoy on its way to assist the victims of a chemical attack, provoking the organization to file a formal protest with the Nasser government.

The Cyrillic markings on bomb fragments suggested that the Soviet Union had furnished chemical weapons to Egypt. If that was the case, the Soviets may have intended to use Yemen as a testing ground to assess the military utility of nerve agents (possibly including the new Soviet V agent R-33) and to gauge the reaction of foreign governments, the United Nations, and world public opinion to the use of such weapons in a remote Third World conflict. Because the 1925 Geneva Protocol banned only the use in war of chemical arms and not their development, production, or transfer, the Soviet Union would not technically be in violation of the treaty if it shipped chemical weapons to Egypt for use in Yemen.

It is also possible that by 1967, Egypt had acquired an indigenous capability to manufacture nerve agents. Four years earlier, the Nasser government had opened a chemical weapons production facility called Military Plant No. 801 at Abu Za'abal, an industrial zone ten kilometers northeast of Cairo. The Egyptian Ministry of Defense operated this plant under the cover of a commercial entity called the Abu-Za'abal Company for Chemicals and Insecticides. Appearing at a congressional hearing on April 30, 1969, Harvard biochemistry professor Matthew Meselson testified, "I asked a British chemist who had spent time in Cairo whether he thought that his Egyptian chemist colleagues could have produced nerve gas in Egypt, and he said without doubt yes."

Despite the credible reports of chemical warfare in Yemen, the international community appeared largely indifferent to Egypt's flagrant violation of the Geneva Protocol. Neither U.N. Secretary-General U Thant nor any of the major powers demanded an official inquiry or referred the matter to the Security Council or to the International Court of Justice in The Hague. Although the United States and Britain both criticized the alleged Egyptian chemical attacks, they did so in a low-key manner. Tellingly, the U.S. diplomatic protest was delivered not by Secretary of State Dean Rusk but by an official from the Arms Control and Disarmament Agency, a little-known bureaucracy. Behind the scenes, however, President Lyndon B. Johnson was quite concerned about Egypt's apparent use of nerve agents. A secret briefing memorandum prepared for Johnson's meeting with Soviet Premier

Alexei Kosygin on June 16, 1967, included the following "contingency talking points":

1. The International Committee of the Red Cross in Geneva has confirmed the use of poison gas recently in Yemen.
2. US has evidence that supports the ICRC release, specifically, we know that UAR [United Arab Republic = Egypt] used lethal nerve gas in Yemen.
3. The use of lethal chemical agents by either party during recent crisis would have seriously increased risk of escalation of conflict.
4. Continued existence and use of such weapons of mass destruction in Middle East could lead others in area to decide they must acquire these or other weapons of mass destruction.
5. Express desire [to] discuss with Soviets ways of preventing acquisition and further use [of] such CW agents by states in area.

Although it is not known if President Johnson used the talking points during his meeting with Kosygin, they clearly indicate the U.S. government's belief that Egypt had employed nerve agents in Yemen, most likely with Soviet assistance. Nevertheless, Washington declined to press these charges in public. On July 27, 1967, the State Department spokesman Robert J. McCloskey told reporters that the United States was "deeply disturbed" by reports of chemical warfare in Yemen, but he refused to confirm the allegations. "I think it's fair to say that we have drawn no specific conclusions based on what we know," he said. A few months later, however, the journalist Seymour Hersh interviewed an unnamed State Department official and asked if the U.S. government had concrete evidence that nerve agents were militarily effective. The official replied bluntly, "We know that nerve agent works. It worked in Yemen."

Even though Egypt was the Soviet Union's most important ally in the Middle East and U.S. relations with Nasser were poor, the Johnson administration was muted in its criticism of Egyptian chemical warfare in Yemen. Why? A possible explanation is that the White House was concerned that strongly condemning the Egyptian chemical attacks would provoke a backlash against the ongoing American use of tear gas and the herbicide Agent Orange in the Vietnam War, which had become a topic of bitter denunciations by the Soviet Union and its allies. Although the United States had not yet ratified the 1925 Geneva Protocol, Washington claimed that the treaty did not ban the combat use of tear gas or herbicides because these agents did not normally cause death or prolonged incapacitation. Few countries, however, supported the U.S. interpretation of the Geneva Protocol. Given this controversy, the Johnson administration may have decided to play down its public denunciations of the Egyptian chemical attacks. The unfortunate result was that several hundred Arab tribesmen in a remote desert country were abandoned to a cruel fate, and the international norm against chemical warfare was weakened.

At the same time that Egypt was escalating its war in Yemen, the Nasser regime launched a series of provocative actions that increased tensions with neighboring Israel. On May 15, 1967, Israeli intelligence learned that Egypt had massed large numbers of ground forces in the Sinai Peninsula. President Nasser accompanied this military buildup with two other threatening steps: he demanded the departure of U.N. peacekeepers patrolling the Sinai border between Israel and Egypt, and on the night of May 22, the Egyptian Navy blocked the straits at the end of the Gulf of Eilat to prevent the passage of Israeli ships. Other Arab states in the region also prepared for war. On May 30, Jordan joined the Egyptian-Syrian military alliance and placed its army under Egyptian command. Iraq quickly followed suit and agreed to send reinforcements.

In late May 1967, an Israeli military intelligence unit conducting a reconnaissance mission in the Egyptian-controlled Sinai Desert came across a reinforced concrete bunker near El-Arish containing six 105 mm chemical artillery shells. An Israeli mobile laboratory attached to the Sharon Division took samples from the shells for analysis and determined that they contained Sarin. Although the projectiles had fuses and were ready to fire, the small number of rounds raised doubts about their military purpose.

In response to the discovery of the Egyptian shells, the Israeli government launched a frantic effort to obtain enough chemical protective gear for the civilian population of Israel. As the threat of war loomed, Israeli agents purchased 20,000 gas masks from U.S. manufacturers and flew them to Tel Aviv aboard a chartered Boeing 707 jet. Israel also requested another 20,000 masks from West Germany. Chancellor Kurt Georg Kiesinger and Defense Minister Gerhard Schröder debated whether gas masks should be considered "war matériel," which would be banned for export under West German law. In the end, Kiesinger overruled his defense minister and authorized the immediate shipment of gas masks to Israel on June 2, only days before the war began. Israeli agents also purchased thousands of autoinjectors containing nerve agent antidotes. At the same time that the government of Israel stockpiled defensive equipment, it sought to deter any Arab use of chemical weapons by making veiled threats that it had the capability to retaliate in kind. Ever since 1960, when Israeli scientists had visited the French chemical weapons testing site in the Algerian desert, Israel had been suspected of pursuing a gas warfare capability, at least as a stopgap deterrent until it acquired a nuclear arsenal.

The rising tensions between Israel and its neighbors finally reached the breaking point on June 5, 1967. Shortly before the Arab armies attacked, Israel launched a preemptive air strike against the Egyptian Air Force and destroyed most of its planes on the ground. During the ensuing ground war, the Israel Defense Forces defeated the combined armies of Egypt, Syria, Jordan, and Iraq in only six days. Fortunately, mutual deterrence and the short duration of the war prevented any use of chemical weapons. A few weeks later, Israel returned the 20,000 gas masks it had purchased from West Germany, unused.

After Egypt's humiliating defeat in what became known as the Six-Day War, President Nasser could no longer sustain his intervention in Yemen. He therefore decided to withdraw his troops and pursue better relations with Saudi Arabia. Cairo's last use of chemical weapons in northern Yemen took place in July 1967, shortly before the departure of the intervention force. The Egyptian attacks were severe and may have constituted a final, unsuccessful attempt to achieve a military victory.

In 1968 the United States shut down its VX production program, and one year later the Newport Army Chemical Plant was officially mothballed. One reason for this decision was a severe shortage of empty bombs and artillery shells, nearly all of which were being filled with high explosives for the U.S. war in Vietnam. The Pentagon even raided its existing stockpile of chemical artillery shells for impact fuses to use in conventional weapons. Given the lack of empty munitions to fill with VX, large amounts of bulk agent continued to be stored in tanks and one-ton containers on the grounds of the Newport facility.

The U.S. Army Chemical Corps also faced a growing crisis over the M55 artillery rocket, which had a warhead that contained either Sarin or VX. M55 rockets had been filled with Sarin at Rocky Mountain Arsenal from 1961 to October 1965, and with VX at Newport from 1964 to 1965. Within two years of their production, many of the Sarin-filled rockets (but not the VX-filled ones) began to leak, creating a huge headache for the Army. In February 1966, the Chemical Corps established an M55 Action Team to investigate the problem. This committee found that some batches of Sarin-filled rockets had a higher rate of leakage than others because the nerve agent fill varied considerably in purity. Although laboratory studies had suggested that Sarin was compatible with the rocket's thin aluminum shell, these experiments had been performed with pure agent synthesized in a pilot plant and not with the industrial-grade product.

The Sarin used in the M55 warheads had been manufactured at Rocky Mountain Arsenal between 1953 and 1957 in a series of 431 lots. The first 241 batches had gone through two distillation steps, yielding a product that was 92 percent pure and had a low level of acidity. For the remaining 190 lots, however, the chief engineer of the Chemical Corps had decided to expedite production and save money by dropping the second distillation step. These so-called round-out lots of Sarin had a purity specification of only 88 percent. The M55 Action Team determined that heavy-metal impurities and high concentrations of acid in the round-out lots had caused the nerve agent to corrode the rocket's thin aluminum shell, resulting in numerous pinhole leaks.

The M55 investigation also found a second explanation for the leaking rockets. In 1965, the Chemical Corps had begun to dispose of the obsolete M34 Sarin cluster bomb, of which some 10,000 units had been manufactured in the late 1950s. Because the Sarin in the cluster bomblets was about 90 percent pure, it was recovered and reused in some of the early lots of M55 rockets. During its residence in the M34 bomblets, however, the nerve agent had picked up copper impurities from the welds. Two or three years after the M34 Sarin was recycled into the M55 rockets, the copper impurities reacted with the aluminum shell in a process known as bimetallic corrosion, causing tiny pits that eventually turned into pinhole leaks.

By 1968, the defective M55 rockets had begun to leak like sieves, leaving puddles of Sarin on the floor of the storage igloos. Because of the extreme safety hazard, inspectors had to enter the affected bunkers wearing full-body protective suits, even in the midsummer heat. The final report of the M55 Action Team, submitted in March 1968, recommended the prompt disposal of the leaking rockets. About 50,000 of the defective rounds were secretly destroyed by open-pit burning at Dugway Proving Ground. Chemical ordnance specialists dug twelve large trenches and placed the M55 rockets nosedown, covered them with dunnage and gasoline, and set them on fire. The results were disastrous: many of the rocket motors ignited, causing the weapons to fly erratically through the air, spewing their deadly contents and contaminating the ground.

After this misadventure, the Army decided to dispose of its leaking M55 rockets under Operation CHASE (an acronym for "cut holes and sink 'em"), an existing program for the ocean dumping of obsolete ordnance. Initially limited to conventional weapons, CHASE had begun accepting chemical munitions in 1967. Thirty leaking M55 rockets were stacked inside massive steel containers called "coffins," which were then filled with concrete and welded shut. Each filled coffin weighed 6.4 tons. In a 1968 mission

Open-air burning pits at Dugway Proving Ground in Utah were used in 1968 to destroy about 50,000 defective M55 rockets that were leaking Sarin.

called CHASE 5, several ton containers of mustard agent and hundreds of M55 coffins were loaded onto an aging Liberty ship, the S.S. *Corporal Eric Gibson.* The Navy towed this vessel from Colt's Neck Naval Pier in New Jersey to a spot 200 miles east of Atlantic City and scuttled it in 7,200 feet of water.

THROUGHOUT THE 1960S, as recommended by the Project 112 report, the Deseret Test Center conducted an extensive series of field trials involving nerve agents and chemical simulants, which were held at sea in the Pacific Ocean and on land in Alaska, Hawaii, and the Panama Canal Zone. Of 134 planned trials, 50 were actually carried out (19 at sea and 31 on land), some of them jointly with Britain and Canada under the Tripartite Agreement. Roughly five thousand U.S. soldiers took part in the ship-based trials and about five hundred in the land-based ones.

A subset of the Project 112 operation, called the Shipboard Hazard and Defense (SHAD) program, involved six sea trials with chemical agents in the Pacific Ocean from 1964 to 1968, three with live nerve agents and the others with simulant chemicals. The Deseret Test Center's naval fleet consisted of five light tugboats and two converted Liberty ships, the U.S.S. *George Eastman* and the U.S.S. *Granville S. Hall.* These vessels were "citadel" ships, meaning that the portholes and hatches could be sealed and an advanced wash-down system eliminated all traces of nerve agent after a test.

In addition to assessing the vulnerability of Navy warships to nerve agent attack, the SHAD trials tested procedures for detection and warning, crew protection, and ship decontamination while maintaining a warfighting posture. The Navy feared that an enemy attack with a few VX-filled bombs or missile warheads might put an entire warship out of action, including an aircraft carrier operated by approximately 5,500 servicemen. Seawater could not be used to decontaminate the carrier deck because it would damage the insides of the aircraft irreparably.

The first SHAD test series, code-named "Flower Drum," extended over several months in 1964 off the coast of Hawaii. A gas turbine mounted in the bow of the *George Eastman* generated a cloud of vaporized Sarin that wafted over the ship, while crew members wearing gas masks took air samples from various parts of the vessel to measure contamination levels. The second phase of Flower Drum involved spraying a towed barge with a solution of VX mixed with a fluorescent dye to assess the effectiveness of a water wash-down system for protection and decontamination. In another SHAD trial called "Fearless Johnny," which took place southwest of Hawaii in August–September 1965, a Navy A-4B aircraft took off from an airfield on the island of Kauai and sprayed the *George Eastman* with VX (or a simulant chemical) mixed with fluorescent dye. The purpose was to measure the extent of exterior and interior contamination caused by aerial spraying of VX under various levels of shipboard readiness, to assess the impact of VX contamination on military operations, and to determine the effectiveness of the ship decontamination system.

The Deseret Test Center also conducted numerous open-air releases of Sarin and VX on land to study the dispersion and persistence of nerve agents under various climatic conditions. A test series in April–May 1967, code-named "Red Oak," involved the detonation of Sarin-filled artillery shells and M55 rockets in Upper Waiakea Forest Reserve, a dense rain forest on the island of Hawaii. In addition, from 1964 to 1968, the Army Tropic Test Center, near Fort Clayton in the Panama Canal Zone, conducted "environmental" tests of nerve agent munitions to determine the effects of tropical climate on long-term storage. Over a two-year period, mines, rockets, and artillery shells filled with VX, and rockets filled with Sarin, were placed on pallets outdoors under ventilated covers and periodically tested for leaks, pressure, corrosion, and agent purity.

The Army performed live-agent trials under arctic conditions at the 1,200-square-mile Fort Greely Military Reservation, about a hundred miles southeast of Fairbanks, Alaska. This vast military reservation encompassed mountains, glaciers, forests, tundra, rivers, and lakes, and its remoteness made it ideal for arctic warfare exercises and cold-weather trials. In winter, temperatures plummeted to minus 50 degrees for days on end. Between 1962 and 1967, the Deseret Test Center conducted hundreds of open-air trials at the Gerstle River Test Site, about thirty miles south of Fort Greely, to measure the effects and persistence of nerve agents or chemical simulants under subzero weather conditions. These tests were code-named "Elk Hunt," "Whistle Down," "Night Train," "Sun Down," "Devil Hole," "Swamp Oak," "West Side," and "Dew Point." In some cases, the Army fired artillery shells and rockets containing Sarin or VX into spruce or aspen forests or open terrain with snow cover to study the behavior of the agent cloud.

Other trials involved the dispersal of nerve agents from mines, bombs, rockets, torpedoes, and spray tanks.

During Elk Hunt, Phase I, Sarin-filled munitions were detonated in place by remote control while cameras filmed the effects of the vaporized agent on sheep or goats tethered nearby. Elk Hunt, Phase II, held in 1965, involved a series of trials to determine how much VX would be picked up by military vehicles and troops passing over contaminated terrain. Army vehicles ran over and triggered VX-filled mines, after which soldiers wearing gas masks and protective suits washed down the vehicles and themselves with decontaminating solution. The vehicles and protective suits were then tested for residual traces of VX.

Although the nerve agent trials in Alaska were shrouded in secrecy, an accident at Gerstle River provided a rare glimpse into the testing program. In February 1966, Army personnel left 200 artillery shells filled with VX and three M55 rockets with Sarin warheads on the frozen surface of Blueberry Lake, a body of water about a thousand feet across in a remote portion of the testing ground. The munitions had been prepared for demolition and then apparently forgotten. When the ice melted in the spring thaw, the shells and rockets sank to the bottom of the lake. In August 1968, more than two years later, the incoming head of the special projects division at the Arctic Test Center heard about the missing weapons and ordered Blueberry Lake pumped dry in the spring of 1969. Workers recovered the munitions from the lake bottom and destroyed them by chemical neutralization.

In January 1971, the Gerstle River incident became public thanks to investigative reporting by a Fairbanks journalist named Richard A. Fineberg, whose articles sparked concern in Congress over the Army's cavalier handling of deadly nerve agents. Fineberg determined that the 203 shells and rockets left on the frozen lake had been part of a large stockpile of surplus chemical arms that the Arctic Test Center had begun to destroy in 1964. In response to probing questions from lawmakers, Army officials explained that the missing weapons had been "aggregate leftovers from a number of tests in the past" and had not been destroyed immediately because of "the priority of test operations over disposal operations." This explanation implied that the lost chemical munitions were a tiny fraction of the total number tested in Alaska. The massive release of nerve agents into the pristine arctic wilderness may have caused serious and lasting environmental damage. In July 1972, animal protection experts at Fort Greely reported the mysterious deaths of fifty-three caribou in the vicinity of Blueberry Lake, where the sunken chemical munitions had been recovered three years earlier.

In the spring of 1968, the Army was implicated in another accident involving nerve agents, this one involving thousands of deaths. Although the victims were sheep and not people, the incident was to have profound implications for the future of the U.S. chemical weapons program.

CHAPTER ELEVEN

INCIDENT AT SKULL VALLEY

On March 13, 1968, Chemical Corps technicians wearing gas masks and protective suits were preparing for an open-air release of VX nerve agent at Dugway Proving Ground. The Army's chemical and biological testing site covered 1,315 square miles of northwest Utah, an area one quarter larger than the state of Rhode Island. It encompassed a varied landscape of barren salt flats, sand dunes, and rugged cliffs, surrounded on three sides by mountain ranges. The base also included an airfield with a 13,000-foot runway that could accommodate aircraft of any size.

Security at Dugway was tight, and armed guards and military aircraft patrolled the 210-mile perimeter. The main entrance gate was on the eastern edge of the proving ground, an eighty-mile drive from Salt Lake City on U.S. Highway 40 through twisting mountain passes and across the flat desert floor. Visitors encountered a military checkpoint with a sign clearly designed to intimidate: WARNING: DANGEROUS INSTRUMENTALITIES OF WAR ARE BEING TESTED ON THIS POST. CAUTION: DO NOT HANDLE ANY UNIDENTIFIED OBJECTS. REPORT THEIR LOCATION TO SECURITY.

About one thousand civilians and five hundred servicemen and their families lived in the residential section of the post, near the southern tip of the snowcapped Cedar Mountains. The officers and civilian scientists had detached houses like those in a modest suburban subdivision, while the troops were housed in barracks. Despite the arid climate, the lawns were watered and wild horses often grazed on the lush grass. The scientists at Dugway were mainly college graduates who had majored in chemistry or biology, along with a few Ph.D.s. Most had been attracted by the generous salaries, which compensated for the isolation and harsh weather of the Utah desert. The post had a PX and a movie theater, but the closest town was Tooele, thirty miles away. On weekends, staff members drove their RVs into the hills to explore abandoned gold mines or follow the tracks of the old wagon trains. During the week, they focused on the technical aspects of their work and spent little time ruminating about its broader political or moral significance.

Roughly fifteen miles west of the quiet residential streets of Dugway, the proving ground began on the salt flats and continued up into the rugged hills. Marked out on the desert floor were circular grids ranging in size from 1,800 square feet to 150 square miles. Tower Grid, for example, had a 75-foot artillery tower and a 300-foot rocket tower on which chemical munitions could be mounted and detonated, releasing cone-shaped plumes of toxic agent that floated downwind over the two-mile-long range. The thirty acres in the center of the grid were instrumented with 3,200 battery-powered air samplers mounted on poles, which measured the concentration of chemical agent at multiple points in space and time. These data could be converted into density maps of the toxic plume and later into application tables and mathematical models of chemical weapons effects, including downwind transport, dilution, and deposition.

The principle underlying the safety of nerve agent testing at Dugway was that the vast majority of droplets in an agent plume would settle to the ground within a mile of the point of release, while the rest would be diluted to harmless levels by the time they reached the border of the proving ground. Under unusual atmospheric conditions, however, toxic clouds were known to travel long distances. Every month or so, Dugway officials placed a confidential call to the sheriff of Tooele County asking him to patrol Highway 40, about thirty-five miles north of the proving ground, and tell people who had stopped by the side of the road to get back into their cars and keep driving. The sheriff never asked the reason for this request, and Dugway officials never volunteered any information.

THE LIVE-AGENT TRIAL planned for March 13, 1968, involved the spraying of VX from a high-performance aircraft. Over the previous fifteen years, Dugway had conducted some 1,200 tests in which roughly a million pounds of nerve agents had been released. This particular trial was the third in a series of three to evaluate the TMU-28B spray tank being developed for the Air Force. Exhaustive testing of the full configuration—two spray tanks loaded with VX and mounted under the wings of a fighter aircraft—was necessary so the developers could assess the weapon system's strengths and limitations under various meteorological conditions.

Although the morning dawned clear and sunny, in late afternoon the skies over the proving ground clouded over, the temperature dropped, the wind began to gust, and the distant flash of a thunderstorm could be seen on the horizon. The weather conditions were far from ideal for an open-air release of nerve agent, but Colonel James H. Watts, the commanding officer of Dugway, gave the go-ahead.

At 5:30 p.m., an Air Force jet fighter flew over the proving ground. Mounted under its wings were two pressurized spray tanks filled with 320 gallons of VX solution, weighing a total of 2,600 pounds. The nerve agent had been mixed with a dark red dye to make it easier to observe the agent cloud and measure the droplet sizes. With a crackling roar, the fighter leveled off at an altitude of 150 feet above the desert floor, the height of a fifteen-story building. A series of burning smoke pots marked the flight line, a half-mile upwind of the target grid. According to the test plan, the pilot would release a linear cloud of VX, jettison the empty spray tanks, and then climb to a higher altitude.

After conducting two practice runs to make sure he was flying in the correct pattern, the pilot received the order to start the release. He opened the valves on the pressurized spray tanks and the nozzles began to discharge two parallel sprays of VX-dye mixture, forming pinkish contrails behind the fighter.

Five seconds later, at the end of the run, the ejection equipment malfunctioned and one of the tanks failed to drop. The pilot had no way of stopping the pressurized flow of agent, and as the plane climbed rapidly, the roughly twenty pounds of VX left inside the tank continued to spray out up to an altitude of about 1,400 feet. Although the Dugway technicians were aware of this problem, they assumed that the prevailing winds would carry the toxic cloud along the west side of the Cedar Mountains, allowing it to dissipate harmlessly over the barren flats of the Great Salt Desert. Their work done for the day, the team members packed up their equipment and headed home to dinner.

Shortly after the cloud of VX droplets had been released, some unusual weather conditions developed. A weak cold front passing over Dugway generated cumulus clouds with strong updrafts that sucked up and retained the oily droplets. The clouds then merged into a broad crescent formation that was carried by the wind beyond the edge of the proving ground. About an hour and a half after the VX test, the wind made a 180-degree shift in direction and began to blow from the west, gusting at up to thirty-five miles per hour. Because of this change in wind direction, the VX-tainted clouds did not remain on the western side of the Cedar Mountains but were carried over the high ridge and into Skull Valley, a desert rangeland twenty-seven miles northeast of the test grid.

Although the federal government managed most of the land in Skull Valley, some of it was owned by private ranchers. Thousands of horses and cattle pastured in the valley throughout the year and flocks of sheep grazed there from November to May, eating native plants such as cheatgrass and bud sage. As night fell, intermittent showers of snow and rain developed and continued until morning. The precipitation washed the oily droplets of VX out of the air and deposited them on several flocks of sheep grazing in Skull Valley and on the slopes of the Stansbury and Onaqui Mountains that formed its eastern edge. The wind also carried VX droplets through a pass in the Onaqui Mountains and into Rush Valley beyond, some forty-five miles from the test site, where they were rained out onto another flock of sheep. Exposure to moisture degraded the VX into a chemical derivative that, while still highly toxic, was not absorbed as readily through the skin. Nevertheless, the sheep ingested the poisonous substance by eating contaminated vegetation and licking snow, their primary source of moisture.

ON THE MORNING of March 14, shepherds from the Hatch Ranch in Skull Valley noticed that many of the roughly 2,800 ewes grazing near White Rock, on the slope of the Cedar Mountains, had begun to act "crazy in the head." The sheep had a profuse nasal discharge and appeared dazed, responding with a delay to noise or rapid movement. They made frequent attempts to urinate and walked in a stilted, uncoordinated manner, often falling when they attempted to leap. Some held their head tilted down and to the side at an odd angle, indicating a weakness in the muscles of the neck. The most seriously affected animals staggered and dropped to the ground in apparent exhaustion, unable to rise. By afternoon, hundreds of sheep at

White Rock had begun to die. A few hours later, another flock grazing on the plain below the Stansbury Mountains on the east side of Skull Valley began to develop identical symptoms.

On the evening of March 14, the foreman of Skull Valley Ranch called two local veterinarians, Dr. Marr Fawcett and Dr. Richard Winward, who arrived the next morning. They examined the sick ewes but were unable to diagnose the cause. Particularly baffling was the fact that sheep were the only animals in the area that showed signs of illness; horses and cattle grazing among them were unaffected. Because sheep are susceptible to several viral and bacterial infections, the ranchers feared that an outbreak of infectious disease might be spreading down the valley.

By March 15, sheep were dying in five separate flocks in Skull Valley, and the local ranchers were seriously concerned. Three days later, two experts from the U.S. Department of Agriculture—Dr. Lynn James, a veterinarian, and Dr. Kent Van Kampen, a veterinary pathologist—performed field autopsies on sixteen dead sheep and found no lesions suggestive of infectious disease or the ingestion of poisonous plants. Van Kampen urged Alvin Hatch, the manager of the Anschutz Land and Livestock Company, to contact Dugway Proving Ground and find out if there had been a recent test with a dangerous chemical or biological agent. Hatch reported the sheep deaths to a Dugway official, who denied that any activities at the proving ground could have been responsible.

On March 19, Utah Governor Calvin Rampton appointed Dr. D. A. Osguthorpe, a veterinarian and consultant to the state department of agriculture, as his special representative to investigate the Skull Valley outbreak. That afternoon, Osguthorpe flew in his private plane to the White Rock area, accompanied by state Livestock Commissioner David R. Waldron. Suspecting that the animals may have been poisoned, the veterinarian drew blood from ten sick ewes that could still walk. Laboratory analysis showed that the animals' cholinesterase levels were depressed, suggesting exposure to an organophosphate compound, possibly a pesticide.

Over the next few days, experts from the Agricultural Research Service and the Animal Health Division of the U.S. Department of Agriculture, the Utah State Department of Health, the University of Utah, and Dugway Proving Ground examined the stricken sheep and took tissue samples. One by one, they ruled out toxic plants, heavy metals, parasites, and viruses as the cause of the mysterious illness and narrowed the search to a toxic chemical. Although organophosphate pesticides were used routinely in Skull Valley to treat sheep for skin parasites and to spray alfalfa fields, it was still too early in the season to apply them.

On March 20, one week after the VX test at Dugway, sheep on the western slopes of the Onaqui Mountains began to manifest symptoms similar to the flocks in Skull Valley. All of the sick animals were located within an area extending east-northeast from the proving ground and shaped roughly like an isosceles triangle with an apex at the test grid, sides fifty miles long, and a base twenty-five miles wide. Within this triangle, the highest levels of illness and death were in the flocks closest to the proving ground. Although Dugway officials continued to insist that "no tests which could be harmful to animals" had been conducted recently, Dr. Osguthorpe remained suspicious. He returned to White Rock and injected several sick sheep with atropine, using the recommended therapeutic dose. Seeing no response, he repeated the treatment at increasingly higher levels. When he administered several times the normal therapeutic dose of atropine, some of the stricken animals were able to walk, although their recovery was only temporary.

By now, Dr. Osguthorpe was convinced that the sick sheep were suffering from organophosphate poisoning. He and Livestock Commissioner Waldron paid a visit to Dugway Proving Ground and met with Colonel Watts, the commanding officer, and Dr. Mortimer A. Rothenberg, the scientific director. After Osguthorpe had laid out his concerns, the Dugway officials assured him that the Army had not tested any toxic organophosphate compounds since July of the previous year.

The next day, the veterinarian accompanied Utah's Governor Rampton on a helicopter tour of the five affected flocks. All of the available evidence still pointed to chemical poisoning. Blood samples from sheep owned by the Deseret Livestock Company and pastured north of the White Rock area had significantly depressed levels of blood cholinesterase. Subnormal levels of the enzyme were also found to a lesser extent in cattle, horses, wild rabbits, rodents, and birds in the affected area.

Although Dugway officials continued to deny responsibility, their position was undercut on March 21, when the Army Testing Command in Washington, D.C., provided a technical report to the office of Utah Senator Frank E. Moss. According to this report, on March 13, the day before the first sheep in Skull Valley had fallen ill, Dugway had conducted three separate operations involving nerve agents: the release of Sarin from a 155 mm artillery shell, the destruction by burning of 160 gallons of persistent nerve agent, and the spraying of VX from a jet aircraft. Although the first two activities were quickly ruled out as possible causes of the sheep illness, the third remained under suspicion.

The Army Testing Command had given the report to Senator Moss in confidence, apparently intending it "for official use only." Because the document bore no restrictive markings, however, Senator Moss's press secretary, Dale Zabriskie, freely distributed copies to Utah-based reporters on March 21. After receiving a flurry of press calls, the Army staff in Washington declared that the information released by Moss's office was wrong.

On March 22, a frustrated Governor Rampton called a meeting at his office in the Utah State Capitol in an attempt to resolve the controversy. Attending were Dr. Osguthorpe; Colonel Watts and Dr. Rothenberg of Dugway Proving Ground; Brigadier General John G. Appel, the commander of the Deseret Chemical Center; and Brigadier General William W. Stone, the head of the Army Matériel Command, who had been sent from Washington to investigate the incident. Speaking on behalf of the Army, General Appel admitted that the information released by Senator Moss's

A cloud of VX released during an open-air trial at Dugway Proving Ground in March 1968 killed thousands of sheep, which were buried in trenches in Skull Valley, Utah.

office was correct and that Dugway had conducted tests with nerve agents on March 13. Governor Rampton replied that if Dugway was responsible for the sheep deaths, it should compensate the affected ranchers. But General Appel disputed the existence of a cause-and-effect relationship between the VX trial and the sheep deaths in Skull Valley.

After the meeting, Colonel Watts arranged for Dr. Osguthorpe to be granted an interim security clearance so that he could be briefed on the secret release of VX at Dugway on March 13. On learning of the live-agent test, the veterinarian was incensed at the Army's stonewalling. Had Dugway acknowledged the accident early on, the sick sheep could have been treated with antidotes, greatly reducing the number of dead and injured. The Army's belated admission also raised concerns about the health of the people who lived and worked in Skull Valley. If the VX-tainted rain and snow had been toxic enough to injure or kill thousands of sheep, it might have affected humans as well. To assess the possible public-health impact, officials from the U.S. Communicable Disease Center (CDC) in Atlanta and the Utah State Department of Health assembled a team of experts in medicine, veterinary medicine, and epidemiology. They arrived in Skull Valley on March 23 and surveyed 110 individuals but found no illnesses or complaints that might be linked to the sheep deaths. Blood samples drawn from forty-three people showed cholinesterase levels within normal limits.

On March 26, the Army made backhoes and operators available so that the ranchers could bury the sheep carcasses in long trenches. Although Dugway officials continued to deny that the VX test had been responsible for the die-off, feeding experiments provided strong evidence of persistent toxic contamination in Skull Valley. On April 1, more than two weeks after the nerve agent trial, researchers from the USDA's Poisonous Plant Research Laboratory collected forage plants from the White Rock area. When the plants were dried, ground, and fed to healthy sheep through a stomach tube, the animals developed low blood cholinesterase levels and clinical symptoms characteristic of "Skull Valley disease." In another experiment, healthy sheep were transferred to graze in Skull Valley. After four to six days, these animals developed neurological abnormalities similar to those of the sick sheep, indicating that the forage was still tainted. A group of "control" animals, which were moved to the affected area but kept muzzled and fed only hay and water brought in from outside, showed no signs of illness.

On April 4, after a long delay, the Army reluctantly provided a small sample of VX, the composition of which remained secret, to the CDC in Atlanta for comparison studies. Eight days later, the center sent a telegram to the Utah State Department of Health reporting that its scientists had isolated a compound "identical" to VX from snow and grass in the White Rock area and from the liver, blood, and stomach contents of sheep that had died there. For the CDC, this evidence proved beyond a reasonable doubt that the Dugway test had been the cause of the sheep deaths. Forage collected in Skull Valley in June 1968, three months after the incident, no longer caused illness when fed to healthy sheep, indicating that the toxic agent had finally broken down.

One mystery remained to be solved: Why were sheep the only animals in Skull Valley to be seriously affected by low-level VX contamination? The explanation turned out to be that other mammalian species, such as cattle, horses, and humans, have a second form of cholinesterase (called butyrl-cholinesterase) that circulates in their blood serum. This reservoir of the enzyme absorbs and sequesters some of the nerve agent that enters the body, limiting its harmful effects on the cholinesterase in the nervous system. Sheep, in contrast, have almost no butyrl-cholinesterase in their blood. Without this natural buffering mechanism, they are exquisitely sensitive to nerve agents and can be injured or killed by less than a milligram of VX, far below the lethal dose in other animals or humans. The sheep grazing in Skull Valley had also been exposed to higher levels of VX than the other animals because they had consumed large amounts of tainted vegetation and snow.

The sheep die-off had now been explained to the satisfaction of everyone but the Army, which continued to call the findings "inconclusive." Although Dugway refused to accept responsibility for the sheep deaths, the Army agreed to compensate the ranchers for their losses under the Military Claims Act. The total number of sheep affected by the incident was determined to be 6,249, of which 4,372 were dead and 1,877 presumed exposed and hence not marketable. (About half of the dead animals had been shot by ranchers.)

On July 5, 1968, in response to the Skull Valley incident, Secretary of the Army Stanley R. Resor established an Interagency Ad Hoc Advisory Committee for Review of Testing Safety at Dugway Proving Ground, chaired by U.S. Surgeon General William H. Stewart. In November 1968, this committee issued its final report. Without officially acknowledging the accidental release of VX, the Stewart committee suggested a number of measures to minimize the risks to public health from open-air testing. Among the report's twelve recommendations were that high-speed fighter aircraft maintain "positive control" over the release of lethal agents; that no releases be made at heights of more than 300 feet or wind speeds greater than fifteen miles per hour; that trials be designed so that the toxic cloud would not cross Highway 40 until three hours after a release; and that open-air tests be forbidden if thunderstorms were present within a hundred miles. Secretary Resor accepted all of the committee's recommendations and also required Dugway to improve its air-sampling and atmospheric-modeling capabilities so that it could predict the downwind behavior of toxic clouds over a distance of several tens of miles.

In early 1969, nearly a year after the Skull Valley incident, the U.S. Army Claims Service paid the affected ranchers a total of $376,685. This sum was based on an average price per sheep of about $55.00, considerably above the purchase price of $30.00. The reason for the high price was that the dead and injured animals had all been ewes that were either pregnant or likely to lamb in the next few months; they had also been heavily laden with wool, which would have provided additional income. Despite the generous compensation, the Army continued to deny responsibility for the incident, creating an enduring legacy of distrust toward Dugway officials on the part of local residents.

THE SKULL VALLEY incident focused a great deal of negative publicity on the U.S. chemical warfare program. Not only was the massive sheep kill voted Utah's number one news story of 1968, but it became the subject of two network television documentaries, including the main segment of the popular NBC newsmagazine *First Tuesday*. Broadcast at 9:00 p.m. on February 4, 1969, the segment documented aspects of the British, Canadian, and U.S. chemical weapons programs and showed images of the dead sheep in Skull Valley. Among the millions of Americans who watched the *First Tuesday* broadcast was Representative Richard D. McCarthy, a Democratic congressman from Buffalo, New York, who was surprised and outraged by what he saw. During the Skull Valley segment, his wife, Gail, turned to him and asked, "You're a congressman. What do you know about this?" "Nothing," he replied, his indignation rising when he realized that, without his knowledge, he had voted to appropriate large sums of money for nerve agent production and testing.

The Pentagon kept the sensitive issue of chemical weapons hidden from most members of Congress by classifying the relevant information and releasing it on a strictly "need to know" basis. Only five members of the House Appropriations Committee, and no more than 5 percent of the entire House of Representatives, were cleared for information on chemical and biological weapons. As a result, a small clique of senior congressmen was able to allocate money for these programs in secret session and then bury the line items in massive appropriations bills that were brought to the floor for a vote with little advance notice, so that few members had time to read them.

To shed more light on the issue, Representative McCarthy requested a briefing for House members on U.S. chemical and biological warfare policy. When Army officials said that the briefing would be secret and conducted behind closed doors, the Buffalo congressman insisted that the first part be held in open session. The Army reluctantly agreed, and the two-part briefing took place on March 4, 1969. Brigadier General James A. Hebbeler, the briefing officer, denied that Dugway had been responsible for the sheep kill in Utah, warned that the Soviet Union had a chemical warfare capability "seven to eight times" that of the United States, and requested more funding from Congress. For Representative McCarthy, the Army's presentation raised more questions than it answered.

Whereas McCarthy soon became a leading critic of the Chemical Corps, its top booster was Representative Robert Sikes of Florida. Not only was he a major general in the Army Chemical Reserve and the chairman of the Defense Appropriations Subcommittee, but his congressional district included Eglin Air Force Base, the home of the Air Force chemical weapons development laboratory and testing station. Representative Sikes continually stressed the Soviet chemical threat and the need for a credible U.S. deterrent.

IN MAY 1969, McCarthy and Representative Henry S. Reuss of Wisconsin, who chaired a subcommittee of the House Government Operations Committee, held two days of hearings on the environmental and health hazards associated with open-air testing of lethal chemical agents. On July 11, at the prodding of Reuss's subcommittee, Army Secretary Resor disclosed that in addition to the hundreds of trials at Dugway, the Chemical Corps was releasing live nerve agents during training exercises at the Army Chemical School at Fort McClellan, Alabama, and at Edgewood Arsenal in Maryland. Although the testing in Alabama was relatively minor, the program in Maryland was extensive: over a three-month period, the Army planned to conduct 239 weapons tests at Edgewood, compared to 358 at Dugway. After this information was released, more than a hundred people demonstrated outside the gates of Edgewood Arsenal. Within weeks, the Pentagon gave in to public pressure and announced a moratorium on the open-air testing of lethal chemical agents.

Also during the summer of 1969, the Army faced a crisis over the presence of U.S. chemical weapons on Okinawa, an island off the coast of Japan that had been under effective American control since the end of World War II. Roughly 45,000 U.S. troops were stationed at various military installations on Okinawa, including storage depots, training areas, barracks, communications centers, ammunition bunkers, and fuel tanks. In addition, Kadena Air Base in the center of the island, with its two 12,000-foot runways, was one of the largest and busiest U.S. Air Force bases in Asia.

In 1961 and 1963, the Kennedy administration had secretly authorized the deployment of mustard and nerve agent weapons to Okinawa without informing either local or Japanese officials. At a depot run by the 267th Chemical Company, hidden in a pine forest a few miles from Kadena Air Base, hundreds of sod-covered concrete igloos were surrounded by three electrified fences and guarded by a sentry-dog platoon. Inside the bunkers were wooden pallets holding stacks of bombs and artillery shells filled with Sarin or VX. The munitions were painted gray and marked with three rings to signify that they contained nerve agents. Rabbits and goats wandered freely among the igloos, serving as living "sentinels" to provide early warning of a nerve agent leak.

On July 8, 1969, twenty-three soldiers and one civilian employee from the 267th Chemical Company were conducting a maintenance operation inside one of the igloos, sandblasting paint from Sarin-filled aerial bombs in preparation for repainting them. During this operation, a small leak developed next to the fill plug of a Sarin bomb, resulting in the release of toxic fumes. All those inside the bunker developed mild symptoms of nerve agent exposure and immediately donned their gas masks. As a precaution, the exposed individuals were evacuated to an Army hospital and placed under medical observation for six hours, after which they were released and returned to full duty. Meanwhile, ordnance teams decontaminated the affected igloo, and the Army Technical Escort Team was flown from Edgewood Arsenal to Okinawa to assist with disposal of the leaky bomb. Army officials were relieved that no one had been seriously injured or killed; because no Okinawan nationals had been involved, it was assumed that the incident would remain secret.

Ten days later, however, on July 18, 1969, *The Wall Street Journal* published a front-page story about the Sarin leak with the dramatic headline NERVE GAS ACCIDENT: OKINAWA MISHAP BARES OVERSEAS DEPLOYMENT OF CHEMICAL WEAPONS. By revealing that the United States had secretly stationed nerve agents in Okinawa since the early 1960s, the article sparked a political firestorm. The news came as a profound shock to Okinawans of all political persuasions. Chobyo Yara, the chief executive of the government of the Ryukyus, of which Okinawa was the main island, said that he was "flabbergasted. If the report is true, this is a serious problem. The presence of horrible nerve gas weapons jeopardizes our lives and thus it is absolutely unforgivable." Yara demanded that United States remove the stockpile immediately.

In Tokyo, the conservative, pro-American government of Premier Eisaku Sato was deeply embarrassed by the Sarin incident. At a news conference, Foreign Minister Kiichi Aichi said that he had asked the United States "not to cause uneasiness" on Okinawa by continuing to store chemical weapons there. In a commentary on the incident published on July 20, 1969, the columnist James Reston observed, "The trouble is not that the Pentagon is wicked but that it seems to be clumsy; it is constantly being caught doing things that embarrass the Government and complicate the conduct of American foreign and even internal policy."

Indeed, the controversy threatened to disrupt the delicate negotiations between Washington and Tokyo over the renewal of the U.S.-Japan security treaty and the planned return of Okinawa to Japan in 1972. The Japanese government insisted that Washington would have to ask permission before redeploying chemical weapons to Okinawa in a crisis or war. As a Japanese official told *The New York Times*, "We used to say, 'No nukes on Okinawa.'

Now we will have to say, 'No nukes and no gas.' It's going to become a national demand."

On July 22, the Okinawan legislature met in special session and adopted a resolution requesting the removal of all U.S. chemical weapons. Hours later, a Pentagon spokesman admitted that such weapons existed on the island and would be withdrawn. Although the United States had planned the removal operation "for some time," he said, it would now be accelerated. Washington was also forced to admit that it had secretly deployed a stockpile of chemical weapons in West Germany. Opposition parties in Bonn angrily called for a government investigation.

In addition to creating a foreign policy crisis, the Sarin leak on Okinawa underscored the hazards of storing and testing chemical weapons. Coming on the heels of the Skull Valley incident, Okinawa heightened Congressional concern over the open-air testing of lethal chemical agents. On November 19, 1969, while the Army's testing moratorium was still in effect, Congress passed the FY 1970 Defense Authorization Act (Public Law 91-121), imposing strict controls on the deployment, storage, and disposal of chemical weapons, both within the United States and outside the country. The new legislation also established an elaborate approval process for open-air releases of lethal chemical agents. A test could take place only if the Secretary of Defense certified that it was essential for national security, the Secretary of Health, Education, and Welfare concurred that it was safe, and Congress was given thirty days' advance notice. Although the new law did not impose an outright ban on open-air testing, it created so many bureaucratic hurdles as to make such trials effectively impossible.

The Chemical Corps suffered another major blow in late 1969. In May of that year, four months after the inauguration of President Richard M. Nixon, National Security Adviser Henry Kissinger had requested an in-depth policy review of issues related to chemical and biological weapons, with a reporting deadline in the fall. On November 25, 1969, this review culminated in the issuance of National Security Decision Memorandum 35, in which the United States officially renounced its offensive biological warfare program and pledged to destroy all existing stocks of biological weapons and to limit future research and development to strictly defensive measures, such as vaccines and antibiotics. In announcing these decisions, President Nixon declared, "Mankind already carries in its own hands too many of the seeds of its own destruction. By the examples we set today, we hope to contribute to an atmosphere of peace and understanding between nations and among men."

With respect to chemical warfare, Nixon restored the "retaliation-only" policy and said that there would be no more production of unitary chemical weapons; any future modernization of the U.S. chemical arsenal would be considered only after binary munitions had been developed. In the meantime, the United States would maintain the existing stockpile of unitary weapons as a deterrent, while striving to negotiate an international treaty to ban chemical arms. Nixon also declared that he would resubmit the 1925 Geneva Protocol to the U.S. Senate for its consent to ratification.

MEANWHILE, the disposal of obsolete chemical weapons remained in limbo. In 1968, Operations CHASE 11 and 12 had dumped bulk nerve agent and thousands of leaky M55 rockets off the U.S. East Coast, but since then no further sea-dumping operations had been carried out. Now the Army wanted to conduct CHASE 10, which was out of numerical sequence because it had been scheduled earlier and then put on hold. This plan called for the long-distance transport by rail and disposal at sea of some 27,000 tons of obsolete chemical weapons. Most were stored at Rocky Mountain Arsenal, including 12,000 tons of M34 cluster bombs containing Sarin-filled bomblets and 9,000 tons of mustard agent in one-ton containers. In addition, 2,600 tons of leaking M55 rockets, encased in steel-and-concrete coffins, were held at Anniston Army Depot in Alabama and Bluegrass Army Depot in Kentucky.

According to the Army proposal, the weapons would be transported by rail from the three storage sites, passing through major cities such as Indianapolis and Dayton and terminating at Elizabeth, New Jersey, where they would be loaded onto old Liberty ships and scuttled off Long Island. When the Army's plan was made known, it provoked public protests and concern in Congress. Representative McCarthy organized a series of hearings on the risks to local communities along the proposed train route and the harmful effects of chemical weapons dumping on the marine environment.

Representative Cornelius Gallagher, through whose New Jersey district the trains would pass, chaired one of the hearings and invited Pentagon officials to testify. Assistant Secretary of the Army Charles L. Poor stated that after studying several alternatives, he had concluded that train transport was "the preferred method in terms of safety, contamination of the environment, time, and cost." But an academic expert, Professor Matthew Meselson of Harvard, countered that if one of the poison gas trains derailed and exploded, and the wind was blowing at ten to twenty miles an hour, thousands of people might be killed in a densely populated area such as Indianapolis. Indeed, not long after the hearing, an ammunition train carrying tear gas and explosives for Vietnam blew up accidentally in Nevada.

In view of these risks, Congress asked the National Research Council (NRC), the policy analysis branch of the National Academy of Sciences, to set up an expert committee to evaluate the Army plan. Chaired by George B. Kistiakowsky, a professor of physical chemistry at Harvard and former science adviser to President Eisenhower, the panel inspected the proposed train route, the stockpiled weapons, and the Liberty ships that were to be loaded with them. The committee also heard testimony from two experts from Edgewood Arsenal: Colonel Sam Bass, the chief of development, and Sigmund Eckhaus, a chemical engineer. They warned that the leaking M55 rockets embedded in coffins were becoming unstable and that it was just a matter of time before Sarin seeped into the rocket propellant and triggered an explosion that could fracture the blocks and release a cloud of lethal vapor. But other witnesses countered that sea-dumping posed a serious and potentially catastrophic risk to civilians. According to one frightening scenario, an explosion on board one of the chemical weapons ships could release a massive cloud of nerve gas that would be blown over densely populated areas of the East Coast.

After weighing the various options, the NRC panel issued a final report in which it recommended the on-site destruction of the obsolete weapons stored at Rocky Mountain Arsenal: the M34 Sarin bomblets by means of disassembly and chemical neutralization, and the bulk mustard agent by incineration. The Army accepted this recommendation and in October 1969 launched Project Eagle, which over the next seven years disposed of thousands of tons of mustard and then Sarin. Because of the special hazards associated with the M55 coffins, however, the NRC advised that they should be disposed of in a final sea-dumping operation, which was designated CHASE 10.

During Operation CHASE, bulk containers of mustard agent and steel-and-concrete "coffins" containing defective Sarin-filled M55 rockets were loaded onto a rusting Liberty ship in preparation for disposal at sea (top). The ship was then towed off the East Coast and scuttled in deep water (bottom).

Beginning on August 10, 1970, 418 steel-and-concrete coffins containing 12,500 M55 rockets and weighing 2,675 tons were loaded onto trains at Anniston Army Depot in Alabama and Bluegrass Army Depot in Kentucky.

The coffins filled a total of thirty-nine railroad boxcars. For security reasons, the Army did not release the exact route the trains would follow. Each train was restricted to a speed of thirty-five miles per hour and preceded by a "pilot" train to make sure the tracks were clear. Accompanied by military police, medical personnel, and chemical ordnance specialists, the trains took more than a day and a half to travel 1,400 miles, passing through twenty-one small towns in seven states while avoiding major cities such as Atlanta. Hospitals along the route were told to stock up on atropine in case of an accident, and the train crews carried their own antidote injectors.

When the trains finally reached Sunny Point Military Ocean Terminal near Cape Fear, North Carolina, cranes loaded the massive steel coffins onto a rusting Liberty ship, the S.S. *LeBaron Russell Briggs*. Accompanied by a destroyer escort and a Coast Guard cutter, a tugboat towed the 442-foot hulk for two and a half days to a location 283 miles east of the Florida coast, beyond the edge of the U.S. continental shelf. The depth of the water at the dump site was 16,000 feet, more than twice that of the previous location off the New Jersey coast.

Lieutenant A. A. Schiavone led an eight-man team of Army ordnance specialists who boarded the *Briggs* and checked on the ship's only passengers: six white rabbits. The animals appeared in good health, indicating that no Sarin had leaked from the coffins during the voyage. After removing the rabbits and all salvageable equipment, the team rigged the ship with hydrophones and depth gauges to monitor its sinking rate. At 11:45 a.m., the crew opened seven flood valves deep inside the hold. Over the next four hours, the freighter sank slowly until it was a little more than half submerged. Then the holds suddenly flooded and the stern went down, followed by the rest of the ship. At precisely 3:53 p.m., the rusty bow of the *Briggs* slipped beneath the waves, followed by a geyser of white foam.

THE PENTAGON ALSO faced the challenge of what to do with the 13,000 agent-ton stockpile of chemical munitions it had promised to remove from Okinawa, including 2,865 tons of mustard, 8,322 tons of Sarin, and 2,057 tons of VX. One proposal was to transfer the weapons to Umatilla Army Depot in Hermiston, Oregon, but this idea elicited strong protests from the citizens and governors of Oregon and Washington State. Alaska was also briefly discussed as a possible storage site until Congress passed a law prohibiting the transfer of chemical weapons from Okinawa to any location on the U.S. mainland.

After considering several alternatives, the Pentagon announced on December 4, 1970, that it would transfer the Okinawan stockpile to Johnston Island, a tiny American-owned atoll in the middle of the South Pacific, about 800 miles southwest of Hawaii. A half-mile wide, two miles long, and surrounded by a coral reef, Johnston Island had no indigenous population and was inhabited only by U.S. military personnel and civilian laborers. The atoll had been unclaimed until 1858, when the United States set it aside as a bird refuge. In the 1930s, the War Department had converted the island into a military base, and in 1941 the Navy had constructed an airfield that took up nearly its entire length. Viewed from the air, Johnston Island resembled a giant aircraft carrier.

The transfer of chemical weapons from Okinawa, code-named Operation Red Hat, began in the summer of 1971 and lasted two and a half months. Five ships loaded with chemical munitions made the eleven-day voyage, the last arriving at Johnston Island on September 21, 1971. The weapons filled with Sarin and VX were stored in igloo-style bunkers until a

Johnston Island in the Pacific Ocean; an airfield runs down the center of the island. Originally a wildlife sanctuary, the island was heavily utilized by the U.S. military during World War II and the Cold War.

During Operation Red Hat in 1971, U.S. chemical weapons that had been based in Okinawa were shipped to Johnston Island for storage and eventual destruction.

safe method could be devised to destroy them. Because the Weteye (Sarin) bombs, M55 rockets, and M34 cluster bombs had not been designed to be dismantled, the Army faced a major disposal problem.

Storage conditions on Johnston Island were far from ideal: shells and bomb casings corroded in the humid, salty air of the atoll, worsening their tendency to leak, and occasional hurricanes tore the roofs off the storage facilities. The hundred or so U.S. troops guarding the chemical weapon stockpile lived on the upwind side of the island and deployed caged rabbits near the storage bunkers to provide early warning of a leak. In 1972, Congress passed the Marine Protection, Research and Sanctuaries Act (Public Law 92-532), which banned any further sea dumping of chemical weapons. Until an alternative disposal technology became available, the leaking M55 rockets were sealed inside steel tubes for long-term storage.

CHAPTER TWELVE

NEW FEARS

DURING THE EARLY 1970S, as the U.S. war in Vietnam wound down and military spending declined, the Army Chemical Corps faced multiple challenges to its organizational survival. The incident at Skull Valley and the rise of the environmental movement had ended the open-air testing of lethal agents; President Nixon's decision to halt the production of unitary chemical weapons had led to the closing of the VX plant in Newport, Indiana; and Congress had voted to slash funding for the Chemical Corps. The United States also faced strong international criticism for its employment of toxic chemicals in Vietnam, including the use of tear gas to flush enemy soldiers out of caves and tunnels, and the spraying of the herbicide Agent Orange to defoliate large expanses of jungle and deprive the Viet Cong guerrillas of cover. Agent Orange contained only trace amounts of the highly toxic substance dioxin as a synthetic by-product, but because vast quantities of the herbicide were sprayed over Vietnam, both U.S. troops and Vietnamese civilians were exposed to concentrations of dioxin high enough to cause chronic illnesses and birth defects. The legacy of Agent Orange would haunt Vietnam for generations.

Although the United States did not consider tear gas and herbicides to be true chemical weapons, their use provoked international and domestic protests and created an enduring stigma that made any plan to modernize the U.S. chemical arsenal highly unpopular. In addition, inspired by President Nixon's unilateral renunciation of the U.S. offensive biological weapons program in November 1969, the Conference of the Committee on Disarmament (CCD), a multilateral arms control forum based at the

United Nations Office in Geneva, negotiated a treaty banning the development, production, and stockpiling of biological and toxin weapons that was opened for signature in 1972. Diplomats participating in the talks expressed the clear intent to follow up the Biological Weapons Convention with a similar treaty to outlaw chemical arms.

In January 1973, the Department of the Army announced plans to downsize the Chemical Corps and eventually eliminate it as a separate branch of the service. As a first step, the Army Chemical School at Fort McClellan, Alabama, was mothballed and its library sold off. For a time, serious consideration was given to establishing a "Chemical Regiment." Instead, the Army decided to merge the Chemical Corps into the Ordnance Corps. Because the missions and cultures of the two organizations proved to be incompatible, however, the Chemical Corps was able to survive as a separate entity.

In response to these multiple challenges, the Chemical Corps staked its future on accelerating the development of binary chemical weapons, while keeping this effort shrouded in secrecy to avoid political controversy. Although President Nixon's 1969 executive order had ruled out further production of unitary chemical weapons, it had left the door open to the possible procurement of binary munitions after their development was complete. By making chemical weapons safer to handle, transport, and deploy, binary technology promised to enhance their value as a deterrent and make them more acceptable politically. The production of binary weapons, if it occurred, would assure the Chemical Corps a multibillion-dollar acquisition pipeline extending over several years. Accordingly, the share of the Corps's research-and-development budget devoted to binary weapons increased from a few percent in 1969 to roughly a quarter in 1970, half in 1971, and two-thirds in 1973.

In addition to developing the Bigeye VX bomb for the Navy, Bill Dee and his team at Edgewood Arsenal were working on a binary Sarin artillery shell for the Army. Called the M687 projectile, it had entered advanced development in 1967. The main technical challenge associated with this munition was to ensure that the two precursor chemicals mixed and reacted efficiently to form Sarin inside the shell during its flight to the target. In order to reduce the weight of the projectile, one of its binary components was reformulated. Instead of the mixture of dichlor and difluor (known as "di-di") that was used to manufacture Sarin at Rocky Mountain Arsenal,

A mock-up of the M687 binary Sarin 155 mm artillery shell, which contained two canisters filled with the relatively nontoxic precursors DF and OPA. When the projectile was fired from a howitzer, the setback forces ruptured thin diaphragms between the canisters, allowing the precursors to mix. They reacted in seconds during the projectile's flight to the target to yield Sarin, which was dispersed on impact.

the M687 projectile contained pure difluor (DF). The other binary component, known as OPA, was a mixture of isopropyl alcohol, a stabilizer, and a catalyst.

The DF and OPA used in the binary artillery shell were stored in separate canisters. Prior to firing, the two canisters would be inserted into the M687 projectile, one behind the other. The facing ends were each covered with a polymer plate and a very thin steel diaphragm known as a "burst disc." When the projectile was fired from a howitzer, the intense setback forces would rupture the discs, causing the DF and OPA to mix and react inside the shell to form Sarin. The spinning of the projectile in flight would facilitate the blending of the chemicals, so that the reaction would take only about four seconds to go to completion. When the shell reached the target, an impact fuse and burster charge would disperse the newly synthesized Sarin as an aerosol of tiny droplets.

During field testing of prototypes of the M687 projectile at Dugway Proving Ground in 1971, Dee and his colleagues observed that about one shell in every eight suddenly became unstable and dropped out of the air. To diagnose the source of the problem, the development team tracked the shells in flight with high-speed cameras and Doppler radar. They concluded that the motion of the liquid precursors inside the spinning shell was causing the instability. By developing a computer model of the projectile in flight—a difficult task given the limited processing power available at the time—they managed to modify the design of the shell and fix the defect. For this accomplishment, Dee received the Army Science Award. The M687 projectile was now considered a technical success, and in 1972 it entered the engineering development phase.

Nevertheless, a major obstacle to field testing of the binary artillery shell was the FY 1970 Defense Authorization Act (Public Law 91-121), passed in the wake of the Skull Valley incident, which had effectively halted all open-air releases of lethal chemical agents. Thus, although the M687 projectile was fired more than two thousand times in the course of its development, only one test (on September 16, 1969) involved the use of live Sarin. All the other trials were performed with chemical simulants, raising questions about their realism. The binary artillery shell also proved to have a number of technical limitations. First, the target had to be far enough away from the point of firing to allow the reaction of DF and OPA to go to completion. Second, because the binary reaction created unwanted chemical by-products, the weapon would deliver less Sarin to the target than a standard unitary shell. Third, although the use of pure DF saved weight and improved reaction efficiency, it had the drawback of generating hydrogen fluoride (HF) gas as a by-product. Whereas Sarin itself was odorless, HF gave off a noxious odor that would warn enemy troops of the approaching agent cloud.

On September 18, 1973, a few months after a major withdrawal of U.S. troops from Vietnam, the Secretary of the Army, Bo Calloway, requested funding to build a new production facility at Pine Bluff Arsenal in Arkansas for DF, the primary chemical component of the binary Sarin projectile. This trial balloon elicited little support in Congress, where chemical weapons remained politically unpopular. One month later, however, a new Arab-Israeli war broke out in the Middle East. By transforming U.S. perceptions of the chemical warfare threat, the conflict gave the Army Chemical Corps a new lease on life.

FOR SEVERAL YEARS, Presidents Anwar Sadat of Egypt and Hafez al-Assad of Syria had plotted to avenge their nations' humiliating defeat in the 1967 Six-Day War. The two Arab leaders devised a plan for a joint surprise attack against Israel on the festival of Yom Kippur, the holiest day in the Jewish calendar, in the hope of catching the Israel Defense Forces off guard. During secret preparations for the invasion, both countries equipped their forces with chemical weapons. Egypt had at least one Air Force wing armed with Sarin-filled bombs and a supply of Soviet-made Scud-B missiles that could deliver chemical warheads over a range of 185 miles. Cairo also sold Syria an arsenal of chemical weapons for $6 million, including Sarin-filled artillery shells, Scud missile warheads, and spray tanks for tactical aircraft.

In September 1973, the Egyptian and Syrian armies began to mass troops near their respective borders with Israel. Egypt called up thousands of reservists and prepared antiaircraft and artillery positions along the west bank of the Suez Canal, facing the Israel-occupied Sinai Peninsula. Because the Egyptian Army had previously mobilized nineteen times without going to war, Israeli intelligence had been lulled into complacency and assumed that this latest deployment was another false alarm. On the afternoon of October 6, however, Egypt and Syria launched a ground invasion on two fronts. Egyptian combat engineers breached the Bar Lev Line, the fortified defensive wall along the Suez Canal that Israel had built after the Six-Day War, enabling columns of Egyptian tanks and armored personnel carriers to advance into the Sinai desert. Meanwhile, on the northern front, Syrian tanks broke through the thin Israeli defenses on the Golan Heights.

After a few panicky days, the Israel Defense Forces managed to mobilize and regroup, and over the next two weeks, the tide of war shifted in Israel's favor. Fear of retaliation in kind deterred the Arab armies from resorting to chemical warfare. Although Israeli intelligence intercepted alert messages to Egyptian and Syrian units equipped with chemical arms, these formations were never ordered into action. On October 24, after eighteen days of fighting, U.S. Secretary of State Kissinger intervened diplomatically to impose a cease-fire.

In the aftermath of the Yom Kippur War, Israeli military intelligence experts inspected Soviet-made tanks and BMP-1 armored personnel carriers that had been captured from the Egyptian and Syrian forces. The Israelis were surprised to discover that the armored vehicles had sophisticated seals, filters, and air pressurization systems to keep out poison gas. The Soviets had also equipped their Arab allies with gas masks, rubber capes, automatic detector alarms, chemical identification kits, autoinjectors filled with nerve agent antidote, portable shelters, and wash-down equipment for decontaminating tanks and planes. In principle, such protective systems could be used for either offensive or defensive purposes. Because Egypt had no reason to believe that Israel would use poison gas first, however, it seemed likely that the intent was to shield Egyptian troops from exposure to their own chemical weapons. Indeed, Egyptian prisoners of war later admitted under Israeli interrogation that their side had possessed munitions loaded with nerve agents.

Israeli leaders perceived Egypt's burgeoning chemical arsenal as a serious threat, particularly when coupled with ballistic missiles that could deliver nerve agents against Israeli cities. Under optimal atmospheric conditions, a few dozen Scud warheads filled with Sarin could generate a toxic plume that would blanket a large area, potentially inflicting thousands of casualties. To prepare for the worst, the Israeli government quietly ordered the manufacture of more than 3 million gas masks of indigenous design for the civilian population. For infants too small to wear a mask, an enclosed crib-type air filtration system was developed. To avoid causing undue public anxiety, the government decided not to distribute the gas masks widely until the imminent threat of war returned.

ISRAELI MILITARY INTELLIGENCE allowed U.S. analysts to inspect captured Soviet equipment from the Yom Kippur War, including BMP-1 armored personnel carriers that incorporated chemical-protective systems. These vehicles had entered development in 1961 and gone into production in 1967, the period when the United States was mass-producing VX. Although the air filtration unit was a standard feature of the BMP-1, it was usually not included in the export versions that the Soviet Union sold to allied countries. Some military analysts speculated that Moscow, in its haste to deliver arms to the Middle East, had mistakenly shipped armored vehicles intended for domestic use. But the fact that the Soviets had also supplied Egypt with gas masks and other chemical defensive gear suggested that the special vehicles had not been provided in error.

The intelligence windfall from the Yom Kippur War caused the CIA to revise its assessment of the Soviet chemical threat, including the possibility that the Red Army might initiate the use of nerve agents in a future conflict.

Other information suggested that the Soviet military was upgrading its chemical warfare capabilities and integrating them into the structure, equipment, and training of its ground, sea, and air forces. According to CIA estimates, more than 100,000 Soviet troops had offensive or defensive chemical warfare missions. Furthermore, the Soviets had about 20,000 vehicles devoted to chemical decontamination, compared to about 600 for NATO. One Soviet vehicle, the TMS-65, resembled a giant blow dryer on wheels; it used a jet engine to decontaminate tanks and other armored vehicles with a blast of superheated air and fluid.

In March 1974, several committees of the U.S. Congress held hearings on the military lessons of the Yom Kippur War. Testifying before a House subcommittee, General Creighton W. Abrams, the Army chief of staff, said that he had been "impressed . . . with the comprehensive CBR [chemical-biological-radiological] defense in the Soviet-equipped forces of the Arab armies. It was comprehensive, sophisticated, complete, and detailed, on every vehicle and for all equipment and for all men. Our forces are not equipped in that fashion."

Commenting on the general's testimony, Representative Sikes of Florida, an outspoken supporter of the Chemical Corps, said that the Soviet Union's sophisticated chemical defenses suggested that their offensive capabilities were equally strong. "The thing that disturbs me greatly about this matter," he added, "is that our capability in the field of CBR . . . has been deteriorating as a part of national policy for several years, and that we are becoming more and more dependent upon a laboratory capability rather than an inventory of weapons or even an adequate defense against them."

ANOTHER CONSEQUENCE of the chemical warfare intelligence derived from the Yom Kippur War was a major effort by the U.S. Army to improve its medical defenses against nerve agents. Among the captured Egyptian items that Israel had passed along to the CIA was a nerve agent antidote kit for crews of armored personnel carriers. Because this kit had writing in Cyrillic letters, the CIA assumed that it was the latest Soviet antidote. When agency chemists analyzed its composition, two of the ingredients were familiar: atropine and a PAM-like compound known as TMB4. The third component, however, was surprising. It was a drug called benactyzine, which Western scientists had never considered using as a nerve agent antidote.

The CIA dubbed the Egyptian antidote "TAB," after the initials of its three components: TMB4, atropine, and benactyzine. In a top secret project code-named "Grand Plot," CIA researchers tested TAB on laboratory monkeys exposed to nerve agents and concluded that the mixture was highly effective as an antidote for Sarin, Soman, or VX. Benactyzine appeared to act extremely rapidly, whereas atropine provided longer lasting protection. In view of the apparent advantages of the new antidote, the Army discarded its old autoinjectors containing atropine and PAM and purchased 7 million copies of the new TAB kits.

Over the next five years, the Army realized that it had made a serious mistake. Not only did benactyzine cause florid hallucinations, incapacitating soldiers and reducing their ability to fight, but the TAB mixture contained too little atropine to be effective. It turned out that the captured antidote kits had not been manufactured by the Soviet Union after all. CIA analysts who did not speak Russian, seeing the Cyrillic writing on the captured kits, had jumped to the conclusion that they were Soviet-made, forgetting that other languages, such as Bulgarian and Serbian, are also written in Cyrillic. In fact, when the Egyptian Army's Soviet-supplied kits had expired in 1969, Cairo had replaced them with the TAB kits, which were made in Bulgaria. Due either to incompetence or inadequate testing, the Bulgarian antidote was defective. On November 4, 1980, the U.S. Army chief of staff quietly ordered the destruction of the 7 million TAB autoinjectors, which fortunately had never been used in combat.

ALTHOUGH THE U.S. intelligence community was correct in assessing that the Soviet Union was modernizing its chemical arsenal, the true size and scope of that effort were unknown in the West. In 1972, the Soviets opened a manufacturing plant for R-33 at the vast chemical industry complex in Novocheboksarsk, a suburb of Cheboksary, the capital of the Russian Chuvash Republic. The V-agent production facility was six stories high, a hundred meters long, and fifty meters wide, and was constructed of reinforced concrete and cement block. Inside, the process equipment was equally overdesigned, with large numbers of glass-lined reactors and rotary driers made of titanium. Additional plants were built nearby to make the precursor chemicals aminomercaptan and chloroester and to produce and fill munitions. Although the Novocheboksarsk factory never operated at full capacity, it manufactured a total of 15,000 tons of R-33. On April 28, 1974, a fire broke out near the munitions-loading line and flames consumed a wooden warehouse containing aviation bombs filled with R-33. About fifty bombs were damaged and leaked, requiring an extensive cleanup by workers in protective suits. Although the accident caused no immediate deaths, several workers later died of chronic illnesses associated with low-level exposure to the Soviet V agent.

In order to keep pace with and eventually surpass the chemical warfare capabilities of the United States, the Soviet Union maintained an active program of research and development. In May 1971, the Central Committee of the Communist Party and the Soviet Council of Ministers approved a new effort to create "fourth-generation" chemical weapons. (World War I agents such as phosgene and mustard were considered the first generation, G agents the second, and V agents the third.) The Kremlin sought to acquire a new class of nerve agents with greater toxicity, stability, persistence, ease of production, and other militarily relevant properties. Code-named "Foliant," this research-and-development program was shrouded in the highest level of secrecy. All related documents were stamped with the classification "Top Secret—Series F" and access was restricted to those with the proper clearance and a need to know.

The main scientific organization involved in the Foliant program was the State Scientific Research Institute of Organic Chemistry and Technology. Known by the Russian acronym GosNIIOKhT (pronounced "gos-ni-ockt"), it was headquartered in an industrial district of eastern Moscow, four and a half miles due east of Red Square. The institute was considered a prestigious place to work, with superb laboratory equipment, generous salaries, and access to imported foods and consumer goods not available in public stores. Only the Soviet Union's most outstanding young chemists, chemical engineers, and physiologists were offered positions there.

From the outside, GosNIIOKhT appeared unimpressive: a cluster of nondescript concrete buildings crowded onto a small triangular plot at No. 53 Highway of the Enthusiasts, with steel steam conduits suspended over the narrow alleyways. The windows of the institute were obscured by faded curtains and a medical clinic occupied a corner of the ground floor. Except for an elaborate gate-pass system that sealed off the closed portion of the building, nothing gave away its real purpose. In fact, GosNIIOKhT was what the Russians called a *yashik,* or box—a classified facility within the Soviet military-industrial complex. Formerly known as Scientific Research Institute No. 42 (NII-42), it was the country's leading center for the development and testing of chemical warfare agents. Like many weapons-related institutes in the Soviet Union, GosNIIOKhT had a postal address for secret correspondence: Post Office Box M-5123.

The Moscow institute reported to the Ministry of Chemical Industry and employed about 3,500 people, of whom about 500 were Ph.D. scientists. Although GosNIIOKhT did some research on toxins (poisonous compounds derived from living organisms), its primary focus was the synthesis of new chemical warfare agents and the development of manufacturing processes. Security at the institute was extremely tight: KGB guards continually patrolled the laboratory complex and members of the staff had to obtain a special pass to go from one floor of the building to another. In addition to its Moscow headquarters, GosNIIOKhT operated three satellite laboratories in Volgograd, Shikhany, and Novocheboksarsk. The Volgograd branch employed between 500 and 700 people; the Shikhany branch about 1,500 (600 of them scientists); and the Novocheboksarsk branch about 300.

Leading the effort to develop fourth-generation chemical weapons was Dr. Pyotr Petrovich Kirpichev, a brilliant young scientist who worked in the Shikhany branch of GosNIIOKhT. Later known as the State Institute for Technology of Organic Synthesis (GITOS), the Shikhany branch was located in the closed city of Volsk-17, twelve miles from the Central Military Chemical Testing Site of the Red Army. In 1973, drawing on some new ideas circulating among Soviet military chemists, Kirpichev synthesized a nitrogen-containing organophosphorus nerve agent that was initially designated K-84 and later renamed A-230. Although the new compound was highly toxic and stable, it was a viscous liquid that crystallized at 14 degrees Fahrenheit and thus presented disadvantages for use in cold weather.

In 1975, a young chemist named Vladimir Ivanovich Uglev joined Kirpichev's research team. At first Uglev felt uncomfortable working on chemical weapons, but he gradually accepted the rationale that they were necessary for the nation's defense. Without a credible deterrent, he was repeatedly told, the U.S. imperialists would unleash the horrors of gas warfare against the Soviet homeland. By the time Uglev arrived at GITOS, the preliminary development work on A-230 had been completed. Over the next few years, Kirpichev, Uglev, and their colleagues synthesized more than a hundred structural variants of A-230 and subjected them to systematic testing in laboratory animals. Most of the analogues were so unstable that they rapidly lost potency, but five were sufficiently toxic and stable to be of potential military interest. These compounds were therefore subjected to intensive study, and the most promising turned out to be A-232.

The molecular structure of A-232 was nearly identical to that of A-230, but with an important difference. Whereas A-230 was a *phosphonate* containing a direct carbon-phosphorus bond, A-232 was a *phosphate,* meaning that the carbon and phosphorus atoms were linked through an oxygen atom. Although phosphonates have only a few civilian applications in the production of certain pesticides and fire retardants, phosphates are used for a wide variety of legitimate industrial and commercial purposes. Because A-232's precursors and breakdown products did not contain a carbon-phosphorus bond—the telltale "signature" of nerve agents such as Sarin, Soman, and VX—its production would be far easier to conceal from foreign spies and arms control inspectors. A-232 had some disadvantages, however: it was less toxic than A-230 by a factor of two or three and tended to degrade rapidly in the presence of moisture.

After the invention of A-230 and A-232, Kirpichev and his colleagues tried to keep their findings quiet so that they could continue their work without interruption. But Victor Petrunin, the deputy director of GITOS, was eager to send news of the breakthrough to his superiors in Moscow. A few days later, the GosNIIOKhT director, Ivan Martinov, arrived at the Shikhany institute and asked to be briefed on the new compounds. Impressed by what he heard, he allocated high priority to the Foliant program and supplied Kirpichev with top-quality laboratory equipment. All research on A-230 and A-232 was henceforth classified "Top Secret—Of Special Importance," the highest level of secrecy at the time. Research reports on the new compounds were no longer circulated for internal review but were sent directly to Moscow, often in the form of handwritten manuscripts.

In 1976, a pilot plant at Volgograd produced a few kilograms of A-230 and A-232 for testing purposes. The first battery of tests involved measuring their physiochemical properties and toxicity in laboratory animals by injection, inhalation, and skin application. Soviet scientists then conducted a series of open-air trials at the Red Army's Central Military Chemical Testing Site at Shikhany using rabbits, guinea pigs, dogs, horses, and monkeys. In the field tests, the new agents turned out to be five to eight times more lethal than VX. This extraordinary potency had not been predicted from laboratory studies and appeared to result from the fact that A-230 and A-232 passed rapidly from the bloodstream into the central nervous system by penetrating the blood-brain barrier. The new agents also inactivated cholinesterase irreversibly in minutes. Indeed, the results of the field trials were so impressive that the GITOS scientists hesitated to report their findings to Moscow for fear of being accused of exaggeration.

DURING THE EARLY 1970s, the technical service of the French Army carried out a series of trials involving live Sarin and nerve agent simulants at the Camp de Mourmelon near Reims. These trials were given a series of code names beginning with the letter "C," including "Citronelle" and "Canelle." Also during this period, the French government decided to produce a stockpile of chemical weapons, although this decision was reversed a few years later.

In 1972, France negotiated with the Algerian government for an additional five-year lease of the chemical weapons testing site at B2-Namous. Under the 1967 agreement, the French had merely been required to notify the Algerian authorities in advance of each test campaign. This time, however, the Algerian side was more demanding. The field trials could no longer be organized and controlled exclusively by the French authorities. Instead, the Algerians demanded the right to authorize each individual experiment and insisted that their own observers be present. Despite these oppressive conditions, the French chemical weapons trials continued for several more years. In 1977, an extensive testing campaign was carried out, but it proved to be the last.

Beginning in 1978, the French dismantled and decontaminated the testing site at B2-Namous, and in 1981 it was finally handed over to the Algerian authorities. From then on, small-scale trials of nerve agents continued in metropolitan France at the Camp de Mourmelon and Bourges. These test campaigns supported the development of offensive chemical weapons as well as defensive equipment for detection, protection, and decontamination.

In the United States, the binary program continued to be plagued with technical and bureaucratic problems. The Navy, which had been ambivalent about chemical weapons from the start, became disenchanted with the slow progress and political controversy surrounding the development of the Bigeye bomb and tried to cancel the program in 1972. After the Air Force expressed interest in taking over the development effort, the Pentagon reinstated the Navy as lead service in February 1974, with the Air Force in the role of "participating service." The Navy, however, delegated the mission of delivering the Bigeye to the Marine Corps, whose vertical-takeoff Harrier fighter-bombers could fly from land bases as well as carriers.

In July 1974, one month before Richard Nixon resigned the presidency over the Watergate scandal, he met with Soviet Premier Leonid Brezhnev in Moscow. The two leaders issued a communiqué on July 3 stating that their governments intended to launch bilateral negotiations aimed at eliminating the "most dangerous" chemical weapons in their respective stockpiles. The relevant paragraph read: "Both Sides reaffirmed their interest in an effective international agreement which would exclude from the arsenals of States such dangerous instruments of mass destruction as chemical weapons. Desiring to contribute to early progress in this direction, the USA and the USSR agreed to consider a joint initiative in the Conference of the Committee on Disarmament with respect to the conclusion, as a first step, of an international Convention dealing with the most dangerous, lethal means of chemical warfare."

In the fall of 1974, the U.S. Senate finally gave its consent to ratification of the 1925 Geneva Protocol, nearly fifty years after it had been negotiated. Gerald Ford, who had assumed the presidency after Nixon's resignation in August, signed the instrument of ratification on January 22, 1975. According to the widely accepted interpretation of the Geneva Protocol, the United States was now legally bound not to use chemical weapons in war except in direct retaliation for a chemical attack.

In Geneva, the Conference of the Committee on Disarmament (CCD), the United Nations' arms control negotiating forum, began to discuss a global ban on chemical weapons. The initial talks were exploratory and soon bogged down over basic questions such as the definition of a chemical weapon and the scope of the future convention. Another major challenge facing the negotiators was the need to verify that countries followed through on their commitment to eliminate existing stocks of chemical weapons and to halt further production. Because nerve agents could theoretically be produced at commercial chemical plants, intrusive on-site inspections would be required at both military facilities and industry sites. Although the participating countries agreed in principle on the need for verification, the Soviet leadership was deeply opposed to on-site inspections, which it viewed as tantamount to espionage. Accordingly, most of the verification proposals allowed each nation to police its own compliance, an approach of dubious effectiveness and credibility.

Arms control experts warned that if the Pentagon began to procure binary weapons, other countries participating in the Geneva talks would conclude that the United States was not negotiating in good faith. Responding to these arguments, Congress denied the administration's funding requests for binary weapons in fiscal years 1975 and 1976. Although the Army did not seek funds for this purpose in FY 1977, Congress passed the Defense Authorization Bill with a provision blocking funds for the production of binary weapons until the president certified that they were "essential to the national interest." At the same time, however, growing U.S. concerns about the Soviet chemical warfare threat led the Secretary of the Army to reverse his earlier decision to abolish the Chemical Corps.

EVEN AS THE UNITED STATES and the Soviet Union were developing next-generation chemical weapons, several other countries, including China, Egypt, Syria, North Korea, Yugoslavia, and Iraq, were pursuing a basic nerve agent production capability. The Iraqi chemical weapons program dated back to the early 1960s, when a group of army officers had traveled to the Soviet Union for training in chemical defense and, after their return, had founded the Iraqi Chemical Corps on January 14, 1964. Seven years later, impressed by Egypt's extensive use of chemical weapons during the Yemen war, the Iraqi leadership launched its own chemical warfare program. In 1971, the Iraqi Chemical Corps built a small research laboratory near the village of Al-Rashad on the northeast edge of Baghdad, where organic chemists, many of them educated in the West, synthesized small amounts of mustard, Tabun, and Sarin.

In 1974, responding to perceived military threats from Iran and Israel, the Iraqi government established a new organization for scientific research and development called the Al-Hazen Ibn Al-Haitham Institute. Although this entity reported to the Ministry of Education, it actually conducted clandestine R&D on advanced military technologies. Directed by a Chemical Corps officer, the Al-Hazen Institute was organized into four divisions or centers. The First Center, focusing on chemical weapons development, was based at Al-Rashad and masqueraded as the Center for Medical Diagnostics. To oversee Iraq's unconventional weapons programs, the Baghdad regime established a three-man Strategic Planning Committee chaired by Saddam Hussein, then chief of internal security and vice president of the Revolutionary Command Council. Saddam did not assume the presidency until 1979, but in the mid-1970s he was already the power behind President Ahmad Hassan al-Bakr.

In January 1975, the Iraqi State Company for Construction Contracts hired a Baghdad design firm to plan a chemical production facility. Later that year, the Iraqi government retained a Beirut-based consulting firm called Arab Projects and Development to recruit scientists and technicians from throughout the Arab world with attractive salaries and perks. This effort was successful, and more than four thousand foreigners came to Iraq to help build chemical plants. In addition, Iraqi procurement teams disguised as commercial representatives traveled to Europe and the United States in search of chemical manufacturing equipment and know-how.

Because Baghdad had broken diplomatic relations with Washington during the 1967 Arab-Israeli war, Iraq hired the U.S. subsidiary of a French engineering firm to find an appropriate American partner. The French intermediary recommended the Pfaudler Company of Rochester, New York, a leading manufacturer of glass-lined reactors and other corrosion-resistant chemical production equipment. Iraqi agents contacted Pfaudler and asked the company to prepare detailed plans for a factory that could manufacture 1,200 tons per year of the organophosphate pesticides Amiton, Demeton, Paraoxon, and Parathion. The planned location of the plant was at Rutbah, in the western desert of Iraq just south of Akashat, the site of a phosphate mine that could supply critical raw materials.

Pfaudler sent two chemical engineers to Baghdad to meet with Ministry of Agriculture officials and discuss the construction of Iraq's first pesticide factory. The Americans then returned to Rochester and began to prepare blueprints and equipment specifications. Because of the high toxicity of the organophosphate pesticides that Iraq planned to produce, Pfaudler considered it advisable to build a pilot plant to train the local workers about safety. But when company representatives presented this proposal to an Iraqi government delegation on January 24, 1976, the Iraqis made clear that they were not interested in a pilot plant and wanted to begin full-scale production as soon as possible. After Pfaudler balked at this request, the Iraqis broke off the negotiations. By this time, they had obtained drawings and specifications for the pesticide plant, although the production equipment was still lacking.

In late 1976, Iraqi officials from the State Organization for Technical Industries approached several West European companies seeking corrosion-resistant reactor vessels, pipes, and pumps for the Rutbah plant. Executives at Imperial Chemical Industries (ICI) in Britain were suspicious of the Iraqi request because the pesticides they planned to manufacture were highly toxic. Convinced that the Iraqis were seeking a plant that could be readily converted to the production of nerve agents, ICI officials rejected the request and informed the British government of their concerns. The Iraqis then turned to Montedison of Milan, a giant Italian chemical company that at the time was in serious financial trouble. According to Iraqi sources, a subsidiary of Montedison called Tecnimont agreed to construct the pesticide plant in nine months for $52 million, but Tecnimont executives later claimed that Iraq had broken off the contract negotiations. Reportedly, the Rutbah plant was finally built in the early 1980s by Klöckner Industrie, a West German chemical company, and was subsequently used to produce nerve agents.

In August 1976, bilateral U.S.-Soviet talks on chemical arms control resumed after a two-year hiatus. Diplomats expressed hope that by the following spring, Washington and Moscow would be able to break the deadlock and develop a joint proposal for a global chemical weapons ban. Although the two superpowers agreed in principle on the scope of the treaty, they remained far apart on the verification measures needed to ensure that countries would not secretly cheat and acquire a militarily significant stockpile of chemical weapons.

In parallel with the arms control talks, the U.S. development of binary weapons continued. Based on extensive field testing with simulant chemicals, the Army Armament Command concluded in November 1976 that the M687 Sarin artillery shell was ready for production, and the Marine Corps submitted a formal procurement request. But five months after President Jimmy Carter took office in January 1977, the new administration took stock of the binary program and its potential impact on the Geneva negotiations. On May 19, 1977, National Security Adviser Zbigniew Brzezinski wrote a memorandum directing a group of high-level officials called the Special Coordination Committee to review U.S. chemical warfare policy. The committee was asked to assess the impact of various chemical disarmament options on the security of the United States and its allies, U.S.-Soviet relations, and other foreign policy interests; the requirements for verification; and the chances of successfully negotiating an arms control treaty. This review concluded that a U.S. decision to procure binary weapons might spur chemical weapons proliferation worldwide and provoke the Soviets into either blocking a disarmament treaty or insisting on a poorly verifiable one.

In a letter to Defense Secretary Harold Brown dated October 23, 1977, Secretary of State Cyrus Vance wrote that in the bilateral negotiations with the Soviet Union, the Carter administration sought to prohibit the development, production, and stockpiling of chemical weapons. "I believe that if we were to forgo plans for production," Vance wrote, "we will have achieved a significant psychological advantage over the Soviets. This would force them into a position of having to respond to the U.S. initiative by taking a positive step toward reducing their own CW program." To convey the seriousness of its commitment to chemical disarmament, the administration rejected an Army request for $13.2 million in the fiscal year 1978 defense budget to prepare for DF production at Pine Bluff Arsenal.

The Chemical Corps and its supporters, in contrast, argued for the need to modernize the U.S. chemical stockpile in parallel with the Geneva talks on chemical disarmament. According to Brigadier General Lynwood B. Lennon, the Army's deputy director of strategy, plans, and policy, "Near-term national and collective security requirements need not and should not be sacrificed to the allure of an elusive arms control agreement. The history of such negotiations seems clearly to reinforce the common sense notion that deterrence must continue until an enforceable, verifiable agreement can be reached."

The most vocal advocates of chemical modernization were a small number of right-wing "defense intellectuals" who worked at Pentagon-funded think tanks and wrote for military journals, including Amoretta M. Hoeber and Stephen Douglass, Jr. They argued that the U.S. stockpile of unitary nerve agent munitions was deteriorating and no longer provided a credible deterrent against the Soviet Union, which continued to upgrade its own arsenal. Douglass postulated that the Soviets did not intend to devastate Western Europe but rather to occupy it and exploit its assets for postwar economic recovery. For that reason, he wrote, the Soviets would prefer to employ weapons that killed people without causing the wholesale destruction of buildings, factories, and other infrastructure. Nerve agents, Douglass argued, "are ideal for Europe if one is interested in minimizing physical destruction."

NATO's Supreme Allied Commander in Europe, General Alexander Haig, Jr., claimed that the capability of the Western alliance to counter a Soviet chemical attack was "very weak." Moscow, he warned, might use missile warheads filled with persistent V agents to knock out NATO airfields and nuclear weapon storage sites early in a war, and nonpersistent G agents to attack allied troop concentrations. In 1979, U.S. military commanders played a war game called "Wintex" that simulated a conflict between NATO and the Warsaw Pact. During this fictional scenario, the Red Army launched a chemical attack with nerve agents, forcing NATO commanders to decide whether to retaliate with tactical nuclear weapons and risk escalation to full-scale nuclear war. This outcome suggested the need for a modernized chemical arsenal capable of deterring the Soviets from the first use of such weapons. But U.S. military strategists were split between those who sought to bolster deterrence by acquiring a credible stockpile of binary artillery shells, and others who wished to procure long-range ballistic missiles with chemical warheads for "deep strikes" against enemy airfields and staging areas inside Warsaw Pact territory, as part of a new NATO war-fighting strategy.

Although the Army reestablished the Chemical School at Fort McClellan in 1979 and began returning chemical units to active duty, the Carter administration deleted the Army's funding request for binary-weapons production from the defense budgets in fiscal years 1979, 1980, and 1981. The U.S. national security agencies were divided over the issue of chemical modernization. Although the Joint Chiefs and the civilians in the Office of the

Secretary of Defense were in favor of binary weapons, the State Department objected that the NATO allies had not been consulted, and the Arms Control and Disarmament Agency and the National Security Council staff worried that procuring the new weapons would undermine the chemical disarmament talks with Moscow. The result of this bureaucratic standoff was that the Carter administration requested funds for improved chemical defenses, including detectors, masks, and protective suits, but not for binary weapons.

For its part, the Chemical Corps was increasingly impatient with the lack of procurement. Maintaining the existing chemical stockpile was far easier and cheaper than destroying it, but the continued viability of the old weapons came to be seen as an impediment to the acquisition of new ones. Once the Chemical Corps became committed to binary munitions, it failed to maintain the unitary stockpile adequately. By neglecting to clean rusty shells, restore identifying markers, and replace fuses, the corps hastened the deterioration of the existing stockpile and thereby increased pressures for its replacement.

MEANWHILE, the détente in U.S.-Soviet relations had begun to sour. Rumors were filtering out of Southeast Asia that the Soviet military was helping the Communist government of Laos to launch attacks with an unknown chemical agent against the remote mountain strongholds of the H'mong tribesmen, who the CIA had recruited to fight on the American side during the Vietnam War. The U.S. intelligence community also concluded that a major outbreak of human anthrax in the Soviet city of Sverdlovsk in April 1979 had resulted from the accidental release of processed anthrax bacterial spores from a clandestine biological weapons factory.

In December 1979, superpower tensions rose dramatically when the Soviet Union sent five divisions into Afghanistan to prop up the shaky Communist government that had come to power in the 1970s and to protect their interests in Central Asia. The Soviets quickly occupied Kabul and installed Babrak Karmal as president. Afghan resistance fighters called *mujahidin* proclaimed a "jihad," or holy war, against the Soviet invaders and were soon joined by militant Islamic volunteers from several parts of the

Arab world. Although Soviet troops controlled the major Afghan cities, they encountered fierce resistance whenever they ventured into the mountainous hinterland.

The Carter administration condemned the Soviet invasion of Afghanistan in the strongest terms. Because of the new tensions between Washington and Moscow, the bilateral talks on chemical disarmament came to an end after twelve inconclusive rounds. Meanwhile, refugees fleeing Afghanistan alleged that Soviet aircraft and artillery had begun using chemical weapons against the *mujahidin* guerrillas. Although the reports were anecdotal, the U.S. intelligence community considered them credible.

In May 1980, Representative Richard Ichord, a conservative Democrat from Missouri who served on the House Armed Services Committee, argued that Congress should respond forcefully to the Soviet Union's invasion of Afghanistan and its alleged use of chemical weapons by voting to modernize the U.S. chemical deterrent. Binary munitions were desirable, in his view, because they would "depoliticize" chemical weapons and make them more acceptable to the public and to military planners. To this end, Ichord offered an amendment restoring the Army's request for a $3.15 million "down payment" on construction of the DF plant at Pine Bluff Arsenal. The amendment, only two paragraphs long, was buried deep in the FY 1981 Military Construction Appropriations Bill and was not even debated on the floor of the House, which approved the legislation by a large majority (308–19) on June 27, 1980.

The Carter administration was stunned by how easily Ichord's troublesome amendment had slipped through the legislative process in the House. Secretary of State Vance urged the Democrat-controlled Senate to block funding for the Pine Bluff plant, but Senator Henry "Scoop" Jackson, a conservative Democrat from Washington State, agreed with Ichord that the Soviet Union enjoyed a "preponderant advantage" over the United States in chemical warfare capabilities. Jackson therefore introduced a similar amendment authorizing $3.15 million to begin construction of the Pine Bluff facility. Public anger over the Soviet invasion of Afghanistan led the Senate Armed Services Committee to defy the White House and vote 14–3 to recommend the amended bill to the full Senate.

On September 16–18, 1980, the Senate conducted several hours of emotional debate on the Jackson amendment, even though it represented only a tiny fraction of the $160 billion Defense Appropriations Bill. Supporters of the amendment, such as Senator Jackson and Senator John Warner (R.–Virginia), argued that Congress had to respond in a forceful manner to the Soviet invasion of Afghanistan. Approving a small down payment on the Pine Bluff plant would send a clear message to Moscow, even though it would take several years before actual production could begin. Senator Carl Levin (D.–Michigan) countered that for the United States to resume the manufacture of chemical weapons after a twelve-year moratorium would be "premature, unwise, and could ultimately be self-defeating from defense, diplomatic, and arms-control perspectives." He argued that intelligence estimates of Soviet chemical warfare capabilities were uncertain and that the existing U.S. stockpile of unitary weapons constituted a "credible" deterrent. Moreover, a unilateral U.S. decision to deploy binary munitions in Western Europe without consulting the NATO allies would provoke a strong political backlash. Senator Gary Hart (D.–Colorado) proposed an amendment to delay partial funding of the DF production plant until after the completion of a top-level policy review, but his amendment failed by one vote, 46–47. The Senate then approved the Jackson amendment by a vote of 52–38 and the entire bill by 89–3. Given the large margin, President Carter did not attempt to veto the legislation.

Six weeks later, the elections of November 4, 1980, caused a tectonic shift in the political landscape. President-elect Ronald Reagan was a former professional movie actor whose avuncular manner belied his far-right conservative beliefs. In sharp contrast to his predecessor, Jimmy Carter, Reagan was an avowed anti-Communist, a military hard-liner, and a strong supporter of U.S. chemical weapons modernization, which he believed was essential to counter the Soviet military threat. Ironically, because the Republicans had also gained control of the Senate, Mark Hatfield, the senior Republican senator from Oregon and a long-standing opponent of binary weapons, became chairman of the Senate Appropriations Committee, giving him considerable influence over the military budget.

On November 21, the Senate passed a version of the Defense Procurement Appropriations Bill that included $19 million to equip the DF production plant at Pine Bluff Arsenal, but this item was deleted from the House version. When a conference committee convened in December to reconcile the House and Senate bills, Senator Hatfield exercised his new power by threatening a full-scale debate unless the committee excised the $19 million for DF plant equipment, leaving only the $3.15 million for site construction.

Even so, by approving at least some money for the Pine Bluff facility, Congress had taken a small but significant step toward production. President-elect Reagan was determined to win the political battle over binary weapons, but the struggle would prove to be far more intense and protracted than anyone had imagined.

CHAPTER THIRTEEN

BINARY DEBATE

In December 1980, a task force of the Defense Science Board, an elite group of scientific advisers to the Pentagon, issued a classified report on U.S. chemical warfare policy that had been prepared over the summer. Chaired by John M. Deutch, a professor of chemistry at MIT, this panel found that the existing U.S. chemical weapons stockpile was outdated and partly unusable, limiting its deterrent value. Upgrading chemical defenses alone would not be a sufficient remedy because gas masks and protective suits were awkward and uncomfortable for troops to wear, particularly in hot weather, causing a 30 to 50 percent decline in unit performance. If the Soviet Union could impose this handicap on the United States without fear of retaliation, it would gain a significant military advantage. The Deutch committee concluded that because the renewed production of unitary nerve agents would be "politically unacceptable," the best option was to destroy the aging unitary stockpile and replace it with a smaller number of advanced binary weapons.

On February 2, 1982, President Reagan formally requested more than $30 million to begin production of the M687 binary Sarin projectile, but the administration faced an uphill battle in persuading Congress to go along. A majority of the Democrat-controlled House of Representatives was opposed to binary production, and the Republican-controlled Senate was split by the closest of margins. Nerve agents were the only weapon system that many liberals and conservatives found equally unpalatable. Not only did chemical arms lack a political constituency, but the public recognized that poison gas was indiscriminate and more likely to kill civilians than well-protected troops. Critics also had serious doubts about the military utility of binary chemical weapons for the defense of NATO and whether they would work as advertised.

The fact that support for binary weapons correlated poorly with party affiliation and ideology made for some strange political bedfellows. On March 12, 1982, for example, twelve Democratic and Republican senators from across the political spectrum wrote a letter to Senator John Tower (R.–Texas), the chairman of the Senate Armed Services Committee, opposing production of binary weapons. Signatories included liberals such as Gary Hart, Edward Kennedy, and George Mitchell, and conservatives such as Thad Cochran and Nancy Kassebaum. The letter read in part, "The production of binary chemical weapons is not necessary for the national defense, nor is it necessary to deter Soviet first use. Our current stockpiles are adequate for that purpose. Our principal emphasis should be the acquisition of additional protective and defensive equipment for U.S. combat forces to reduce the effects of a chemical weapons attack."

During consideration of the FY 1983 Defense Authorization Bill, the Senate endorsed binary production by a narrow 49–45 vote. In the House, however, a political odd couple—Ed Bethune, an archconservative Republican from Arkansas, and Clement Zablocki, a liberal Democrat from Wisconsin—teamed up to defeat the legislation. When members of both chambers met in conference committee to reconcile the two versions of the bill, they agreed to support the House position against binary weapons.

Despite Congress's refusal to fund production of the M687 Sarin projectile, the Reagan administration moved ahead with site preparation for the DF plant at Pine Bluff Arsenal. Meanwhile, the development of the Bigeye VX bomb continued to be plagued with technical problems. When the two binary components (QL and sulfur) were mixed together, the reaction generated intense heat and pressure that sometimes caused the bomb to "cook off," or explode. In one laboratory test in October 1982, a Bigeye prototype was deliberately tested to failure and ruptured, spewing hot nerve agent in all directions. For subsequent tests, Army scientists installed pressure-relief valves on the prototype bombs, even though the actual weapons would not be so modified. The Bigeye also failed to meet the technical specifications for the purity of VX.

Yet another team at Edgewood's Chemical Systems Laboratory was

A mock-up of the Bigeye bomb, an aircraft-delivered binary weapon that was designed to produce VX by combining two relatively non-toxic precursors (QL and powdered sulfur). The bomb would glide to earth while spraying nerve agent over the target. Because of chronic technical problems, the Bigeye bomb was never produced.

developing a binary warhead for the Army's Multiple Launch Rocket System (MLRS) that would generate a mixture of Sarin and an analogue of Cyclosarin with an added methyl group (CH_3). This mixture was termed an "intermediate-volatility agent" because it evaporated more slowly than Sarin but faster than VX. On March 31, 1982, the Army awarded the Vought Corporation a contract to begin concept definition of the XM135 Binary Chemical Warhead for the MLRS.

IN FRANCE, scientists at the Centre d'Études du Bouchet (CEB) were also working on binary chemical weapons. This effort had been inspired by the major Soviet investment in offensive chemical warfare capabilities, as well as by the U.S. binary weapons program. The CEB focused on developing a binary formulation of an intermediate-volatility nerve agent.

During the mid-1980s, in an effort to work around the ban on live-agent testing at Dugway Proving Ground, Bill Dee arranged with the French Army and the CEB to conduct some small-scale outdoor trials of the American MLRS binary warhead at the Camp de Mourmelon near Reims. In return, the Americans shared with their French colleagues some aspects of U.S. binary technology, such as the warhead fusing mechanism.

IN 1983, CONGRESS resumed deliberations over production of the M687 binary Sarin projectile. Although the local congressman from Pine Bluff supported the binary program because it would create new jobs, Arkansas Senator David Pryor placed his moral opposition to chemical weapons above the economic interests of his state. In February 1983, he wrote a letter to Senator Tower that stated, "My opposition to nerve gas production has never been parochial or entirely budgetary. Therefore, I do not want the activity transferred to another location in the country or the funds reprogrammed. . . . I also welcome the support of those, like myself, who oppose nerve gas production on military, diplomatic, and humanitarian grounds."

When the FY 1984 Defense Authorization Bill came up for a vote, the House decisively rejected funding for binary weapons production 256–161. The Senate split evenly, however, allowing Vice President George H. W. Bush to cast the tiebreaking vote in favor of production. In late July, a House-Senate conference committee met to reconcile the two different versions of the bill. Because the House was represented by hawkish congressmen, the conference voted narrowly to authorize binary weapons funding. The battle lines now moved to the Defense Appropriations Bill, which contained an administration request of $124.4 million for production of the M687 projectile. This legislation reached the Senate in the fall of 1983. During the floor debate, Senator Hatfield called the plan to acquire new chemical weapons "morally and politically indefensible." But Senator Tower countered that the deteriorating unitary stockpile would soon become too dangerous for U.S. troops to handle. "Future adherence to a unilateral moratorium on the production of new weapons risks unilateral disarmament by obsolescence," he said.

The Senate vote came on November 8, 1983. During the roll call, Vice President Bush again took the presiding officer's chair, ready to break a tie if necessary. Once again, the Senate split evenly (46–46), and the vice president cast the deciding vote for production. Bush was clearly uncomfortable with his two tiebreaking votes, which provoked sharp criticism from his mother. As Dorothy Bush told *The Washington Post,* "George knows that I disapprove of it. He knows how I feel. He said that we have to have it to deter other countries from using it. But George knows I would die if this country ever uses it." In the end, the House voted against funding for binary weapons, and the House-Senate conference committee adopted the House position.

AT THE SAME TIME that Congress was debating the production of new binary munitions, Iraq began to employ chemical weapons in its bloody war against Iran. The source of the conflict could be traced back to early 1979, when an aging Iranian cleric named Ayatollah Ruhollah Khomeini had led a popular revolution that toppled the monarchy of Shah Reza Palavi and established a conservative Islamic theocracy in Tehran. After the revolution, tensions escalated between Iran's new rulers and the secular Ba'athist regime of Saddam Hussein in neighboring Iraq. A long-simmering border dispute between the two countries flared up over the Shatt-al-Arab waterway, which provided Iraq's sole access to the Persian Gulf.

In the turmoil that followed the Iranian revolution, Saddam saw a strategic opportunity to eliminate his long-standing rival for regional hegemony. The Iranian Army had been weakened by the loss of its main weapons supplier, the United States, and by a major purge that had removed high-ranking officers who were corrupt or loyal to the shah. Believing that Iran's internal disarray would permit a rapid Iraqi military victory, Saddam launched a surprise attack on September 22, 1980. The Iraqi invasion began with air strikes by MiG fighter-bombers and a blitzkrieg-style tank assault by six armored divisions. Although the Iraqi Army rapidly gained control of the contested Shatt-al-Arab waterway, Khomeini rejected a proposed settlement on Iraqi terms and launched a series of fierce counterattacks. Despite the fact that Iran had fewer tanks and artillery pieces and lacked spare parts for its American-made fighters and armored vehicles, it enjoyed a large advantage in number of troops. Iran's population of 64 million people was more than twice Iraq's 24 million and included many young men of military age. Accordingly, Iran began to use "human wave" infantry tactics to hold the line against Iraq's armored assaults.

By early 1981, the momentum of the Iraqi offensive had stalled. To counter Iran's reliance on superior numbers of troops, Saddam began to consider the use of chemical weapons as a "force multiplier." On June 8, 1981, the development program at Al-Rashad, code-named "Project 922," was transferred to the Iraqi Ministry of Defense and given a high priority. Iraq also signed a strategic cooperation agreement with Egypt and paid Cairo $12 million for technical assistance with chemical weapons production and weaponization. This arrangement included the building of chemical plants in Iraq by the Egyptian branch of the West German company Walter-Thosti-Boswau International, and the use of Egypt as a transshipment point for Iraqi imports of nerve agent precursors, such as hydrogen fluoride, from Western suppliers. In late 1981, the construction of a large pesticide production complex was under way on a stretch of empty, semi-arid grazing land fifty miles northwest of Baghdad. The complex was given a benign cover name: the State Enterprise for Pesticide Production (SEPP).

Over the next year, Iraq suffered a series of military setbacks. On March 17, 1982, the Iranian Revolutionary Guards, or Pasdaran, defeated the Iraqi Third Army at Khorramshar, taking thousands of prisoners. By early summer, the Iranian forces had advanced to within a few miles of Basra, Iraq's second largest city. During a battle near Basra in July, the Iraqi Chemical Corps used tear gas for the first time, panicking an entire Iranian division. Encouraged by this success, Iraq escalated in 1983 to the small-scale use of mustard agent. These chemical attacks, referred to as "special strikes," could be launched only with the explicit authorization of the general commander, Saddam Hussein. Fearing the consequences of violating the 1925 Geneva Protocol, to which Iraq was a party, Saddam moved cautiously at first, limiting the number and scale of chemical attacks, but the lack of a harsh international response emboldened him.

During the summer of 1983, the staff and equipment of Project 922 moved from Al-Rashad to SEPP. Construction at the new site continued at a feverish pace and resulted in five large research laboratories, pilot production plants for mustard and Tabun, and an administrative center. One of the labs had an inhalation chamber for testing chemical agents on experimental animals. Iraqi scientists preferred to use beagles imported from West Germany because the characteristics of the canine respiratory system were well understood.

At first, the Iraqi chemical weapons program was heavily dependent on foreign suppliers. More than thirty Western firms sold specialized production equipment and 800 tons of precursor chemicals to Iraq, including fourteen companies in West Germany, three in the Netherlands, three in Switzerland, and two each in France and the United States. One of the U.S. suppliers was a small chemical company in Nashville, Tennessee, that shipped six and a half tons of potassium fluoride to SEPP without asking how the chemical would be used.

Although Western assistance to the Iraqi chemical warfare program was at times unwitting, some unscrupulous businessmen sold chemicals or equipment to SEPP with full knowledge of the intended purpose. One such individual was Helmut Maier, the managing director of the West German firm Karl Kolb GmbH, a medium-sized chemicals producer based near Frankfurt Airport that had contracted with SEPP to build an entire chemical weapons complex. Working through a front company called Pilot Plant Chemical, Maier advised Iraqi officials on designing and equipping the "pesticide" plants and served as a go-between with other West German suppliers. From 1983 to 1986, Pilot Plant engineers built six production lines at SEPP for mustard and nerve agents. The Iraqi government also purchased from foreign suppliers 40,000 artillery shells, 20,000 artillery rockets, and 7,500 bomb casings that were subsequently modified for chemical delivery, along with machinery and components to produce its own chemical munitions.

To operate the chemical weapons plants, the Iraqi government recruited the best graduates in chemistry and chemical engineering from the University of Baghdad and other leading universities. Young Iraqis who were "invited" to work at SEPP actually had little choice: the alternative was to be drafted and sent to the Iranian front. To preserve the secrecy of the chemical weapons program, scientists and technicians were issued fake work documents and sworn to secrecy. They were also kept under close surveillance by the Iraqi security services. Sometimes a government agent knocked on a scientist's door in the middle of the night and ordered him back to work, either because the military had an urgent need for chemical weapons or simply to reinforce the fact that the scientists were at the beck and call of the authorities.

DURING THE SUMMER OF 1983, the government of Iran sent letters to the U.N. Secretary-General alleging that Iraq had employed chemical weapons several times since the war began, in flagrant violation of the 1925 Geneva

Protocol. Tehran argued that a failure by the United Nations to sanction Iraq would undermine the credibility of the treaty and lead to a chemical arms race in the region.

U.S. government sources secretly verified the Iranian charges. On November 1, the State Department's Bureau of Politico-Military Affairs sent a memorandum to Secretary of State George Shultz recommending that the United States "approach Iraq very soon in order to maintain the credibility of U.S. policy on CW [chemical weapons], as well as to reduce or halt what now appears to be Iraq's almost daily use of CW." Three weeks later, on November 21, American diplomats delivered a démarche, or official diplomatic note, to the Iraqi Foreign Ministry in Baghdad. This message stated that Washington was aware of Iraq's use of chemical weapons and strongly opposed it as a matter of principle. In response to the U.S. pressure, Iraq halted its chemical attacks temporarily, although they resumed several months later after a new Iranian offensive.

Despite its concern over Iraq's illicit use of chemical weapons, the Reagan administration was deeply conflicted because it believed that Iran posed a far greater threat to U.S. interests. Washington remained angry over the hostage crisis of 1979–81, when Iranian militants had held fifty-two U.S. diplomats and citizens captive in Tehran for 444 days, and also resented the vitriolic anti-American rhetoric emanating from the Khomeini regime. In addition, the Reagan administration feared that an Iranian victory over Iraq would destabilize key U.S. allies in the region such as Kuwait, Saudi Arabia, and even Jordan, threatening the vital flow of oil from the Persian Gulf.

Because of these geopolitical considerations, Washington made a strategic decision to abandon its officially neutral position on the Iran-Iraq War and "tilt" toward Iraq, while keeping its support low-profile. In December 1983, President Reagan named former Defense Secretary Donald Rumsfeld as his special envoy to Baghdad and dispatched him to meet with Saddam Hussein and discuss the normalization of U.S.-Iraqi relations. During a ninety-minute meeting with Saddam on December 20, Rumsfeld made no mention of Iraq's repeated chemical attacks. In another session with Deputy Foreign Minister Tariq Aziz, Rumsfeld reassured the Iraqi official that the U.S. government's opposition to the use of chemical weapons was a matter of principle and that Washington's interest in normalizing relations remained "undiminished."

Over the course of the war, Iraq's use of chemical weapons became progressively more effective. Early chemical warfare operations were seriously flawed: pilots dropped their bombs from too high or too low an altitude, and ground troops fired chemical artillery shells and rockets in unfavorable weather conditions or failed to concentrate enough agent at the point of attack. In several instances, Iraqi aircraft inadvertently gassed their own ground forces by dropping chemical bombs near the frontline trenches, generating lethal clouds that were blown by the wind back over the Iraqi lines. As the war dragged on, however, the Iraqi forces learned to tailor chemical strikes to specific tactical situations, maintaining the momentum of a ground assault or denying terrain to the enemy.

IN THE SOVIET UNION, the Foliant program continued to move forward. The Central Committee of the Communist Party and the Soviet Council of Ministers issued a secret decree on March 25, 1983, directing GosNIIOKhT to develop binary versions of the fourth-generation agents. The rationale for this effort was to catch up with the United States, which already had three binary chemical munitions under development. In contrast to the U.S. program, which Congress had debated openly for years, Soviet binary development was conducted under extreme secrecy. One reason was that the Soviets sought to develop binary nerve agents whose components resembled ordinary industrial chemicals, with a view to circumventing the verification provisions of a future treaty banning chemical arms.

The first binary formulation developed by the Foliant scientists was for R-33, the Soviet version of VX. This weapon was given the code name "Novichok" (pronounced *no*-we-shoke), the Russian word for "newcomer." Subsequently, Igor Vasiliev and Andrei Zheleznyakov at GosNIIOKhT in Moscow developed a binary version of A-232 that they termed Novichok-5. Although the unitary form of A-232 was unstable and deteriorated rapidly in storage, limited shelf life was not a problem for the binary version because the agent would be synthesized en route to the target and would have to persist for only a short time to serve its intended military purpose.

Novichok-5 had two binary components, one containing phosphorus and the other nitrogen. Both precursor chemicals had legitimate industrial uses and were relatively nontoxic, so that they could be produced at plants ostensibly designed to manufacture agricultural fertilizers or pesticides. This ambiguity would make it easier to conceal the illicit production of Novichok-5 components from international arms inspectors. Although the Soviets manufactured limited quantities of the two precursors for testing purposes at pilot plants in Volgograd and Novocheboksarsk, the chemicals were not stockpiled. Instead, wartime mobilization plans were developed for ramping up production of the binary components and loading them into munitions.

In the mid-1980s, the Soviet Ministry of Chemical Industry began to construct new military production lines at the Pavlodar Chemical Plant in northern Kazakhstan. This plant had been built in 1965 on the banks of the Irtysh River as a dual-use factory capable of manufacturing both civilian chemicals and chemical warfare agents. The military section of the plant operated under tight security and was headed by a chief engineer who reported directly to Moscow. In order to replace the obsolescent factories at Volgograd and Novocheboksarsk that had been built after World War II, construction began at Pavlodar on a series of corrosion-resistant reactors made of high-nickel steel, in which the Soviet military planned to produce up to six types of Novichok binary precursors.

THE UNITED STATES, the Soviet Union, and France were not the only countries to work on binary weapons during the 1980s. British scientists at Porton Down did collaborative research with their Edgewood colleagues on intermediate-volatility nerve agents that were more persistent than Sarin but less persistent than VX. In so doing, they took a second look at compounds such as the Tammelin esters, which had been rejected because they were too unstable to be stored for long periods but would have military utility if produced in a binary system shortly before use.

Research and development on nerve agents also took place in a number of Warsaw Pact countries, including Czechoslovakia, East Germany, Hungary, Romania, and Bulgaria. Because of Czechoslovakia's exposed position in the center of Europe, it had developed particular expertise in the field of chemical defense. At Research Institute No. 70 in the Czech city of Brno, military chemists did "threat assessment" studies on novel chemical agents being developed by NATO countries. After learning that the United States was developing an intermediate-volatility agent for the MLRS rocket system, Colonel Jiri Matousek, the director of research at the Brno institute, sought to synthesize a similar agent for testing the field detectors and personal protective gear used by the Czechoslovak Army and the civil defense.

In May 1983, Matousek and his team of military chemists synthesized and characterized a new family of intermediate-volatility nerve agents that they termed "GV" because they combined characteristics of both the G-series and V-series agents. Several years later, Matousek published an unclassified paper describing this research. "The results presented," he wrote, "show that a new group of supertoxic lethal organophosphorus compounds exist as candidates for new chemical warfare agents with possible use in binary system, possessing extremely high inhalation toxicity and very high percutaneous toxicity. This means that it will be necessary to include such or similar compounds within the framework of chemical defense and the known list of chemical warfare agents should never be considered as definitively closed."

IN EARLY 1984, the Iran-Iraq War continued to intensify. On February 15, Iran launched a major offensive called Operation Dawn V along a hundred-mile front north of Basra, with the aim of seizing the strategic Basra-to-Baghdad highway and cutting Iraq in two. The Iranian forces consisted of some 500,000 Pasdaran troops and People's Army (Basij) volunteers, who ranged in age from nine to over fifty. Despite their poor equipment and lack of formal military training, the Basij militia were inflamed by propaganda, religious fervor, and the promise of martyrdom. Among them were tens of thousands of young boys who had been roped together in groups of twenty to prevent desertions. Supported by attack helicopters, the waves of Iranian infantry advanced steadily through the marshes north of Basra and approached within a few kilometers of the Basra-to-Baghdad highway. Between February 29 and March 1, the Iraqi Army counterattacked in one of the largest battles of the war. Massed Iraqi tanks and helicopter gunships slaughtered thousands of Iranian troops and pushed them back into the marshes, inflicting heavy losses for no gain in territory.

Meanwhile, on February 21, an Iraqi government radio broadcast warned that Iran was preparing another major military offensive to seize

Iraq's land, violate its women, and colonize its population. To repel the invaders, the Iraqi Army would no longer confine itself to a static defense but would strike deep into enemy territory. The announcer concluded ominously, "The invaders should know that for every harmful insect there is an insecticide capable of annihilating it, whatever their number, and Iraq possesses this annihilation insecticide."

Despite the crushing defeat at Basra, Iran opened a second front farther north, in the Majnoon marshes near the confluence of the Tigris and Euphrates Rivers. *Majnoon,* the Arabic word for "crazy," was an appropriate name for this eerie realm of shallow lakes crisscrossed with large sandbars, where the water level rose and fell with the seasons. The unmarked border between the two warring countries ran through the middle of the swamp. Along the western edge of the marshes, the Iraqis had dug a drainage canal, creating two thin strips of dry land called the Majnoon Islands that controlled the northern approaches to Basra. Iraqi prospectors had discovered a rich oil field on the islands with the potential to produce 800,000 barrels a day for thirty-five years. In preparation for drilling, the Iraqis had built an administrative base camp consisting of seven large blue-and-white bungalow sheds and offices, with two more under construction.

In early March 1984, some 15,000 Iranian troops crossed the marshes at night, traveling on foot and in shallow-bottomed fiberglass boats, and attacked the Majnoon Islands. Catching the Iraqi sentries guarding the levees by surprise, the Iranian forces seized the administrative base camp and the surrounding oil field. By dawn, both islands were firmly under the control of the Iranian troops, who dug reinforced trenches and built a pontoon bridge across the canal to bring in supplies and reinforcements. A few days later, they replaced the temporary bridge with a dirt causeway linking the island to the mainland, and expanded the defending force to 30,000 men.

Because of the strategic and economic importance of the Majnoon Islands, the Baghdad regime was determined to dislodge the Iranian defenders and drive them back across the border. The Iraqi Army's Abu Nawar Brigade launched an assault on the islands with amphibious tanks, but the marsh reeds fouled the vehicles' propellers and made them vulnerable to Iranian antitank fire, forcing a retreat. After this setback, the Iraqi commander requested permission to use chemical weapons against the Iranians' fortified positions to break their morale, and Saddam Hussein gave his authorization. Once the special strike had been approved, the order was passed down the operational chain of command from the minister of defense to the chief of staff, the Army general headquarters, and finally the commanders in the field.

Although it was early spring, the heat on the Majnoon Islands was already of furnacelike intensity. Gusts of dry, dusty wind blew over the desolate expanse of marsh and sand. Advancing from the west, Iraqi commandos and helicopter gunships assaulted the dug-in Iranian positions. In the midst of the battle, four Iraqi fighter-bombers thundered over the Iranian trenches at a height of 750 feet and dropped a dozen 250-kilogram bombs. The weapons burst on impact with a muffled thump, some releasing a dirty-white vapor, others spewing a yellowish smoke with a sharp garlicky odor. Carried on the brisk wind, the toxic cloud enveloped the Iranian defenders before they had time to run. The Basij militia lacked any chemical protective gear, and although the Pasdaran carried gas masks, their heavy beards precluded an airtight seal. Finding shelter in bunkers and trenches was impossible because the poisonous vapor was heavier than air and pooled in low-lying areas.

Within minutes of exposure to the toxic cloud, the Iranian troops began to sweat profusely and gasp for breath. Their noses ran with thin, watery mucus and the pupils of their eyes narrowed to pinpoints, darkening and blurring their vision. Those who had absorbed a lethal dose began to twitch uncontrollably and fell to the ground, convulsed by violent spasms. Dark patches of urine and feces soaked through their uniforms. Finally they stopped breathing, although their hearts continued to beat for a few minutes before ceasing. In the aftermath of the attack, between fifty and a hundred Iranian dead lay scattered over the battlefield in grotesque postures, some arched in rigid contraction, others sprawled in flaccid paralysis.

The surviving Iranian troops, stunned and traumatized, retreated from the western half of the Majnoon Islands. Many of those on the periphery of the cloud had jabbed themselves in the thigh with an atropine autoinjector to counteract the lethal effects of the nerve agent. But because atropine also elevates core body temperature and dries out the skin and mucous membranes, dozens of soldiers who had survived the chemical attack later succumbed to dehydration and heatstroke. The casualties were taken to the chemical emergency unit at a nearby Iranian field hospital. From there, the seriously injured were evacuated to a recovery unit in the city of Ahwaz, a few hours' drive from the front, where they were given artificial respiration and intravenous infusions of atropine.

On March 5, 1984, the White House issued an official statement to the press that publicly criticized Iraq's renewed use of chemical weapons. This statement read, "The United States has concluded that the available evidence substantiates Iran's charges that Iraq has used chemical weapons. The United States strongly condemns the prohibited use of chemical weapons wherever it occurs. There can be no justification for their use by any country." Although senior Iraqi officials were surprised by the harshness of the U.S. statement, the Reagan administration never followed up its rhetoric with action.

IN RESPONSE TO the Iranian government's urgent pleas for assistance, U.N. Secretary-General Javier Pérez de Cuéllar dispatched a small team of international experts to Iran on March 13–19, 1984, to investigate the alleged chemical attacks. This team consisted of four medical and military specialists from Sweden, Spain, Australia, and Switzerland. After flying to Tehran, the U.N. experts traveled to two sites within five kilometers of the Iraqi border, east of the Majnoon Islands combat zone. Although they saw no fighting, they could hear the distant rumble of artillery fire.

Iranian officials showed the U.N. experts fragments of Iraqi chemical bombs and rocket casings recovered from the battlefield, as well as captured Iraqi gas masks and other protective gear that had been manufactured in Eastern Europe and bore Arabic script. The U.N. team also visited a military field hospital where some forty Iranian chemical casualties were being treated and examined six of the injured soldiers. Some of them had pinpoint pupils indicative of nerve agent exposure, while others had reddened and blistered skin caused by mustard agent. Finally, the U.N. experts inspected an unexploded bomb that Iranian officials claimed had been dropped by an Iraqi aircraft. Donning gas masks and rubber gloves, they drilled a hole in the bomb casing and extracted a sample of amber-colored liquid, which was sent for analysis at laboratories in Sweden and Switzerland. On March 26, 1984, the U.N. team reported that the Iraqi bomb had contained Tabun.

In response to the expert-group report, the U.N. Security Council voted unanimously to condemn the use of chemical weapons in the Iran-Iraq War, although it declined to mention Iraq by name. The Iranian government then sought to bolster its case in the court of international public opinion by arranging for the treatment of several young soldiers with severe mustard injuries at hospitals in Sweden and Austria. Covered with disfiguring burns and massive, fluid-filled blisters, the Iranian casualties were extensively photographed and interviewed by the European press. Although Tabun had actually inflicted far more deaths than mustard, the nerve agent casualties were not as "photogenic" for propaganda purposes. Iranian soldiers exposed to nerve agents were either dead or comatose or had recovered without visible injuries.

Much to Tehran's dismay, its propaganda campaign failed to arouse much moral indignation from the international community, let alone a concerted response. Fearing the consequences of an Iranian victory, Western countries declined to back up their criticism of the Iraqi chemical attacks with political or economic sanctions against Baghdad. Although Iran had not possessed a chemical arsenal when the war began, the tepid international response to its repeated pleas for help led the Iranian government to conclude that the only way to halt the Iraqi chemical attacks was to acquire the capability to retaliate in kind. In 1984, Tehran launched a crash chemical weapons development program. Much as Iraq had done earlier, Iranian officials hired Western European firms to build dual-use pesticide plants and purchased precursor chemicals, equipment, and technical expertise from a variety of commercial suppliers.

The United States did make a modest effort to stanch the flow of chemical weapons precursors to both Iraq and Iran. On March 2, 1984, U.S. customs officials learned that a chemical company in Nashville, Tennessee, was about to make a second large shipment to Iraq of potassium fluoride, a key ingredient in the manufacture of Sarin and other nerve agents. Alerted by Customs, FBI agents found and confiscated the shipment in a cargo warehouse owned by the Dutch airline KLM at New York's John F. Kennedy International Airport. The seventy-five drums of potassium fluoride were addressed to SEPP in Baghdad. At the end of March, the U.S. government introduced special licensing requirements for the sale to Iraq and Iran of dual-use industrial chemicals that could be diverted to make chemical weapons. West Germany also came under pressure to crack down on illicit exports. In August 1984, after *The New York Times* published a leaked CIA report describing the shipment by West German companies of entire pesticide plants to Iraq, Chancellor Helmut Kohl intervened to stop any further deliveries.

As MORE NATIONS joined the "chemical club," others felt vulnerable to attack, increasing the risk of further proliferation. In June 1984, Kenneth Adelman, the director of the U.S. Arms Control and Disarmament Agency, testified before a Senate hearing, "All too often in the past, the nuclear issue has so overshadowed as to drive out concerns on chemical weapons. I personally put the threat of a nuclear war very low, very low. I put the increasing use of chemical weapons around the world very high."

The Reagan administration's response to this threat was to seek funding from Congress for production of binary chemical munitions. After two failed attempts, the administration changed tactics in 1984 by linking the binary program to an arms control initiative, a traditional gambit to win congressional support for a controversial weapon system. At an international meeting in Stockholm in January, Secretary of State George Shultz announced a "two-track" policy that combined chemical modernization and disarmament. The United States, he said, would soon present a draft treaty text at the U.N. disarmament forum in Geneva (renamed the Conference on Disarmament), which had created a working group in March 1980 to negotiate a multilateral convention banning chemical weapons. Secretary Shultz stressed the importance of the binary weapons program as a source of negotiating leverage with Moscow. Because the Soviets were tough negotiators who rarely made concessions unless it was in their interest to do so, a U.S. decision to move forward with binary production would create a powerful "bargaining chip" to pressure the Soviets to negotiate in good faith.

President Reagan conveyed the same message at a press conference on April 4, 1984, when he observed, "We must be able to deter a chemical attack against us or our allies. And without a modern and credible deterrent, the prospects for achieving a comprehensive ban would be nil." The NATO supreme commander, General Bernard Rogers, used a more concise formulation: "Rearm now, so as to be able to disarm later." But critics of the binary program argued that resuming chemical weapons production after fifteen years would sacrifice an important U.S. propaganda advantage.

On April 18, 1984, Vice President Bush presented a draft of the Chemical Weapons Convention (CWC) to the Conference on Disarmament in Geneva. A striking feature of the U.S. draft was that it provided for "anywhere, anytime" inspections of facilities suspected of the illicit development, production, or storage of chemical arms. In fact, the Reagan administration was not prepared to accept such intrusive inspections on its own territory and made the proposal only because it knew that Moscow was certain to reject it, giving the Soviets a black eye in the struggle for international public opinion. Nevertheless, the U.S. draft became the basis of the "rolling text" of the CWC, which underwent a process of continuous modification over the course of the negotiations. Language that had not been adopted by consensus was placed in square brackets, sometimes with an explanatory footnote. Soon the rolling text was peppered with hundreds of brackets.

Meanwhile, Congress continued to address the binary weapons issue. The 1984 debate was largely a replay of the earlier debates in 1982 and 1983. The House considered a bill, backed by the Armed Services Committee, authorizing $95 million to buy production equipment for the DF plant at Pine Bluff Arsenal. But Representative Bethune of Arkansas introduced an amendment to delete the funding that passed by a vote of 247 to 179. Several factors contributed to the Reagan administration's large margin of defeat in the House. First, the fact that an archconservative like Bethune opposed the binary program made it easier for rank-and-file Republicans to vote against it. Second, the vote on binaries came one day after the House had endorsed the Reagan administration's MX nuclear weapons program, leading several congressmen who had supported the MX missile in the face of constituent opposition to cast a pro–arms control vote on nerve gas. Finally, the White House was distracted by the concurrent political battles over the MX missile and antisatellite weapons and did not devote enough time and effort to lobby House members on binaries.

The lopsided rejection of the binary weapons authorization in the House had a major impact on the subsequent legislative process in the Senate. On May 24, 1984, during the markup of the FY 1985 Defense Authorization Bill in the Senate Armed Services Committee, Chairman Tower reluctantly stripped the $95 million line item for binary production equipment from the bill before it went to the Senate floor, sparing his fellow

Republicans an unpopular vote in an election year. Senator Tower's action was a tacit admission that voting for binary weapons had become a political liability.

ON NOVEMBER 26, 1984, the Reagan administration quietly restored diplomatic relations with Baghdad after a hiatus of more than seventeen years, and proceeded to support the Iraqi regime with military intelligence, bank loans, and other forms of assistance. At the same time, Iraq continued to manufacture chemical weapons and employ them on the battlefield. SEPP produced sixty tons of Tabun in 1984, although the agent was about 50 percent pure and tended to deteriorate rapidly. Because impure Tabun was ineffective unless used in combat within four to six weeks of production, the filled munitions were shipped directly from the factory to frontline airfields and artillery units.

By 1985, SEPP, now called the Muthanna State Establishment, was a sprawling industrial complex of more than twenty manufacturing plants for chemical weapons intermediates and final products, along with filling lines for bombs, shells, rockets, and missile warheads. The Sarin production facility was five stories high and incorporated powerful ventilation and air filtration systems. To make aerial targeting more difficult, the research laboratories and production buildings were dispersed over an area of twelve square miles. In addition, the entire complex was surrounded by a double perimeter fence and defended by SA-2 antiaircraft missile batteries. Adjacent to the main production zone, West German companies had built eight large underground storage bunkers of reinforced concrete, with roofs thick enough to resist a direct hit with an aerial bomb.

The fact that Iraq had acquired most of its chemical production equipment and precursors from foreign suppliers made evident the need for stricter controls on the international chemical trade. Several Western countries tightened their export regulations, which required companies to apply for a license to ship dual-use chemicals, production equipment, or knowhow to suspected proliferators. Nevertheless, national export controls were often rendered ineffective when other suppliers undercut them. In April 1985, at the suggestion of Australia, fifteen Western industrialized countries founded an informal coordinating body called the Australia Group to "harmonize" their national chemical export regulations. The founding members of the group met for the first time in June 1985 at the Australian Embassy in Brussels and subsequently developed a common "control list" of dual-use precursor chemicals and production equipment. Members of the Australia Group also shared intelligence about countries suspected of seeking chemical weapons.

AFTER FAILING for three consecutive years to persuade Congress to fund the production of binary weapons, a senior Reagan administration official seemed ready to concede defeat. "Three strikes and you're out," he said. "We won't try again for binary weapons." Contrary to expectations, however, the White House did try again. In early 1985, the Reagan administration launched a full-court press to obtain $170 million in the FY 1986 defense budget for procurement of the M687 Sarin projectile and the Bigeye VX bomb. Administration officials calculated that because Congress was so closely divided, it would not take many votes to turn the situation around.

A series of Pentagon officials—binary program manager Robert Orton, Undersecretary of the Army James Ambrose, his deputy Amoretta Hoeber, and Chemical Corps chief Major General Gerald Watson—testified before the armed-services committees in support of the binary program. In addition, two more blue-ribbon panels weighed in. General Frederick J. Kroesen, who had recently retired as commander-in-chief of U.S. Army forces in Europe, chaired a classified study by twenty-one retired generals and admirals on the chemical warfare threat to NATO. This panel found that the chemical component of the alliance's "flexible response" strategy, which had been adopted in 1967 and remained in effect, had eroded to the point that the Soviets would be "militarily foolish" not to initiate the use of nerve agents in a future European war. Moreover, the billions of dollars being spent to upgrade NATO's conventional and nuclear forces were "hostage to the absence of a companion program modernizing our ability to survive and fight in a chemical environment." The Pentagon was pleased with the Kroesen committee's findings and paid an outside contractor $70,000 to prepare a declassified version of the report for distribution to all members of Congress.

The second blue-ribbon panel, the Presidential Chemical Warfare Review Commission, issued its final report on April 1, 1985, after several months of deliberation. Chaired by Walter J. Stoessel, Jr., a former Deputy Secretary of State, this committee found that although the 120,000 U.S. chemical weapons deployed in Europe would remain "serviceable" well into the 1990s, their replacement with binaries was warranted. The commission's main findings were that "modernization of the U.S. chemical weapon stockpile would not impede and would more likely encourage negotiations for a multilateral, verifiable ban on chemical weapons; that only a small fraction of the current stockpile has deterrent value; that the proposed binary program will provide an adequate capability to meet our present needs and is necessary; and that any expectation that protective measures alone can offset the advantages to the Soviets from a chemical attack is not realistic."

During a hearing of the Senate Armed Services Committee, critics attacked the credibility of the Stoessel report, noting that the supposedly impartial commission had included no known opponents of binary weapons and several outspoken advocates, such as former Secretary of State Alexander Haig, Jr., and John C. Kester, a senior aide to former Defense Secretary Harold Brown. Because of this apparent bias, Senator Carl Levin said that the Stoessel report was "not going to inspire a lot of confidence."

To provide some context for the Kroesen and Stoessel reports, Representative John Porter (R.–Illinois) asked the CIA to give classified briefings for House members on the Soviet chemical threat to NATO. Surprisingly, the available intelligence appeared to support the critics' position. The CIA briefers said that although the Red Army had accumulated large stockpiles of chemical arms, it had reduced its reliance on these weapons after recognizing that little military advantage could be gained by using them against NATO forces equipped with personal protective gear. This assessment differed sharply from that of the Kroesen report, which had found that the threat of Soviet chemical weapons was "serious" and the potential for their use in war "likely."

Opponents of the binary program also challenged the recommendations of the two blue-ribbon panels. On June 17, 1985, Representatives Porter and Dante Fascell (D.–Florida) published an opinion piece in *The Washington Post* titled "New Nerve-Gas Weapons That We Don't Need." The article began, "Would you support a new Pentagon program that adds billions of dollars to the $200 billion deficit, that has never been field tested because it has failed 80 percent of its controlled laboratory tests, that has been rejected by our closest allies in NATO, that if put into effect would kill civilians in droves while leaving protected enemy soldiers unharmed, and that makes chemical weapons proliferation and terrorist use more likely and arms control less so? Of course not."

When the FY 1986 Defense Authorization Bill was debated on the Senate floor, Senator John C. Danforth (R.–Missouri), an ordained Episcopal minister, argued that chemical weapons were immoral because they could not be aimed and would inevitably kill large numbers of innocent civilians. "The whole concept is abhorrent . . . to what Western values have stood for since Thomas Aquinas," he said.

Prior to the floor vote, Vice President Bush again took his seat as president pro tem of the Senate, prepared to break another tie as he had done twice in 1983. Much to his relief, he was not called upon to do so. By a margin of 50 to 46, the Senate rejected an amendment introduced by Senator Pryor to delete $163 million for production of the M687 projectile and the Bigeye bomb. The Stoessel report, combined with intense lobbying by the politically influential National Guard, had managed to persuade five senators who had previously opposed binary production to change their votes. As Vice President Bush left the Senate chamber, he grinned and said that his mother would "rest easy."

The action now shifted to the House, where the Reagan administration launched an intense lobbying campaign. On June 19, 1985, Representatives Ike Skelton (D.–Missouri) and John Spratt (D.–South Carolina) introduced an amendment to the Defense Authorization Bill allocating $124 million for binary weapons production. The recent retirement of Representative Bethune had weakened the opposition, and an international incident also influenced the vote. Less than a week earlier, on June 14, 1985, Hizbollah terrorists had hijacked TWA Flight 847 carrying 153 people from Athens to Rome and ordered the pilot to fly to Beirut, where they demanded the release of 766 prisoners held in Israel. Among the 104 American hostages were five U.S. Navy divers. The terrorists severely beat two of the divers and executed one of them, Robert Dean Stethem, with a shot to the head. They then dumped his lifeless body onto the tarmac. On June 18, Stethem's remains were flown back to the United States.

The next day, when the Skelton-Spratt amendment authorizing binary production came up for a vote, the House was in a defiant and angry mood. Fifty members changed their vote from the previous year, enabling the amendment to pass 229–196. The legislation placed two conditions on the release of FY 1986 funds: production of binary weapons could not begin until October 1987, and then only if NATO's supreme governing body, the North Atlantic Council, formally agreed to station the new U.S. chemical weapons on European soil. Representative Porter, who had led the failed opposition, predicted that the second condition would be "changed or put in the hands of conferees and lost."

Final passage of the Defense Appropriations Bill by both houses meant that after several years of acrimonious debate, Congress had "crossed the Rubicon" and approved the production of binary chemical weapons.

CHAPTER FOURTEEN

SILENT SPREAD

ALTHOUGH CONGRESS had finally approved production funds for the binary artillery shell, it had imposed a tough political condition on their release: President Reagan first had to obtain the agreement of the NATO foreign ministers to station the new U.S. weapons on Western European soil. Because this hurdle seemed nearly insurmountable, the Reagan administration persuaded Congress to modify the legislative condition so that the president was merely required to certify that the North Atlantic Council, NATO's highest political body, had endorsed the U.S. chemical modernization plan. Still, given the extreme unpopularity of chemical weapons in Western Europe, even this watered-down condition would be difficult to satisfy.

Seeking to avoid a fractious debate in the North Atlantic Council, the U.S. ambassador to NATO, David M. Abshire, decided to raise the issue of binary weapons in a lower-level body called the Defense Planning Committee (DPC), composed of the permanent representatives to NATO of the fifteen nations participating in the alliance's military structure. (This committee had been created to discuss NATO military issues without the participation of France after Paris withdrew from the integrated military command.) The U.S. gambit was successful: on May 15, 1986, the DPC endorsed the NATO "force goal" for chemical weapons, and on May 22, the allied defense ministers meeting as the DPC simply "noted" the permanent representatives' action without debate.

On July 29, 1986, President Reagan sent a letter to the Speaker of the House stating that the administration had satisfied the legislative condition set by Congress and that production of the M687 binary projectile would proceed. But congressional opponents of the binary program protested that Ambassador Abshire's "end run" around the North Atlantic Council had subverted the will of Congress. Representative Fascell and Senator Pryor prepared a legal brief arguing that because the "wrong" NATO body had approved the U.S. chemical modernization plan, the political condition in the appropriations bill had not been met. When this tactic failed, the binary opponents in the House tried to amend the FY 1987 Defense Authorization Bill to block production of the weapons. But Representative Les Aspin, the Democratic chairman of the House Armed Services Committee, caved to Senate negotiators in conference committee and allowed the binary program to proceed.

Despite the DPC's nominal endorsement of the NATO "force goal," the Western European allies viewed the U.S. chemical modernization plan with extreme distaste. Ironically, the price for their acceptance of the binary program turned out to be a U.S. pledge not to station the new weapons on European soil in peacetime. Instead, the binary components would be stored at bases in the continental United States and airlifted to the host countries during a crisis or war. Once arrived at their destinations, the binary shells would be assembled and deployed, either for deterrence or to force the enemy to halt his first use of chemical weapons.

West German Chancellor Helmut Kohl imposed another condition on Bonn's willingness to accept U.S. binary weapons in wartime: he demanded the withdrawal of the roughly 120,000 unitary chemical weapons that the United States had stationed for decades on West German soil. When President Reagan agreed to this demand, conservative policy analysts excoriated the administration for having "shot itself in the foot." Removing all U.S. chemical weapons from West Germany without replacing them would effectively create a "chemical-weapons-free zone" in Western Europe, weakening NATO's ability to deter a Soviet chemical attack.

Administration officials insisted that even if binary chemical weapons were stationed on U.S. soil in peacetime, they would still constitute a credible deterrent. But critics pointed out other problems with the wartime deployment plan. First, air-lifting the stockpile of binary weapons to Europe would require sixteen days of transatlantic flights by the entire U.S. fleet of C-141 Starlifter transport planes, displacing more urgent cargo such as troops and conventional munitions. Second, a massive airlift of U.S.

binaries to Western Europe during a crisis might lead Moscow to believe that Washington was preparing a chemical attack, causing the Soviets to use their own weapons preemptively.

Over the next few years, congressional opponents tried to cut appropriations for binary weapons production and worked with the General Accounting Office (GAO), the investigative arm of Congress, to expose technical and managerial problems with the program. In June 1986, Representative Fascell made public a GAO report stating that recent tests of the Bigeye bomb had left numerous operational problems unresolved. The analysis concluded that the weapon "should either move back to developmental and chemical testing . . . or should be abandoned in favor of newer concepts."

France, meanwhile, decided in 1986 to develop a binary chemical warhead for its multiple-launch rocket system under a program called ACACIA, an acronym for Armement Chimique Adapté pour Contrer les Intentions Agressives ("chemical weapon designed to counter aggressive intentions"). The proposed French military budget for the five-year period 1987–91, released in early November 1986, included funds for procurement of the binary warhead, which was justified with the argument that "France cannot renounce definitively those categories of armament that other nations claim the right to possess." For many years, the French armed forces had pressured their political leaders to procure a "minimal" stockpile of nerve agent weapons as a deterrent. France's lack of a chemical retaliatory capability, they argued, created a military imbalance in the European theater that risked lowering the nuclear threshold in the event of war between NATO and the Warsaw Pact. In 1987, however, French President François Mitterrand suspended the ACACIA program in the development phase because his government intended to sign the Chemical Weapons Convention, which was nearing completion in Geneva.

THROUGHOUT MOST OF THE 1980s, the Iran-Iraq War continued to rage. In December 1986, Saddam Hussein authorized the Iraqi Third and Seventh Army Corps and the Iraqi Air Force to employ chemical weapons without obtaining his prior approval. When the field commanders balked at this unexpected change in policy, Saddam traveled to the front in January 1987 to confirm the order in person. Once he had done so, the Iraqi commanders took full advantage of their new authority, and their chemical attacks against the poorly protected Iranian forces became increasingly effective. The Iraqi Chemical Corps integrated chemical shells, bombs, and rockets into the fire plan for large military operations, using nerve agents to attack Iranian reinforcements, forward defenses, command posts, artillery positions, and logistics facilities.

Despite Saddam Hussein's flagrant violations of the Geneva Protocol, the Reagan administration viewed the secular Iraqi regime as a necessary bulwark against Iran's militant Islamic ideology and began to provide Baghdad with military advice and logistical support. Washington also encouraged friendly Arab states to sell military equipment to Iraq. The goal of this policy was to ensure a protracted stalemate between Iraq and Iran, so that neither country would pose a threat to Israel or to U.S. oil interests in the Persian Gulf. An eventual victory by a weakened Iraq was also seen as an acceptable outcome. In a highly classified program, more than sixty officers from the Pentagon's Defense Intelligence Agency (DIA) provided Iraqi commanders with detailed tactical intelligence on Iranian troop movements, including reconnaissance satellite images, and assisted the Iraqi military with tactical planning for land battles, air strikes, and bomb damage assessment. The American officers participating in the DIA assistance program were not particularly disturbed by Iraq's extensive use of chemical weapons, which they saw as simply another way of killing the enemy.

After 1986, Iraq halted the manufacture of Tabun at the Muthanna State Establishment and concentrated exclusively on Sarin. Egyptian chemical weapons experts provided assistance, enabling Iraq to increase its Sarin output from only 5 tons in 1984 to 200 tons in 1987 and 390 tons in 1988. The nerve agent was loaded into aerial bombs and 122 mm artillery rockets with a range of 25 kilometers. Because Iraq had never mastered the process of distilling Sarin, however, the purity of the agent ranged between 60 and 70 percent, resulting in its rapid deterioration. After three months in storage, the purity level usually dropped below 40 percent, the cutoff for filling munitions. Iraqi Sarin therefore had to be consumed on the battlefield almost as fast as it was produced. As Muthanna's output of Sarin increased, so did the amount Iraq employed in the war against Iran. The annual consumption of Sarin-filled artillery rockets, for example, jumped from 1,200 rounds in 1986 to 15,000 rounds in 1987.

In addition to Iranian troops, the victims of Iraq's chemical weapons included political prisoners, mostly Kurds and Shi'ites. A mysterious entity known as Unit 2100, reporting directly to the Iraqi Ministry of Military Industry, ran a secret chemical weapons testing facility near the village of Al-Haditha, in a remote area of Iraq's western desert. Unit 2100 reportedly conducted experiments with human guinea pigs, and no prisoners sent to Al-Haditha ever returned alive. The Iraqi security services also used nerve agent for at least one mass execution in 1987. Troops from the Second Army Corps transported ten truckloads of political prisoners to a remote gulch near the town of Jalula on the Iranian border. As an Iraqi intelligence officer watched from a distance, a misty white cloud rose from the gulch. A few hours later, the trucks rumbled past in the opposite direction, piled with dead bodies; the corpses were unmarked and looked as if they were asleep. When the security agents escorting the convoy saw the intelligence officer, they angrily ordered him to leave the area immediately. By then he understood that what he had witnessed had been no ordinary execution. The dead were reportedly buried in a mass grave near the town of Ba'qubah in east-central Iraq.

In response to tightened international controls on the export of nerve agent precursors to Iraq, Baghdad diversified its foreign suppliers and began to pursue an indigenous capability to manufacture its own chemical intermediates so that it would no longer be vulnerable to cutoffs in supply, a strategy known as "back integration." Between 1986 and 1988, the Iraqi government contracted with two West German companies to build chemical plants at Fallujah, sixty kilometers west of Baghdad, for converting elemental phosphorus from the giant phosphate mine at Akashat into phosphorus trichloride.

Iraq also acquired more persistent types of nerve agents. Because Sarin dissipated rapidly in the intense heat of the Iraqi desert, it was difficult to generate the concentrated clouds of vapor needed to inflict heavy casualties on Iranian troops. Accordingly, Iraqi military chemists at the Muthanna State Establishment began to develop less volatile agents, such as Soman. Iraq's plan to manufacture Soman was foiled, however, by its failure to find a supplier of pinacolyl alcohol. As an alternative, Iraq chose to make Cyclosarin (GF), an analogue of Sarin whose lower volatility made it superior for use in hot climates. The production process for Cyclosarin was identical to that of Sarin except for the replacement of isopropyl alcohol with cyclohexyl alcohol, which was readily available from the Iraqi petrochemical industry. Iraqi military chemists also mixed Sarin and Cyclosarin together to form a "cocktail" that was more toxic than either agent separately. In 1987, Muthanna began to produce Sarin and Cyclosarin together in the same vessel by using a mixture of alcohols.

Another persistent nerve agent developed by Iraq was VX. Although chemists at Al-Rashad had done preliminary research on V agents in 1975–76, serious development work on VX did not begin at Muthanna until 1985, under the direction of Dr. Emad Husayn Abdullah Ani, an Iraqi chemist who had studied at the Timoshenko Academy of Chemical Defense in Moscow. In 1987, the director-general of Muthanna, General Nazar al-Khazarji, wrote a top secret letter to senior Iraqi officials in which he compared VX to a nuclear weapon. Two metric tons of the nerve agent delivered by aircraft, he claimed, could kill as many people as the atomic bomb that destroyed Hiroshima. Accordingly, the general observed, acquiring the ability to mass-produce VX would usher Iraq "into the [field] of armament of advanced countries."

A pilot plant at Muthanna produced a total of 2.4 tons of VX in five production runs between late 1987 and the end of May 1988, although the purity of the agent was only about 50 percent. Because VX could not be distilled, agent that was less than 90 percent pure was unstable and had a limited shelf life: Iraqi VX stored in bulk containers or filled munitions deteriorated to the point of nonusability within a few weeks. Toward the end of the Iran-Iraq War, Iraq loaded some bombs with VX but apparently did not use them.

Iran, for its part, had decided in 1984 to acquire its own chemical warfare capability after enduring numerous Iraqi chemical attacks and attempting without success to persuade the international community to enforce the Geneva Protocol. With assistance from Western European companies, Tehran built factories for the production of mustard, phosgene, and hydrogen cyanide, which were loaded into bombs and artillery shells. Even after Iranian leaders had acquired a stockpile of chemical weapons, they debated whether to employ them. Ayatollah Khomeini was opposed to chemical warfare on religious grounds, noting that the Koran forbade the use of poisoned weapons. But as Iraq's chemical attacks intensified, the Iranian military put increasing pressure on Khomeini to authorize retaliatory strikes, and in 1987 he finally relented and issued a secret order. Although the Iranians employed chemical weapons sporadically in 1987 and 1988, their lack of training and experience prevented these attacks from having any real military impact. Toward the end of the war, Khomeini decided to halt the use of chemical weapons and deny any previous attacks so as to regain the moral high ground vis-à-vis Iraq. Nevertheless, the Iranian leadership did not abandon its pursuit of chemical weapons but sought to intensify its efforts and acquire a militarily significant stockpile for future use.

ON APRIL 10, 1987, Soviet President Mikhail Gorbachev, who had taken office in 1985, declared publicly that the Soviet Union was ending all production of chemical weapons and would henceforth convert its existing military chemical facilities to civilian purposes. Despite Gorbachev's pledge, however, the secret development and testing of Novichok binary agents continued. In 1986, the Soviets had built the Chemical Research Institute at a closed military complex in Nukus, Uzbekistan. This facility employed 300 people and included laboratories, a pilot production plant for Novichok agents, a munitions filling line, and a large test chamber in which dogs were exposed to toxic agents through gas masks placed over their muzzles.

In May 1987, a serious accident occurred at GosNIIOKhT in Moscow. Andrei Zheleznyakov, an experienced military chemist at the institute, was conducting an experiment under a fume hood in which he combined the binary precursors of Novichok-5 inside a small stainless-steel reactor and measured the reaction temperature. Previous experiments had shown that the higher the temperature, the greater the purity of the end product. Using a syringe connected to a flexible metal tube, Zheleznyakov drew samples from the reactor for analysis. Suddenly overcome by a spell of intense dizziness, he saw that the tube had become disconnected from the syringe and was leaking invisible fumes into the air. He quickly sealed the leak, but his ears were ringing and orange spots flashed before his eyes. Paralyzed with fear, he murmured, "Guys, I think it's got me."

Zheleznyakov's coworkers quickly took him outdoors for some fresh air and gave him a shot of vodka. When he returned to the laboratory, he looked pale and drawn, and his section chief told him to go home and rest. A colleague escorted him to the bus stop. As he waited for the bus, Zheleznyakov experienced a hallucination in which the onion-domed church across the way suddenly glowed brightly and broke up into a thousand swirling pieces. He fainted and collapsed on the sidewalk, and a friend brought him back to GosNIIOKhT. There the KGB security detail called an ambulance and followed in a car as the emergency vehicle, sirens blaring, carried the stricken chemist to the Sklifosovsky Institute for Emergency Care, the leading poison center in the Soviet Union.

By the time Zheleznyakov arrived at the hospital, his breathing was labored, his heart was barely beating, and his nervous system was shutting down. The KGB escorts told the admitting physician, Dr. Yevgeny Vedernikov, that the patient was suffering from food poisoning caused by the ingestion of bad sausage. They then made the doctor sign a security form pledging never to discuss the case in public. Although Dr. Vedernikov did not know the exact cause of the poisoning, a blood test revealed that Zheleznyakov's cholinesterase level was close to zero. By treating the chemist symptomatically with atropine and other antidotes, Vedernikov managed to save his life.

For the next ten days Zheleznyakov remained in critical condition, oblivious to his surroundings, and only gradually regained consciousness. After another week of intensive care in Moscow, he was transferred to the Institute of Labor Hygiene and Occupational Pathology in Leningrad. This hospital had a top secret ward for the treatment of nerve agent injuries known as the Special Department for Foliant Problems. (The Leningrad institute, together with its affiliates in Volgograd and Kiev, was part of a closed system of classified medicine under the Third Main Administration of the Soviet Ministry of Health.) Unable to walk, Zheleznyakov remained at the Leningrad clinic for three months. Although he gradually improved, he suffered from chronic weakness in his arms, a toxic hepatitis that gave rise to cirrhosis of the liver, epilepsy, spells of severe depression, and an inability to read or concentrate that left him totally disabled and unable to work. He died five years later, in July 1992. The devastating consequences of Zheleznyakov's exposure to a whiff of Novichok-5 demonstrated the extraordinary toxicity of the Foliant nerve agents.

GORBACHEV'S "PEACE OFFENSIVE" included a dramatic new initiative in the field of chemical arms control. In January 1986, the Soviet leader had agreed to all of the basic elements of the CWC, with the sole exception of "challenge" inspections of suspect sites. On August 6, 1987, Soviet Foreign Minister Eduard Shevardnadze addressed the Conference on Disarmament in Geneva. Much to the surprise of the other delegations present, he said that the Soviet Union could now accept the principle of mandatory challenge inspections without the right of refusal. In effect, Gorbachev was calling the Reagan administration's bluff by adopting the earlier U.S. proposal for "anywhere, anytime" inspections of suspect sites. This provision had been included in the U.S. draft treaty that Vice President Bush had presented in Geneva on April 18, 1984. Because the Pentagon refused to accept "anywhere, anytime" inspections of its own facilities, the sole purpose of the U.S. proposal had been to embarrass the Soviet Union. Now, however, Gorbachev's act of "diplomatic jujitsu" turned the tables and put the United States in the awkward position of having to back away from its own proposal. U.S. diplomats scrambled to weaken the challenge-inspection provisions in the draft treaty, much to the irritation of Britain and other countries.

Gorbachev followed up this public relations coup with another stunning gesture. In a demonstration of the Soviet leader's new policy of *glasnost*, or "transparency," the Red Army held an "open house" at the Central Military

Foreign observers examine a static display of Soviet chemical munitions during an "open house" at the Central Military Chemical Testing Site in Shikhany, Russia, on October 4, 1987.

Soviet technicians demonstrate the operation of a mobile nerve-agent-destruction unit at the Shikhany military base in Russia on October 4, 1987. Former chemical weapons scientist Vil Mirzayanov claims that the process did not work and the demonstration was faked.

Chemical Testing Site at Shikhany on October 3 and 4, 1987. The invited guests, including diplomats and military observers from forty-five countries, U.N. representatives, and journalists from Soviet and foreign publications, were flown to a military airfield 900 kilometers southeast of Moscow in the vast emptiness of the Russian steppe. They were then transported by bus to the nearby town of Shikhany, a cluster of modern, five-story apartment blocks that housed the five thousand people who worked at the testing site.

Hosting the open house were Colonel General Vladimir Karpovich Pikalov, the commander of the Soviet Chemical Troops, and his deputy, Lieutenant General Anatoly Kuntsevich. A hard-line Communist and heavy drinker, Kuntsevich was an old-school Soviet bureaucrat with a reputation for thuggishness. His face was sallow and deeply furrowed, and the sides of his mouth were turned down in a permanent frown. Like many Soviet officials of his generation, Kuntsevich was accustomed to double-dealing: at the same time that he was overseeing the development of new chemical weapons at Shikhany, he served on the Soviet delegation to the chemical disarmament talks in Geneva.

At the Shikhany testing site, the international guests were escorted through a series of military checkpoints to the demonstration area, where they toured a static display of the nineteen types of chemical munitions in service with the Soviet Chemical Troops. The exhibit included mockups of artillery shells and rockets, missile warheads, bombs, aircraft spray tanks, and even chemical hand grenades. As the foreign diplomats and journalists filed past the display, Soviet military briefers calmly described the specifications and function of each munition, including caliber, chemical fill, and detonator type. According to an article in *International Defense Review,* "the visit was very carefully orchestrated and the visitors were shown a well arranged array of hardware and given precisely measured amounts of information. All supplementary questions were greeted with a polite but firm refusal to add any further detail." Despite the tightly controlled nature of the event, some of the Soviet officers were visibly uncomfortable with the presence of foreigners and journalists at the formerly top secret facility. One general admitted to a reporter that the experience was "rather like taking one's trousers off in public."

When General Pikalov was asked whether the static display included every type of munition in the Soviet chemical arsenal, he replied, "We displayed *all* our toxic agents and *all* our chemical munitions, with the exception of certain modified types that are not fundamentally different in terms of apparatus or armament from those that were shown." He also insisted that the Soviet Union did not possess "American-type binary" munitions and that Moscow had never transferred chemical weapons to other states. Despite the claims of full transparency, however, the Soviets concealed several aspects of their chemical arsenal. No mention was made during the open house of the Foliant nerve agents or the Novichok binary formulations then under development. Moreover, although only four types of V-agent munitions were displayed at Shikhany, the Novocheboksarsk factory actually manufactured at least fourteen different types, and the display of Soman-filled weapons was also incomplete.

The open house at Shikhany included the demonstration of a new Soviet technique for the chemical neutralization of nerve agents that had been developed for use at a planned chemical weapons destruction facility in Chapayevsk, near Moscow. Inside an airtight enclosure, chemical troops wearing gas masks and protective suits opened the filling port in the side of a 250-kilogram bomb containing thickened Sarin. One soldier carefully drew a small amount of the nerve agent into a syringe and injected it into a laboratory rabbit. Several minutes later, the animal went into convulsions and died. The soldier then removed a larger sample of Sarin, placed it in a chemical reaction vessel, and added a neutralizing solution. After two hours, the vessel was opened and a sample of the mixture injected into a second rabbit. This time the animal remained unharmed, indicating that the nerve agent had been destroyed. Because the Soviet neutralization method did not work effectively, however, the second part of the demonstration had reportedly been faked.

On December 16, 1987, the United States began to manufacture the various hardware and liquid components of the M687 binary Sarin artillery shell. For logistical and safety reasons, the production process involved facilities in three different states. At Pine Bluff Arsenal in Arkansas, the Integrated Binary Production Facility was under construction. The first element to become operational was an open-air plant for the conversion of dichlor into difluor (DF). Because no commercial chemical company was willing to

This manufacturing plant for the binary Sarin component DF was part of the Integrated Binary Production Facility at Pine Bluff Arsenal in Arkansas, which operated from December 1987 to the end of 1990.

supply dichlor to the Army, a contract was awarded in January 1988 to build a dedicated production facility for this key intermediate at Pine Bluff. Until the dichlor plant went on line, enough of the chemical had been stockpiled at Rocky Mountain Arsenal to meet the Army's immediate DF production needs.

In Van Nuys, California, the Marquardt Corporation manufactured two types of plastic-lined steel canisters designated M20 and M21, which were each about the size of a coffee can. The M20 canisters were shipped to Pine Bluff Arsenal, where they were filled with DF and stored on-site. Meanwhile, Marquardt loaded the M21 canisters with the second binary component, code-named "OPA"—a mixture of isopropyl alcohol, a stabilizer, and a catalyst—and shipped the filled canisters to the Louisiana Army Ammunition Plant near Shreveport, which manufactured the steel artillery projectiles. There the M687 shell bodies were packed with only the M21 canisters installed, and shipped to Tooele Army Depot in Utah and Umatilla Army Depot in Oregon. Filling and storing the M20 and M21 canisters separately precluded the accidental mixing of the binary precursors.

In wartime, the artillery shells and the binary canisters would be airlifted directly from their storage sites in the United States to deployment areas in Western Europe for possible use. To prepare the M687 projectile for firing, the M21 OPA canister would be removed from its storage position inside the shell and the two canisters reinserted in the correct sequence: the M20 DF canister in front and the M21 OPA canister behind, with the thin steel burst discs in contact. Then a soldier would seal the back of the artillery shell and screw an impact fuse into the nose. When the projectile was fired from a howitzer, the intense setback forces would rupture the burst discs between the canisters, allowing DF and OPA to mix and react to form Sarin during the several seconds it took for the shell to fly to the target. The nerve agent would then be dispersed explosively on impact.

IN LATE 1987, Saddam Hussein launched a brutal military campaign against the Kurdish minority of northern Iraq. The 3.5 million Iraqi Kurds belonged to a nation of some 20 million people who inhabited a broad swath of territory in the Middle East, including portions of Turkey, Syria, Iraq, and Iran. Long victims of military and political repression in their host countries, the Kurds dreamed of an independent homeland that they called

"Kurdistan." During the Iran-Iraq War, groups of Kurdish guerrillas known as *peshmerga* formed a loose alliance with Iran to fight against the Iraqi regime in Baghdad.

Saddam Hussein gave the task of putting down the Kurdish rebellion to his cousin Ali Hassan al-Majid, the commander of the northern region of Iraq. With Saddam's blessing, Ali Hassan launched a program of violent repression against the Iraqi Kurds. Called the Anfal campaign, it involved mass executions and attacks with chemical weapons against hundreds of Kurdish villages, with the aim of terrorizing and depopulating the rural areas controlled by the rebel groups. At a meeting of the ruling Ba'ath Party in Baghdad, Ali Hassan—later nicknamed "Chemical Ali"—was tape-recorded discussing the anti-Kurd campaign. "I will kill them all with chemical weapons," he boasted. "Who is going to say anything? The international community? F—— them!"

The most devastating Iraqi chemical attack against the Kurds took place in mid-March 1988 in the city of Halabja, about 150 miles northeast of Baghdad. (This operation was not technically part of the Anfal campaign, which focused on repressing Kurds in rural areas of northern Iraq.) A bustling market town whose population had been swollen by an influx of refugees to some 50,000 inhabitants, Halabja was a warren of flat-topped buildings and unpaved streets the color of dried mud, set in the green foothills of a snowcapped mountain range. The city's residents had the misfortune of living about eleven miles west of the Iranian border, on the edge of the war zone. In the spring of 1988, they were trapped between the Iraqi forces defending Darbandikhan Lake (whose dam controlled part of the water supply for Baghdad) and the Iranian forces attacking down from the mountains.

On March 13, 1988, a joint force of Iranian Revolutionary Guards (Pasdaran) and Kurdish *peshmerga* guerrillas launched an offensive near Halabja in an attempt to penetrate deep into Iraqi-held territory. Over the next three days, the Pasdaran forces advanced to the eastern edge of Darbandikhan Lake, causing the Iraqi defenders to withdraw to the opposite shore. Heavy Iranian shelling also forced the evacuation of several Iraqi military posts between the border and Halabja, enabling Pasdaran soldiers to infiltrate the city. At this point, Saddam Hussein, seeking to repel the Iranian advance and strike a crushing psychological blow against the Kurdish *peshmerga* and their civilian supporters, authorized a "special strike" against Halabja.

The Iraqi counterattack began on the morning of March 16 with an artillery barrage on the city. Then waves of Iraqi Mirage fighter-bombers flew over Halabja, dropping high-explosive and incendiary munitions that shattered windows and darkened the sky with roiling clouds of black smoke. Thousands of city residents crowded into government air-raid shelters and private basements to wait out the attacks. The air raids continued for several more hours, each new wave of fighter-bombers following immediately upon the last.

After a brief period of calm in early afternoon, six warplanes returned at about 3:00 p.m., flying so low that observers on the ground could make out the Iraqi flags painted on the undersides of the wings. This time the aircraft dropped bombs that burst with a muffled thump and spewed dirty white clouds reeking of garlic, sweet apples, perfume, and gasoline. As the lethal mist spread rapidly through the city streets, birds fell out of trees, livestock collapsed, and families hiding in basements and air-raid shelters began to choke, vomit, and struggle for breath. According to an eyewitness account in the Human Rights Watch report *Genocide in Iraq*:

> In the shelters, there was immediate panic and claustrophobia. Some tried to plug the cracks around the entrance with damp towels, or pressed wet cloths to their faces, or set fires. But in the end they had no alternative but to emerge into the streets. It was growing dark and there were no streetlights; the power had been knocked out the day before by artillery fire. In the dim light, the people of Halabja could see nightmarish scenes. Dead bodies—human and animal—littered the streets, huddled in doorways, slumped over the steering wheels of their cars. Survivors stumbled around, laughing hysterically, before collapsing. Iranian soldiers flitted through the darkened streets, dressed in protective clothing, their faces concealed by gas masks. Those who fled could barely see, and felt a sensation "like needles in the eyes."

After the first wave of Iraqi chemical attacks, thousands of survivors began to flee the city under a freezing rain, following roads that led to the highlands and the Iranian border. Many had lost their shoes and walked barefoot through the snow and mud. Every ten minutes or so, another wave of Iraqi planes flew over, strafing the refugees with machine-gun fire and dropping bombs. Those who had been exposed to sublethal doses of mustard or nerve agent suffered from blurred vision and shortness of breath, and their symptoms continued to worsen. Groups of refugees who had lost their eyesight formed human chains with belts and ropes so that no one would get lost. During the night, many children died of exposure, and their grieving parents abandoned their small bodies by the side of the road.

At dawn, fearing more air raids, the Kurdish refugees left the main roads and dispersed into the mountains despite the grave danger of land mines. When a group of Halabja survivors finally reached the swift-moving river that marked the border with Iran, the Iraqi air attacks continued. As the panicked refugees crossed a narrow pontoon bridge spanning the river, the pressure of the crowd caused dozens of people to fall into the ice-cold water, and the raging current swept several young children away to their deaths. At another point along the Iranian border, some six thousand refugees from Halabja congregated near two ruined Kurdish villages. Iranian doctors arrived in helicopters and administered atropine to the survivors before ferrying them across to safety. The victims were given medical attention and then transported to two refugee camps, where they would remain until the end of the Anfal campaign.

The Iraqi chemical attacks on Halabja killed between two thousand and five thousand people and injured another ten thousand, many of whom suffered chronic medical symptoms and deep psychological trauma. When news of the attack reached the outside world, Iraqi officials blamed Iran for the atrocity, but the evidence pointed clearly to Saddam Hussein. In the course of routine electronic surveillance, U.S. signals-intelligence satellites had intercepted radio communications between Iraqi fighter-bombers and ground controllers coordinating the chemical attack.

The Iraqi Army did not attempt to recapture Halabja immediately, leaving the city under de facto Iranian occupation for about five months. Seeking to exploit the chemical attack for propaganda purposes, the Iranian authorities bused dozens of foreign journalists, photographers, and television crews from Tehran to the stricken city. As artillery fire continued to echo around the hills, the silent streets of Halabja were filled with the sickly sweet odor of decomposing bodies. Scores of corpses—of men, women, children, livestock, and pets—lay sprawled on the earthen streets and sidewalks where they had fallen, sometimes clustered into piles, with no visible wounds. David Hirst, the Mideast correspondent for *The Guardian* of London, filed the following eyewitness report:

The skin of the bodies is strangely discolored, with their eyes open and staring where they have not disappeared into their sockets, a grayish slime oozing from their mouths and their fingers still grotesquely twisted. Death seemingly caught them almost unawares in the midst of their household chores. They had just the strength, some of them, to make it to the doorways of their homes, only to collapse there or a few feet beyond. Here a mother seems to clasp her children in a last embrace, there an old man shields an infant from he cannot have known what.

Many unanswered questions remained. Journalists who went to Halabja found no fragments of chemical munitions, presumably because they had been confiscated by Iraqi or Iranian troops. Moreover, some photos of the dead that appeared in the world press gave the impression that the corpses had been arranged to create a more horrific impression. Also puzzling was the fact that many of the dead had blood running from their ears and noses (normally a sign of blast impact) and bright blue lips suggestive of hydrogen cyanide poisoning. A possible explanation of the latter was that the Iraqis had employed impure nerve agent that was heavily contaminated with

Civilian victims in the streets of the Kurdish city of Halabja after Iraqi aircraft dropped nerve-agent-filled bombs on March 16, 1988.

cyanide. One of the reaction steps in the manufacture of Tabun involves sodium cyanide, and Iraqi military chemists reportedly had difficulty removing the excess cyanide from the final product. Given the nearly instantaneous lethality of the chemical attack, however, some experts believed that the Iraqis had used a more volatile and fast-acting nerve agent, such as Sarin.

At the request of ten countries, U.N. Secretary-General Pérez de Cuéllar asked the government of Saddam Hussein to permit an impartial investigation of the Halabja attack, but a senior Iraqi official refused. "This is a question of sovereignty," he said, "and therefore I do not think we are going to accept that." At the same time, the Reagan administration suggested that Iran as well as Iraq might have been responsible for the gassing of Halabja. This allegation, made on the basis of flimsy evidence, had the effect of reducing the political pressure on Baghdad.

AFTER THE ATTACK on Halabja, the Iran-Iraq War ground on inconclusively. Although the U.N. Security Council had passed Resolution 598 in July 1987 calling on both sides to stop fighting and withdraw to their prewar borders, Tehran had refused to accept a cease-fire in the belief that it could still prevail militarily. During the spring of 1988, however, Iran suffered a series of grave military setbacks. The first came in mid-April, when Iraq launched Operation Ramadan to recapture the Al-Fao Peninsula, which provided access to the Persian Gulf and had been one of Iraq's major oil-exporting ports before the war. Iraqi aircraft and helicopters dropped Sarin-filled bombs on Iranian frontline troops, opening the way for an amphibious attack from the sea and a massive land assault by Iraqi regular forces and Republican Guard units, equipped with two thousand tanks and six hundred heavy guns.

After the Iraqi victory on April 18, 1988, a U.S. Air Force officer named Lieutenant Colonel Rick Francona, who served with the U.S. Embassy in Baghdad as liaison to the Iraqi military intelligence service, visited the Al-Fao battlefield. The once-lush peninsula had been transformed into a cratered moonscape, and areas were cordoned off with colored warning flags to indicate persistent chemical contamination. Scattered over the ground were dozens of atropine autoinjectors that Iranian soldiers had used after an Iraqi Sarin attack.

During the late spring of 1988, Iraq and Iran launched hundreds of conventionally armed ballistic missiles at each other's population centers. Known as the "war of the cities," this campaign involved the most extensive military use of missiles since the German V-1 and V-2 attacks on southern England during World War II. If Ayatollah Khomeini refused to accept a cease-fire, Saddam Hussein planned to attack Tehran with conventional bombs and missiles to shatter windows throughout the Iranian capital. He would then follow up with a barrage of chemical-tipped Scuds, creating clouds of Sarin vapor that would penetrate into homes and offices and kill thousands of civilians. Although Iran did not have its own nerve-agent-filled warheads with which to retaliate, it did have chemical bombs that could be dropped from aircraft. To prepare for an Iranian retaliatory strike, the Iraqi Ba'ath Party organized a large-scale civil defense exercise in Baghdad that involved the evacuation of two entire city districts. Saddam was clearly willing to accept a heavy toll in both Iranian and Iraqi civilian casualties to end a war that threatened his regime.

Although an Iraqi chemical strike on Tehran never materialized, the specter of missiles armed with nerve agent warheads caused widespread panic in the Iranian capital, triggering a mass evacuation in which some two million people abandoned their homes. The pervasive terror inspired by the Iraqi chemical threat, combined with a deep sense of war weariness, led to the collapse of civilian morale and helped to convince the Iranian leadership that the war could not be won. On August 20, 1988, Ayatollah Khomeini finally accepted a humiliating cease-fire under the terms of U.N. Security Council Resolution 598, an act that he compared to drinking from "a poisoned chalice."

After eight years of bloody attrition, neither Iraq nor Iran could claim victory: the border dispute between them was unresolved and both countries had suffered devastating human and financial losses. Iraqi chemical attacks had inflicted some fifty thousand Iranian casualties, of which about five thousand had been fatal. Even so, these figures were only a small fraction of the more than one million military and civilian casualties on both sides.

The Anfal campaign continued even after the end of the Iran-Iraq War, forcing tens of thousands of Kurdish refugees to flee into southeastern Turkey. In July 1988, the Iraqi Army retook Halabja and razed nearly every building in the city with dynamite and bulldozers, leaving the bodies of the dead to rot where they had fallen.

In October 1988, Physicians for Human Rights (PHR), a nonprofit humanitarian organization, sent an investigation team to northern Iraq to interview and examine Kurdish refugees who had witnessed or experienced Iraqi chemical attacks. Refugees from several villages reported that low-flying jets had dropped bombs that produced lethal clouds. The first animals to die had been birds and domestic fowl, followed by sheep, goats, cows, and mules; humans close to the bomb bursts had succumbed in minutes. Based on this testimony, the investigators concluded that the Iraqi chemical strikes had involved mustard and at least one type of nerve agent.

Four years later, in 1992, a forensic team from PHR visited the site of a 1988 attack on the Iraqi-Kurdish village of Birjinni. The team collected soil samples from bomb craters near the village and sent them to the British Chemical Defence Establishment at Porton Down. There, analysis by a sensitive technique known as gas chromatography–mass spectrometry confirmed the presence of chemicals with the unique spectral "fingerprint" of Sarin breakdown products.

THE FAILURE OF the international community to punish Iraq's flagrant violations of the 1925 Geneva Protocol had a deeply corrosive effect on the legal, political, and moral norms constraining the spread of chemical arms. In an article published in September 1988, *New York Times* columnist Flora Lewis criticized the "deafening silence" of Western governments in response to Iraq's extensive use of chemical weapons against Iran and its own Kurdish minority. Lewis wrote that the "complicity of the world community" in this systematic violation of international law would encourage other states in the Middle East region to acquire chemical arms. A year earlier, in an address to retired intelligence officers, deputy CIA director Robert M. Gates had made a similar prediction, warning that Iraq's massive use of mustard and nerve agents during the Iran-Iraq War had "broken the moral barrier" against chemical warfare.

These ominous predictions were soon borne out. Iran, the chief victim of Iraq's chemical attacks, moved quickly to expand its capabilities for the manufacture and delivery of chemical arms. In October 1988, two months after the end of the Iran-Iraq War, the speaker of the Iranian Parliament (and future president) Ali Akbar Hashemi Rafsanjani declared, "Chemical and biological weapons are . . . the poor man's atomic bomb. We should at least consider them for our defense. Although the use of such weapons is inhuman, the war taught us that international laws are only drops of ink on paper." In 1989, Iranian officials signed contracts with the Swiss company Krebs AG of Zurich to build a production facility for phosphorus pentasulfide, a dual-use chemical, and with the German chemical giant Hoechst to build a large pesticide plant near Tehran.

Other states in the Middle East also expanded their chemical warfare capabilities. In 1985, the Egyptian government–owned El Nasr Pharmaceutical Company had built a new factory to manufacture phosphorus trichloride at the production complex in Abu Za'abal, north of Cairo. Egypt had hired Krebs AG to construct the plant and Stauffer Chemicals of Pennsylvania to supply the production equipment. In March 1989, after learning of these contracts, the United States and Switzerland pressured the Egyptian government to certify that the Abu Za'abal plant would be used only for civilian purposes and to declare what chemicals would be produced there. When Cairo refused to comply with these demands, the Swiss Foreign Office sent a letter to Krebs ordering the firm to sever all business ties with its Egyptian partner. Because the Swiss government had been slow in acting, however, it failed to prevent the Egyptian plant from becoming operational.

Syria, too, acquired a sophisticated chemical arsenal. Since 1979, when Egyptian President Anwar Sadat had negotiated the Camp David peace accord with Israel, Syria had found itself without a close regional ally in its confrontation with the Jewish state. After the Israeli invasion of Lebanon in 1982 had again demonstrated the conventional military superiority of the Israel Defense Forces, Syrian president Hafez al-Assad decided to expand his country's chemical warfare capability as a deterrent. The main Syrian organization responsible for developing nerve agents was the Center for Scientific Studies and Research, known by its French acronym CERS. Although Syria had extensive mineral deposits of phosphorus, it relied for chemical precursors and specialized production equipment on commercial suppliers in West Germany, Switzerland, and India. By the end of the 1980s, Syria had built chemical weapons production plants that were reportedly located near the cities of Allepo, Homs, and Hama.

Assad also purchased from North Korea medium-range Scud B and C ballistic missiles, which were capable of delivering nerve agent warheads over hundreds of miles. By threatening cities in northern Israel, these missiles would give Syria a degree of "strategic parity" and constrain Tel Aviv's military options. At the same time, the integration of ballistic missiles and chemical weapons ushered the Middle East region into a frightening new era in which civilian populations were increasingly vulnerable to surprise chemical attack.

CHAPTER FIFTEEN

PEACE AND WAR

AFTER THE IRAN-IRAQ WAR ended in 1988, Baghdad destroyed its old, degraded stocks of chemical weapons and the Muthanna State Establishment switched to the production of pesticides and other legitimate chemicals. But Muthanna remained under the control of the Iraqi military, which continued research and development work on nerve agents under the cover of commercial activity. In August 1989, the director-general of Muthanna wrote to senior Iraqi officials that "research on munitions and chemical weapons are very important in war conditions. We must always be ready and prepared and must follow up every new development in this domain."

One priority task was to improve the stability of nerve agents, the low quality of which had caused them to deteriorate in a matter of weeks during the Iran-Iraq War. Even storing Sarin-filled munitions in refrigerated bunkers had failed to solve the problem. Although a brief shelf life had been tolerable in wartime, when the weapons were consumed almost as fast as they could be produced, in peacetime more stable formulations that could be stockpiled for long periods were required. Iraqi scientists addressed this problem by developing field-mixed nerve agent bombs and warheads, in which the final precursors would be combined by hand one day before use. One of the binary components was DF, which could be stored for years in plastic jerry cans. The other component was a 60:40 mixture of isopropyl alcohol and cyclohexyl alcohol, which was preloaded into bombs and Scud missile warheads. Before a weapon was prepared for delivery, an Iraqi soldier wearing a gas mask and protective suit would pour the DF solution into a bomb or warhead already containing the alcohol mixture. The chemicals would then react vigorously to produce a 60:40 mixture of Sarin and Cyclosarin. Although the resulting nerve agent "cocktail" would be impure, it would still be considerably more potent than a unitary agent after a few months of storage. Iraq later claimed to have produced and stockpiled 1,024 aerial bombs and 34 missile warheads of the field-mixed type.

Iraqi scientists also experimented with dual-chamber artillery shells and 122 mm rockets, in which DF and the alcohol mixture were stored in separate compartments and allowed to combine and react after the munition was fired. About a hundred prototype "true binary" shells were tested in 1989 and 1990 with "encouraging" results, but the technology never worked reliably enough to warrant large-scale production. Another focus of Iraqi chemical weapons research and development was on improving the manufacturing process for VX. The earlier method had resulted in an agent of low purity that deteriorated rapidly and had a shelf life of only one to eight weeks. In April 1988, Iraqi scientists developed a salt form of VX known as "dibis" that remained stable for up to eight months and could be converted into active VX as needed. Although Muthanna produced dibis on a trial basis, Iraqi officials later claimed (without providing hard evidence) that all of the material had been converted into 1.5 metric tons of VX, which had then deteriorated and been destroyed.

At Pine Bluff Arsenal in Arkansas, the U.S. Army Chemical Corps was converting dichlor into difluor (DF), one of the chemical components for the M687 binary Sarin artillery shell. Although the Chemical Corps had built a dedicated plant at Pine Bluff to produce dichlor, it was not yet operational. By August 1989, however, the available supply of dichlor was shrinking and would soon be exhausted. To get the new plant up and running, the Army needed to find a reliable supplier of thionyl chloride, the preferred chlorinating agent used in the production of dichlor. At least 160,000 pounds of thionyl chloride would be needed by June 2, 1990. Since Congress was holding up a $47 million appropriation for the M687 projectile until the dichlor plant was fully operational, the stakes for the Army were high.

The prime contractor for the Pine Bluff facility, Combustion Engineering, ordered thionyl chloride from the two U.S. chemical companies that manufactured it: Occidental Chemical Corporation of Dallas and Mobay Corporation of Pittsburgh. These firms produced thionyl chloride as an intermediate for the synthesis of commercial pesticides, dyestuffs, and plastics. In September 1989, both Occidental and Mobay declined to sell thionyl chloride to the U.S. government, citing legal issues and corporate policies against involvement in chemical weapons production. An Occidental spokesman stated, "As a matter of company policy, Occidental Petroleum will not sell or distribute chemicals that contribute to the production of chemical weapons or illicit drugs." Mobay's parent company, the West German firm Bayer, also refused to supply thionyl chloride to the U.S. Army because of concerns about negative publicity, including possible references to the Nazi era. Observed a Mobay spokesman, "In this day and age, who wants to be involved in providing chemicals that go into chemical weapons?"

The administration of President George H. W. Bush, which had taken office in January 1989, found itself in an awkward position. Given the refusal of the two U.S. manufacturers to cooperate, the Army considered ordering thionyl chloride from foreign suppliers, but doing so would create thorny political problems. In response to U.S. pressure, the West German government had finally begun to crack down on domestic companies that were suspected of selling chemical weapons precursors to proliferators such as Libya, Iraq, and Iran. Indeed, only a few months earlier, the United States had helped block a shipment of thionyl chloride from a West German company to Tehran. If the Pentagon now ordered the same chemical for its own nerve agent production program, Washington would be open to charges of hypocrisy.

In view of the political drawbacks of purchasing thionyl chloride from abroad, the Army Chemical Corps asked the Department of Commerce to force the two U.S. manufacturers to sell the chemical. Under an obscure 1950 law known as the Defense Production Act, the federal government had the authority to compel qualified companies to accept defense production orders if the Pentagon certified that the product was vital to national security; corporate executives who refused to comply could face fines and imprisonment. Even so, taking legal action against the two firms would prevent the Army from meeting the June 1990 Congressional deadline. Going to court would also antagonize the U.S. chemical industry, which was making a good-faith effort to prevent the sale of chemical weapons precursors to proliferators. All told, the issue risked becoming a major embarrassment for the White House.

As a stopgap measure, the Chemical Corps shipped railroad tank cars filled with leftover dichlor from Rocky Mountain Arsenal to Muscle Shoals for redistillation. The purified dichlor was then sent to Pine Bluff for conversion to DF, providing enough feedstock for one month of production. Nevertheless, a shortage of thionyl chloride was not the only problem facing the M687 projectile. Marquardt Corporation, which produced the canisters that held the binary components, had a three-year backlog of orders for parts it was having trouble manufacturing. The company had also come under federal scrutiny for alleged quality-control problems and ties to figures in a defense procurement scandal known as "Ill Wind."

As TECHNICAL DIFFICULTIES slowed the U.S. production of binary weapons, there was some movement on the arms control front. On July 28, 1988, U.S. Ambassador Max Friedersdorf made a speech to the Conference on Disarmament in which he revealed the locations of all U.S. chemical weapons production facilities, outlined plans for their destruction under the future Chemical Weapons Convention (CWC), and called on the Soviet Union and other states to follow suit. Six months later, on February 21–23, 1989, the United States conducted a trial inspection at a DuPont chemical plant in New Jersey to test draft procedures for verifying the chemical weapons ban. The goal of this exercise was to check treaty compliance at a "dual-use" chemical industry site that could potentially be converted to military production, without unduly disrupting commercial production or disclosing industrial trade secrets.

During bilateral talks with Moscow that ran in parallel with the multilateral CWC negotiations, the Bush administration proposed a reciprocal exchange of data on chemical weapons stockpiles and related facilities in the United States and the Soviet Union. Despite lingering suspicions between the two countries, arms control officials from both sides began to negotiate the agreement and developed a degree of mutual trust as they hammered out the details. This effort culminated in a Memorandum of Understanding (MOU), which was signed on September 23, 1989, by U.S. Secretary of State James Baker III and Soviet Foreign Minister Eduard Shevardnadze. The signing ceremony took place at an outdoor table at Jackson Lake Lodge in the heart of Grand Teton National Park, just north of Jackson Hole, Wyoming, where Secretary Baker liked to vacation. Under a deep blue sky, the sweeping panorama of the Grand Teton range, rising almost vertically from the table-flat plain, provided a spectacular backdrop to the event.

Henceforth known as the "Wyoming MOU," the U.S.-Soviet agreement called for a two-phase exchange of data between the two countries on their respective chemical weapons stockpiles and related facilities, plus a series of reciprocal site visits as a confidence-building measure. The first milestone would come on December 29, 1989, when the two countries would exchange data on aggregate stockpile size; types of stockpiled agents; percent of chemical agents in munitions, devices, and bulk containers; locations of storage, production, and destruction facilities; and types of agent and munitions at each storage facility.

Two days after the signing of the Wyoming MOU, President Bush made a speech to the U.N. General Assembly in New York in which he reaffirmed the United States' commitment to negotiating a global ban on chemical weapons. To facilitate progress in the ongoing multilateral talks in Geneva, he proposed a bilateral agreement with the Soviet Union in which both sides would reduce their chemical weapons stocks to equal levels, with strict provisions for verification. The President suggested that the interim level be set at 5,000 metric tons, or less than 20 percent of the existing U.S. stockpile.

Bush also pledged that after the multilateral CWC entered into force and the Soviet Union became a party, the United States would destroy 98 percent of its chemical stockpile within eight years, while keeping the remaining 2 percent—about 500 tons—as a "security stockpile." Once all chemical weapons–capable states had joined the treaty, Washington would destroy its remaining stocks over a two-year period. Another element of the Bush plan was that the United States would retain the option to continue manufacturing binary weapons for the 500-ton "security stockpile" if a sufficient number had not been produced by the time the CWC entered into force. This rather awkward proposal was the result of President Bush's attempt to strike a balance between the conflicting recommendations of the State Department and the Pentagon.

France endorsed the U.S. initiative on the condition that all countries with chemical weapons would be allowed to retain a residual "security stockpile." Although French officials denied possessing chemical weapons at the time, they wanted to acquire a small stockpile of binary munitions and "grandfather" them in under the proposed 500-ton limit. Most other countries strongly criticized the United States for seeking to retain some of its chemical weapons for an indefinite period after the CWC entered into force, despite the fact that the chemical "have-nots" would be prohibited from acquiring them in the first place. Such a two-tier system had already been created by the 1968 nuclear Non-Proliferation Treaty (NPT), which gave special status to the five states that had tested nuclear weapons before January 1969. Because many developing countries viewed the NPT as discriminatory, they had no desire to create a similar arrangement for chemical weapons. Arms control advocates also attacked Bush's "security stockpile" idea on the grounds that other countries would follow the example of the U.S. by continuing to produce chemical arms after the CWC went into effect. According to Representative Fascell, the administration's proposal would have the result of "unwittingly legitimizing the very thing that President Bush and Congress want to halt—chemical weapons proliferation."

In December 1989, at a summit in Malta with Soviet leader Mikhail Gorbachev, President Bush made a small but significant concession: he dropped his insistence that all states capable of producing chemical weapons would have to ratify the CWC before it could enter into force. The two leaders also made some incremental progress on the bilateral negotiating track. Although Moscow was considering the U.S. proposal for both countries to reduce their chemical stockpiles to equal levels, the Soviets were not technically ready to begin destruction.

At a meeting in Moscow on February 9, 1990, Foreign Minister Shevardnadze and Secretary of State Baker agreed to negotiate a Bilateral Destruction Agreement (BDA) that would reduce the chemical stockpiles on both sides to 5,000 metric tons. Since the declared Soviet stockpile of 40,000 tons was considerably larger than the American stockpile of 31,000 tons, reduction to equal levels would require the Soviet side to make a deep and asymmetric cut in its chemical arsenal. Shevardnadze insisted that his government could not accept such a deal if the United States continued to manufacture new binary weapons. Accordingly, he proposed including in the BDA a commitment by both sides to halt all further production of chemical arms.

Seeking to conclude the bilateral agreement in time for a planned Bush-Gorbachev summit in Washington in early June, Secretary Baker returned to Moscow for intensive negotiations with his Soviet counterpart. On May

8, 1990, in a dramatic concession, Baker offered to terminate the U.S. production of binary weapons if the Soviets made deep cuts in their existing chemical stockpile. Given the binary program's growing technical and political problems, Baker's proposal sought to make a diplomatic virtue of necessity. In any event, the Soviets agreed.

On June 1, 1990, during the U.S.-Soviet summit in Washington, Presidents Bush and Gorbachev signed the BDA, whose official title was the "U.S.-Soviet Agreement on Destruction and Non-production of Chemical Weapons and on Measures to Facilitate the Multilateral Chemical Weapons Convention." The key provisions of the BDA were that both sides agreed to halt all production of new chemical weapons, and that destruction of the existing stockpiles would begin by the end of 1992 and reach the agreed level of 5,000 metric tons by the end of 2002. Washington and Moscow also pledged to negotiate a bilateral verification mechanism to check the accuracy of their respective stockpile declarations and monitor the destruction of the weapons. The proposed verification measures would include on-site inspections of chemical weapons storage facilities and the continuous presence of inspectors and monitoring instruments at the destruction facilities. The BDA further committed the two sides to cooperate in developing safe and environmentally sound technologies for destroying chemical arms.

In July 1990, a month after the BDA was signed, U.S. Defense Secretary Dick Cheney notified the Army that he was withdrawing the administration's request for new production funds for binary weapons, canceling planned tests of the Bigeye bomb, and preparing to put all binary manufacturing facilities in mothballs as soon as the BDA was approved by Congress and the Supreme Soviet. Because of the shortage of dichlor, however, production of DF for the binary Sarin projectile had already ended.

DURING THE SUMMER OF 1990, the U.S. Army prepared to withdraw its stockpile of unitary chemical weapons from West Germany. President Reagan had originally promised Chancellor Kohl that the U.S. weapons would be removed by the end of 1992, but the unexpected fall of the Berlin Wall on November 9, 1989, and the rapid pace of German unification had led Kohl to advance the timetable by two years. For political reasons, Kohl wanted to complete the transfer before the all-German parliamentary elections scheduled for December 1990. Although the German Green Party filed suit to delay the removal of the U.S. weapons pending a detailed assessment of the safety and environmental risks, the German courts rejected this legal action and a last-minute appeal, allowing the plan to proceed. Whereas the West German government called the transfer of chemical weapons Operation Lindwurm, the German word for a mythical dragonlike beast, the Americans gave it the more prosaic name Operation Steel Box, a reference to the sealed metal containers used to transport the munitions.

On July 26, 1990, the buzz of police helicopters circling overhead and the idling of heavy trucks broke the early-morning calm in the southwestern German town of Clausen, population 1,600. More than 500 American troops, 500 German troops, and 1,200 German state and local police participated in the first phase of the operation. Despite threats of disruption, no protesters materialized and only a few town residents came out to watch, relieved that the weapons were finally being removed. Less than four months before, they had been shocked to learn that 437 tons of Sarin and VX, contained in 120,000 artillery shells, had been stored for decades in concrete bunkers at the nearby U.S. Army ammunition depot.

At 8:00 a.m., a five-mile-long convoy of seventy-nine vehicles began to snake along the empty roads from Clausen to the rail depot in the town of Miesau, thirty miles away. Twenty U.S. Army tractor-trailers carried the first shipment of 3,500 Sarin-filled artillery shells, packed into airtight steel containers. The tractor-trailers, escorted by armored personnel carriers, decontamination trucks, and West German police, fire, and emergency vehicles, drove at an average speed of eighteen miles per hour on secondary roads through populated areas, and thirty miles per hour on the closed Autobahn. After a journey of two and a half hours, the last vehicles in the convoy arrived in Miesau only a few minutes behind schedule. Similar convoys took place every weekday over the next month, and by the end of August all 437 tons of nerve agents had been moved to the rail depot.

In the second phase of the operation, the steel containers packed with chemical shells were loaded onto special rail cars for a twelve-hour, 600-mile journey to the North Sea port of Nordenham, across from the larger harbor at Bremerhaven. The rail shipments started on the evening of September 12 and continued for the next six nights. Each convoy consisted of two trains carrying a total of eighty steel containers, plus an escort train carrying command-and-control and disaster response personnel.

In the final phase of the transfer, the steel containers were loaded into the holds of two U.S. Navy cargo ships, U.S.S. *Flickertail State* and U.S.S. *Gopher State*. By the afternoon of September 19, all of the chemical weapons were safely onboard. Because of stormy weather in the North Sea, the two ships did not depart until the afternoon of September 22. Escorted by U.S. Navy guided missile cruisers, they crossed the Atlantic, rounded Cape Horn at the tip of South America, and traversed the Pacific, refueling three times at sea. The two cargo ships reached Johnston Island in early November and were off-loaded by November 18, after which the chemical shells were transported to storage bunkers in a secure part of the island. Remarkably, the $46 million "retrograde" operation, involving more than 23,000 U.S. and German personnel and the movement of hundreds of tons of lethal chemicals halfway around the world, was completed without a single accident or injury.

Johnston Island now housed 120,000 chemical munitions from West Germany, along with the 300,000 bombs and shells that had been transferred from Okinawa in 1971. These weapons would eventually be destroyed in a special high-temperature incinerator called the Johnston Atoll Chemical Agent Disposal System (JACADS). Built at a cost of $240 million, the JACADS facility was then undergoing testing. The U.S. chemical weapons "demilitarization" program had begun with Project Eagle in 1969–1976, when the Army had disposed of large quantities of Sarin and mustard agent at Rocky Mountain Arsenal. In the early 1980s, the U.S. Congress, concerned over the hazards associated with leaking M55 rockets and other aging chemical munitions, had assigned the Army the task of destroying these weapons in a safe and environmentally sound manner. (Dumping at sea was no longer an option because Congress had banned it in 1972.)

In 1984, the National Research Council's Board on Army Science and Technology had assessed a variety of chemical weapons disposal technologies and endorsed the Army's choice of high-temperature incineration over chemical neutralization. The rationale was that incineration could decontaminate all parts of a chemical munition (agent, explosive, metal parts, and packing materials) and would be effective for every type of toxic agent. Chemical neutralization, in contrast, was complex, time-consuming, specific to each type of agent, generated large amounts of liquid waste, and could not destroy explosives and metal parts, which would have to be incinerated. The advisory committee also concluded that incineration was as safe as neutralization, an assessment that would later become the focus of intense criticism.

AFTER THE END of the Iran-Iraq War, Iraq's military, intelligence, and security services began to focus on other regional threats, particularly Israel. Saddam Hussein's goals were to counterbalance Israel's undeclared nuclear arsenal and deter a possible attack against Iraq's own nuclear weapons installations. On April 1, 1990, Saddam gave a bellicose speech in which he warned that if Israel launched a preemptive strike, Iraq would retaliate in a way that would "make the fire eat up half of Israel." This statement was widely interpreted as a threat to use chemical weapons against Israeli cities.

Shortly after Saddam's speech, Iraq launched the development of two new chemical weapons. The first was the R-400 aerial bomb, which was designed for low-level release from aircraft and based on the reverse engineering of an imported, parachute-retarded system. Prototypes of the R-400 were field-tested on May 22, 1990, and the weapon went into production soon thereafter. The second project was to develop a chemical warhead for the Al-Hussein ballistic missile, an extended-range version of the Soviet Scud. Iraqi engineers conducted two flight tests of the new warhead, the first with a mixture of oil and water, and the second with degraded Sarin. In June 1990, the first lots of R-400 bombs and "special" missile warheads were delivered to Muthanna.

Saddam's strategic objectives vis-à-vis Israel culminated in the so-called Thunderstrike project, which involved acquiring the capability to attack Israeli cities using Al-Hussein ballistic missiles armed with chemical (and possibly biological) warheads. Saddam chose ballistic missiles rather than aircraft for this strategic mission because of the poor combat record of the Iraqi Air Force and the fact that equipping aircraft with his most potent weapons could create an internal threat to the regime. The Thunderstrike project envisioned a total of sixty fixed-arm missile launchers, configured in groups that took into account the limited accuracy of the Al-Hussein's guidance system. Six to eight launchers grouped together would fire salvos of missiles, ensuring multiple hits on a city-sized target.

On April 12, 1990, during a meeting with a visiting delegation of U.S. senators, Saddam warned that if Israel carried out a surprise nuclear attack on Baghdad that destroyed the Iraqi chain of command, special military units were under standing orders to drive to secret launch sites in the western desert, mate the chemical warheads in their custody with long-range missiles, and fire them at targets in Israel. Twenty-eight of the sixty planned fixed-arm launchers had been completed by the summer of 1990.

Meanwhile, the eight years of war with Iran had devastated the Iraqi economy, which was heavily dependent on oil exports. Baghdad's financial straits exacerbated a long-standing border dispute with Kuwait over a valuable oil field that straddled the border between the two countries. After months of rising tensions over Iraq's claims that Kuwait was "stealing" its oil reserves, Saddam Hussein ordered the surprise invasion of his southern neighbor. In the early-morning hours of August 2, 1990, columns of Iraqi tanks, armored vehicles, and trucks ferrying 100,000 troops sped down the north-south highway and into the oil-rich emirate. By mid-morning the invaders had occupied and looted Kuwait City, and Saddam quickly annexed the country as a new province of Iraq. Three days later, U.S. President George H. W. Bush declared that the Iraqi aggression would "not stand." After obtaining the permission of Saudi Arabia, he ordered a major buildup of U.S. military forces in the region for what he called Operation Desert Shield. The United States also began to assemble a broad international coalition to expel Iraq from Kuwait.

UNDER THE WYOMING MOU, the Soviet Union gave the United States a detailed set of data declarations on its stockpiled chemical weapons and related development, production, and storage facilities. Despite Gorbachev's rhetoric of *glasnost,* the Soviet declaration contained several major gaps, including a failure to mention the Foliant nerve agents and the Novichok binary formulations. In 1989–90, the Soviets conducted trials of Novichok-5 at the Chemical Research Institute in Nukus, Uzbekistan, and its open-air testing site on the Ustyurt Plateau, an expanse of arid desert several hundred miles west of the Aral Sea. The Kremlin intended to keep these top secret activities under wraps in order to preserve its technological advantage. This gambit probably would have succeeded, had it not been for the courageous decision by a senior scientist at GosNIIOKhT to disclose the institute's most explosive secrets to the outside world.

Vil Sultanovich Mirzayanov was born in a small Russian town on the European side of the Ural Mountains. He was a member of the Tatar minority, a Turkic-speaking, mostly Muslim ethnic group that had experienced a long history of persecution. As a young man, he showed an aptitude for math and science and was admitted to the Moscow State Academy of Fine Chemical Technology, where he earned a degree in petroleum-refinery engineering in 1958. He was then hired by a state research institute on synthetic fuels, moving after a few years to another institute involved in the development and production of boranes as rocket propellants. He also began his doctoral studies in chemistry at the Institute of Petrochemistry of the Soviet Academy of Sciences, where he wrote a dissertation on the chromatographic analysis of trace concentrations of chemical compounds.

After Mirzayanov successfully defended his thesis in 1965, one of the members of his doctoral committee recommended him for a job at GosNII-OKhT. Although he had already obtained a security clearance for his work on boranes, the new position required a more rigorous background investigation that took three months to complete. Finally the clearance was approved and Vil joined the technical staff at GosNIIOKhT in 1966, at the age of thirty. His first assignment was to monitor toxic emissions from the laboratory into the Moscow air and water, and to perform chromatographic analyses of the various chemical warfare agents under development.

Mirzayanov enjoyed his work at GosNIIOKhT and received good performance reviews. To further his career, he became a member of the Communist Party, which was considered a prerequisite for high-level positions at the institute. In the late 1980s, after two decades of service, he was promoted to the position of chief of the Department of Technical Counterintelligence, responsible for shielding the highly classified research on chemical weapons from the eyes of foreign intelligence services. In this capacity, Mirzayanov carried a notebook filled with top secret code names and traveled frequently to meetings at Shikhany and Novocheboksarsk. He did his job so well that the U.S. intelligence community remained unaware of the Foliant program.

As the years went by, however, Mirzayanov began to have moral qualms about his work. He was disturbed by the duplicity of Kremlin leaders who continued to invest vast resources in chemical weapons development while paying lip service to disarmament and failing to meet the basic needs of the Soviet people. When President Gorbachev eased restrictions on political speech, Mirzayanov became a bit more outspoken about his personal views. In 1989, he helped to organize the GosNIIOKhT branch of the opposition party Democratic Movement of Russia.

During the spring of 1990, Mirzayanov was told to help prepare the chemical weapons production facilities at Volgograd and Novocheboksarsk for upcoming visits by U.S. experts under the bilateral confidence-building provisions of the Wyoming MOU. At the Khimprom plant in Volgograd, he took environmental samples throughout the Soman production unit. Although the manufacturing line had been shut down, samples from the smokestack contained fifty to a hundred times the allowed concentration of nerve agent. Mirzayanov also found that samples from the wastewater pond were highly contaminated, even though the plant manager claimed that the treated waste did not inhibit cholinesterase in laboratory tests. Puzzled, Mirzayanov did his own analysis and discovered that salts in the waste water interfered with the reaction between Soman and cholinesterase, masking what was in fact a dangerous level of toxicity.

Mirzayanov wrote up his findings and took samples back to Moscow to bolster his case. When he told GosNIIOKhT director Viktor Petrunin that contamination of the Volgograd facility posed serious health risks for the workers and the local population, Petrunin frowned. "You did a good job with the analysis, but these findings are extremely troublesome for us," he said. "I'm sure you would find the same thing at Novocheboksarsk."

Mirzayanov was taken aback by the implications of this remark. "You mean you don't intend to correct the situation?" he asked.

Petrunin shook his head. "Don't be naive," he replied sharply.

Mirzayanov refused to follow the director's advice and simply drop the issue. At an interagency meeting on counterintelligence problems, he reported his findings on the high level of toxic contamination at Volgograd. One senior official, the Deputy Minister of Chemical Production, took strong exception. "Dr. Mirzayanov was not authorized to make this report, which he did strictly on his own initiative," he said. "Accordingly, his information is not trustworthy." The deputy minister went so far as to cast doubt on Vil's loyalty by implying that he might be working under the influence of a foreign intelligence service.

After the meeting, Petrunin told Mirzayanov that he was fortunate not to be living in Stalin's time, when such a remark would have sealed his fate. Although Vil was allowed to keep his job at GosNIIOKhT, it was clear that nothing would be done to clean up the toxic contamination at Volgograd. Deeply disillusioned, Mirzayanov resigned his membership in the Communist Party in May 1990. Petrunin retaliated by denying him access to laboratory equipment and transferring several of his colleagues in an attempt to isolate him. Although Mirzayanov debated whether to go public with what he knew, he hesitated, fearing the consequences for himself and his young family.

MEANWHILE, THE U.S. binary program was winding down. In December 1990, Combustion Engineering completed the dichlor plant at Pine Bluff Arsenal and then, because of the unavailability of thionyl chloride from commercial sources, proceeded to put the plant into mothballs. Construction of a manufacturing plant for QL, the liquid component of the Bigeye bomb, had already been halted in July 1990 when the facility was 65 percent complete. The third binary weapons program, the Army's MLRS rocket warhead, was canceled while still in development. Because the binary artillery shell had taken so long for Congress to fund and had experienced a series of technical and political delays, relatively few M687 projectiles had actually been produced by the time the program came to an end. As a result, the U.S. chemical weapons stockpile consisted of 30,600 tons of unitary agents but only 680 tons of binary components.

Under the BDA, the United States and the Soviet Union were supposed to negotiate a set of bilateral inspection procedures to verify the destruction of chemical weapons, but the talks bogged down and the BDA never entered into force. Another problem that sank the agreement was that the schedule for reducing the chemical weapons stockpiles on both sides to 5,000 metric tons by the end of 2002 proved to be highly unrealistic. Although the Soviet Union had spent $165 million to build a chemical weapons neutralization facility ten miles outside Chapayevsk, a city of 90,000 people south of Moscow, the local inhabitants protested so vehemently when the plant opened in 1989 that the Kremlin finally agreed not to put it into operation. A new destruction facility would therefore have to be built elsewhere, resulting in a lengthy delay.

IN THE PERSIAN GULF, war clouds were gathering on the horizon. The United States, having built up a large invasion force in Saudi Arabia over the previous four months, persuaded the U.N. Security Council to pass

Resolution 678 on November 29, 1990, giving Saddam Hussein an ultimatum: either withdraw all Iraqi forces from Kuwait by January 15, 1991, or face military action to expel them. The Iraqi regime gave no sign of backing down, however, and instead prepared for war by stockpiling large quantities of weapons, including chemical arms. Between December 1990 and January 1991, the Muthanna State Establishment churned out Sarin and Cyclosarin at the rate of one metric ton per day and loaded a mixture of the two agents into 8,320 artillery rockets.

In early January 1991, General Hussein Kamel, a son-in-law of Saddam Hussein who directed all of Iraq's unconventional weapons programs, ordered the Iraqi Ministry of Defense to provide thirty-one trailers to transport the chemical rockets to munitions depots near Ukhaydir and Nasiriyah in southern Iraq. A few days before the January 15 war deadline, Iraqi officials ordered the evacuation of Muthanna. Chemical munitions stockpiled on-site were dispersed to airfields and military bases in the western desert and other remote locations, and Al-Nida mobile missile launchers (which had replaced the vulnerable fixed-arm launchers) were deployed to forward storage and support centers.

At a meeting of the Iraqi leadership shortly before the start of the war, Saddam personally authorized the use of chemical weapons against Israel, Saudi Arabia, and U.S. forces. Although he did not spell out the exact circumstances under which the weapons might be employed, possible triggers included a direct threat to the Iraqi regime. In addition to stockpiles of chemical shells, bombs, and rockets, a special mobile-launcher unit equipped with seven fully fueled Al-Hussein missiles with chemical warheads was reportedly designated as a strategic reserve. Because Saddam did not trust the regular armed forces with such powerful weapons, he placed them under the exclusive control of the Special Security Organization (SSO), his most trusted praetorian guard. To deter a surprise attack by Israel that might aim to "decapitate" the Iraqi leadership, Saddam predelegated to the SSO missile unit commander the authority to launch retaliatory strikes with chemical warheads.

As the U.N. deadline neared for Iraq's withdrawal from Kuwait, the coalition forces prepared for the worst. For the first time since World War

II, American and British soldiers faced an enemy that had already employed chemical weapons extensively in battle and was considered likely to do so again. According to CIA estimates, Iraq possessed more than 1,000 tons of blister and nerve agents loaded into a variety of munitions, including artillery shells, aerial bombs, and rockets. In the face of this threat, scientists at Edgewood (now part of Aberdeen Proving Ground) and Porton Down scrambled to improve the ability of coalition forces to survive and fight on a contaminated battlefield.

Of particular concern was intelligence that Iraq had large stockpiles of Cyclosarin, a nerve agent that had never before been weaponized on a large scale. Chemical defense specialists worried that the handheld chemical agent monitors (CAMs) employed by allied forces to detect nerve agent vapors would fail to recognize Cyclosarin. To eliminate this potential vulnerability, the staff of the Detector Technology Division at Porton Down worked twenty-four hours a day, seven days a week, to reprogram the Erasable Programmable Read Only Memory (EPROM) chips in every CAM in the British stockpile.

U.S. troops deploying to Saudi Arabia were issued the MARK-1 antidote kit containing two autoinjectors, one loaded with atropine to counteract the immediate effects of nerve agent poisoning and the other with 2-PAM chloride to reactivate cholinesterase. However, these standard antidotes were known to be of limited effectiveness against Soman, which the United States suspected was in Iraq's arsenal. Because Soman inactivates cholinesterase irreversibly in minutes (a phenomenon known as enzyme "aging"), the Pentagon decided to augment the MARK-1 antidote kit with a pretreatment drug called pyridostigmine bromide (PB). U.S. soldiers were issued bubble packs of PB tablets and ordered to start taking one 30 milligram tablet every eight hours as soon as they entered combat.

PB works by binding *reversibly* to cholinesterase, converting a portion of the body's supply of the enzyme into a reserve that is protected from permanent inactivation by Soman. Thus, pretreatment of soldiers with PB, followed by postexposure therapy with 2-PAM chloride to displace PB from cholinesterase and reactivate the enzyme reserve, would enable troops to survive exposures to Soman that would otherwise be lethal. Nevertheless, PB was contraindicated in individuals who had already been exposed to nerve agents, including low levels of Sarin, and its use as a pretreatment was known to cause serious side effects in a small number of susceptible individuals. (Only after the war did it become clear that Iraq did not possess Soman.)

Not only did the risk of Iraqi combat use of nerve agents appear high, but some intelligence suggested that Baghdad might sponsor terror attacks with chemical or biological agents against coalition targets at home and abroad. In early December 1990, the British chemical weapons expert Ron Manley and his colleague John Clipson were called into the director's office at Porton Down and told to organize two emergency response teams, which could be deployed on short notice to any country in the world where British interests might be targeted. The response teams were equipped with protective gear, chemical agent detectors, and supplies of antidotes, drugs, and decontamination equipment.

Even as the U.S. and British governments prepared to fight in a toxic environment, they tried to deter Saddam Hussein from ordering the use of unconventional weapons. During an eleventh-hour meeting in Geneva on January 9, 1991, Secretary of State Baker handed Iraqi Deputy Foreign Minister Tariq Aziz a letter addressed to Saddam Hussein. The letter warned that if Iraq launched chemical or biological attacks, the American people would demand "the strongest possible response. You and your country will pay a terrible price if you order unconscionable actions of this sort." Although Aziz refused to accept the letter, he presumably conveyed the gist of its contents to Saddam Hussein. Aziz later said that Saddam had interpreted the U.S. statement as a veiled threat to retaliate with nuclear weapons.

In the morning darkness of January 17, 1991, the first phase of Operation Desert Storm—the coalition air campaign—got under way. Among the priority bombing targets were a number of Iraqi chemical weapons sites, including the production complexes at Muthanna and Falluja and the ammunition storage depots at Muhammadiyat and Ukhaydir. Although the U.S. Air Force used a combination of explosive and incendiary munitions to break open the bunkers and burn the chemical weapons inside at high temperature, some U.S. military officials worried that the air strikes would vent plumes of toxic material that could endanger the health of Iraqi civilians and coalition troops downwind.

Indeed, on January 19, two Czech chemical defense detachments deployed nearly fifteen miles apart in the desert near Hafar-al-Batin in northern Saudi Arabia began to detect low levels of Sarin in the atmosphere with a highly sensitive Soviet-made detection system based on the inhibition of purified cholinesterase. The Czechs were convinced that the nerve agent had been released by the U.S. bombing campaign that had begun two days earlier and was targeting suspected Iraqi chemical weapons facilities. Although the Czech soldiers were ordered to don their gas masks, U.S. troops in the area did not receive a similar order and were not even told that nerve agents had been detected. Later the same day, French chemical troops also reported detecting "infinitesimal amounts of nerve and blister agent" in the atmosphere approximately nineteen miles from King Khalid Military City, a sprawling military base in northern Saudi Arabia. Although senior American commanders were aware of the Czech and French detections, they chose to ignore them because they did not want to create a "panic."

When the coalition ground campaign began on February 24, 1991, U.S. intelligence assessed the threat of Iraqi chemical attacks as high. Although Iraq had not used poison gas during an early skirmish on January 30 at Al-Khafji, a Saudi town near the Kuwait border, General Norman Schwarzkopf, the commander of coalition forces, remained extremely concerned. According to his memoir, "My nightmare was that our units would reach the barriers [along the Saudi-Iraq border] in the very first hours of the attack, be unable to get through, and then be hit with a chemical barrage. We'd equipped our troops with protective gear and trained them to fight through a chemical attack, but there was always the danger that they'd end up milling around in confusion—or worse, that they'd panic."

To bolster the coalition's chemical defenses, the German government donated to the United States sixty FOX nuclear, biological, and chemical (NBC) reconnaissance vehicles. These six-wheeled, lightly armored vehicles were mobile laboratories that could take air, water, and ground samples and analyze them immediately for the presence of toxic agents.

U.S. troops participating in the ground campaign carried a full set of personal protective gear, including a gas mask, helmet cover, battle-dress overgarment (BDO), hood, overboots, and rubber gloves. The BDO is a coat and trousers made of an outer layer of cotton material with camouflage markings and an inner layer of charcoal-impregnated polyurethane foam that absorbs and traps chemical warfare agents. Although the suit provides good protection, wearing it for more than short periods significantly impairs fighting ability. The BDO and hood cause a rapid buildup of body heat, increasing the risk of exhaustion and heat stroke in the desert sun; the gas mask degrades the ability to see, speak, and hear and causes severe claus-

Soldiers from the 24th Mechanized Infantry Brigade undergo chemical defense training in eastern Saudi Arabia in November 1990, prior to the start of Operation Desert Storm against Iraq in January 1991.

trophobia in some individuals; and the rubber gloves limit the sense of touch and the ability to perform delicate manipulations.

To minimize these problems, the U.S. Army employed an operational doctrine during the Gulf War known as Mission Oriented Protective Posture (MOPP), which allowed commanders to adjust the amount of protective gear worn by their troops so that they were adequately protected while minimizing the concomitant loss in fighting effectiveness. In response to changes in the chemical warfare threat, troops were ordered to don or doff various items of protective equipment. The five threat levels were based on strategic as well as tactical intelligence, including reports of enemy weapons deployments, chemical-detector alarms in nearby sectors, and confirmed chemical attacks. Each increase in MOPP level involved adding additional protective gear and resulted in some degradation in combat performance. The alert levels were as follows:

- Level 0: No chemical protective gear worn, but readily available.
- Level 1: BDO and helmet cover worn.

- Level 2: Overboots added.
- Level 3: Gas mask and hood added.
- Level 4: Rubber gloves added.

As it turned out, no Iraqi chemical attacks of any consequence took place during Operation Desert Storm, but low-level exposures to chemical fallout may have occurred. One indication was provided by the M8 portable automatic chemical agent alarms, which were deployed upwind of U.S. units and continually monitored the atmosphere for blister and nerve agents. Throughout the air and ground campaigns, thousands of M8 alarms went off across the battlefield—so many, in fact, that troops started disabling them so that they could get some sleep. The Pentagon claimed that all of the alarms had been false, triggered by chemical "interferents" such as diesel fumes and pesticides. But some of the alarms may have been triggered by nerve agents released by the bombing of Iraqi chemical weapons depots and production facilities and carried downwind over coalition forces.

Another serious threat during the Gulf War was the possibility of Iraqi chemical attacks against Israeli cities, using extended-range Scud missiles armed with nerve agent warheads. To counter this threat, the government of Israel distributed gas masks and antidote kits to the entire civilian population and broadcast public-service announcements on radio and television with detailed instructions for their use. Over the course of the war, Iraq launched thirty-nine Al-Hussein missiles at Israel. U.S. reconnaissance satellites provided a few minutes' warning of each launch, which was relayed directly to the Israeli civil defense authorities. As soon as the air-raid sirens went off, civilians evacuated to a "sealed room" in each house or apartment building that had been built to special standards. They donned their gas masks and waited, sometimes for hours, until the all-clear sounded.

In a brave gesture of solidarity with the beleaguered Israeli people, the American violinist Isaac Stern traveled to Jerusalem in the midst of the Gulf War to perform with the Israel Philharmonic Orchestra. During a rehearsal on February 23, the air-raid sirens blared. Stern donned his gas mask but refused to be cowed and kept on playing. A photograph of the beetle-masked violinist, defending Western culture in the face of barbarism, was published in newspapers around the world. Although none of the Scud missiles that landed in Israel carried a chemical warhead, two Israeli civilians died and nearly a thousand were hospitalized with symptoms attributable to the missile attacks, including heart attacks and severe anxiety. Three quarters of the casualties resulted from the improper use of chemical-defense equipment, such as the failure to remove the plug from a gas mask filter or the panicked self-administration of atropine, which is itself toxic in the absence of nerve agent exposure.

On February 27, coalition troops liberated Kuwait City and the Iraqi Army fled north, suffering heavy losses along what became known as the "highway of death." Although the political imperative to hold the diverse coalition together meant that Saddam Hussein was allowed to remain in power, U.S. officials expected that his ignominious defeat would weaken his power base and lead to his overthrow in a coup d'état.

Iraq's capitulation after exactly one hundred hours of ground combat repudiated the idea that chemical weapons were a potent deterrent, or "poor man's atom bomb," that could intimidate a military superpower. Not only had Iraq failed to prevent the U.S.-led invasion but the coalition forces had minimized the potential impact of Iraqi chemical attacks by conducting rapid, highly mobile ground operations and equipping themselves with personal protective gear. Iraq's decision not to resort to its chemical arsenal during the war was attributed to several factors, including the fear of severe retaliation; the fact that the Iraqi Air Force was quickly grounded or fled across the border into Iran; the U.S. bombing of Iraq's logistical supply lines to prevent chemical weapons from reaching the front; and the direction of the prevailing winds, which blew toward the Iraqi lines for most of the conflict. Even if Saddam had ordered the use of chemical weapons, the speed of the coalition advance would have prevented the Iraqi Army from carrying out the coordinated artillery strikes needed to employ chemical weapons effectively on the battlefield. To saturate a square kilometer of territory with a lethal concentration of Sarin, the Iraqis would have needed favorable winds, precise targeting, and enough time to fire a few hundred chemical shells.

On March 4, 1991, several days after the cease-fire, demolition units with the U.S. Army's 37th Engineering Battalion blew up thirty-seven Iraqi munitions bunkers at the vast Khamisiyah ammunition depot in southeastern Iraq. The explosions shook the ground and sent up huge columns of

A destroyed munitions bunker (lower left) at the Khamisiyah Ammunition Storage Complex in southern Iraq, March 1991. U.N. weapons inspectors later determined that the bunker had contained Iraqi 122 mm rockets loaded with Sarin and Cyclosarin. The blast exposed large numbers of U.S. troops to low levels of nerve agents, with possibly harmful effects.

smoke and dust that were carried away by the prevailing winds. Shortly after the detonations, an M8 chemical agent detector-alarm in the area went off, but subsequent tests were negative.

ON APRIL 3, 1991, the U.N. Security Council adopted Resolution 687, which set out the terms of the Gulf War ceasefire. The Iraqi regime was henceforth banned from possessing chemical, biological, or nuclear weapons, or ballistic missiles with a range of more than 150 kilometers (i.e., capable of reaching Israel). Baghdad was required to declare all of its prohibited weapons, materials, and production facilities and to cooperate with their verification and destruction. To oversee the disarmament of Iraq, the United Nations created a new corps of international inspectors called the U.N. Special Commission on Iraq, or UNSCOM.

Shortly after the adoption of Resolution 687, the Iraqi regime declared about 10,000 chemical munitions filled with mustard and nerve agents, and 1,000 tons of bulk agent in storage containers. In early June, a team of UNSCOM inspectors arrived in Baghdad to verify the accuracy of the Iraqi declarations. Dressed in gas masks and protective suits, they visited the Muthanna State Establishment, which had been heavily bombed during the air campaign. Several production buildings, storage bunkers, and the main administration building lay in ruins. In addition, hundreds of leaking chemical munitions emitted a witches' brew of lethal gases that tainted the air several miles downwind. Fortunately, there were no populated areas nearby.

Meanwhile, the threat of Iraqi chemical warfare during the Persian Gulf War gave strong impetus to the ongoing Chemical Weapons Convention negotiations in Geneva. On May 13, 1991, President Bush announced a new set of steps to improve prospects for the successful conclusion of the CWC. First, the United States would formally forswear the use of chemical weapons against any state for any reason, including retaliation, effective when the CWC entered into force, provided that the Soviet Union was also a party to the treaty. Second, Washington would drop its earlier demand for a "security stockpile" and commit itself unconditionally to the destruction of all of its stocks of chemical weapons within ten years of the CWC's entry into force. Bush also called for setting a target date to conclude the negotiations and recommended that that Conference on Disarmament remain in continuous session if necessary to meet the deadline. A few months later, on July 15, 1991, the United States, Britain, Australia, and Japan tried to break the logjam over procedures for "challenge" inspections of suspect sites by jointly presenting a draft proposal that included provisions for "managed access" to protect confidential business and national-security information unrelated to CWC compliance.

THE WEAPONS INSPECTIONS in Iraq posed a unique organizational challenge for the United Nations, which had never before conducted an operation of this type. UNSCOM Executive Chairman Rolf Ekéus, a veteran Swedish diplomat, recruited a team of experienced scientific and technical experts from the United States, Australia, Canada, the Soviet Union, and Western and Eastern Europe to account for and eliminate Iraq's prohibited chemical, biological, nuclear, and missile capabilities in a verifiable manner. Once UNSCOM certified that Iraq had been disarmed, the inspectors would continue to monitor the country's "dual-use" factories for an extended period to preclude any future Iraqi effort to rebuild its banned arsenals. Because of the unprecedented nature of this task, UNSCOM officials faced a steep learning curve.

In June 1991, British chemical defense specialist Ron Manley was at work in his laboratory at Porton Down when he received a call from the director, Dr. Graham Pearson, ordering him to attend an important meeting the following week at U.N. Headquarters in New York. Although Manley was about to leave with his wife, Jean, on a long-planned vacation in the Canadian Rockies, Pearson told him to postpone the trip. When Manley protested that the plane tickets were nonrefundable, the director arranged to have a refund check issued to him by the end of the day. Even so, Manley had to go home and break the news to his wife. Resigned, she asked whether they should reschedule the holiday or simply cancel it. He told her to wait and see; the meeting in New York was only supposed to last a week.

At the United Nations, Manley met with Brian Barrass, the British commissioner of UNSCOM, and John Gee of Australia, the commissioner responsible for chemical and biological weapons. Gee asked Manley, "How do you fancy chairing an advisory panel on destroying Iraq's chemical weapons?" When Manley tried to turn down the offer, Gee became insistent. "We have a problem," he explained. "For political reasons, the chair can't be an American or a Russian. The Swede and the Frenchman aren't acceptable, and the only other possibility is the Canadian, but his government won't let him do it." Manley finally gave in and agreed to chair the Destruction Advisory Panel. For the next two years, he commuted between London, New York, and Baghdad, while still running his division at Porton Down. Much to his regret, the vacation in the Canadian Rockies was put off indefinitely.

Because many of Iraq's chemical munitions were damaged and leaking, they could not be shipped out of the country. The Destruction Advisory Panel decided to consolidate them at the Muthanna State Establishment where they had been manufactured. Thousands of chemical bombs, rockets, and artillery shells were recovered from various depots and airfields around Iraq and transported to Muthanna, where they were lined up on the desert floor in front of the ruined factories. Early in 1992, Manley established a Chemical Destruction Group (CDG), responsible for the day-to-day supervision of the Iraqi personnel who would do the actual work of destroying the chemical munitions. To staff this oversight body, Gee and Manley persuaded UNSCOM to hire about a hundred military personnel from twenty-five countries who had training in chemical defense and agreed to serve in Iraq for tours of six to nine months. Many of these experts were seconded by their governments. Whenever a new phase of the destruction operation began or a technical problem arose, one or two members of the Destruction Advisory Panel flew to Baghdad to solve the problem. They

United Nations chemical weapons experts seal dozens of leaking Iraqi 122 mm rockets containing nerve agents to prepare them for destruction after the 1991 Gulf War.

then left and allowed the local Iraqi staff to continue their work under CDG supervision.

The chemical weapons destruction operation at Muthanna had two prongs: the construction of an incinerator to burn mustard agent, and the conversion of the Sarin pilot plant into a neutralization facility that transformed nerve agents into relatively nontoxic liquid waste. This material was then sun-dried in large pans or shallow trenches with an impervious lining and mixed with concrete into large blocks, which were permanently sealed in underground munitions bunkers. In some cases, the design of the weapons or their deteriorating condition made it necessary to find expedient solutions for destruction, while managing safety and environmental risks. For example, because 122 mm rockets filled with nerve agents were considered too dangerous to dismantle, the Destruction Advisory Panel instructed the Iraqis to dig deep holes in the desert, place an open-topped storage tank filled with diesel fuel at the bottom of each hole, and lay twenty rockets carefully across the top of the tank. Small explosive charges served to puncture the rocket warheads, allowing the nerve-agent fill to drain into the diesel fuel, which was ignited a few seconds later by an additional small charge. The resulting fireball had a temperature of about 2,000 degrees Celsius—hot enough to break down the molecules of Sarin and Cyclosarin into harmless by-products.

Over the three-year duration of the chemical weapons destruction program in Iraq, UNSCOM inspectors found and eliminated more than 46,000 filled chemical munitions, 30 ballistic missile warheads, and 5,000 tons of bulk agent and precursor chemicals. Even so, the U.N. inspectors were unable to account for the entirety of Iraq's chemical arsenal, and telltale bits of evidence suggested that the Iraqi officials were not telling the whole truth. For example, the inspectors confiscated an Iraqi Air Force document indicating that fewer chemical weapons had been consumed during the Iran-Iraq War than Baghdad had declared. The Iraqis also claimed that during the summer of 1991, they had secretly begun the unilateral destruction of selected filled chemical munitions and bulk agents. Over a period of a few months, they had purportedly eliminated more than 28,000 filled and unfilled munitions, 30 metric tons of bulk precursor chemicals for Sarin and Cyclosarin, and more than 200 metric tons of precursors for VX. Yet Iraqi officials refused to provide documentary or physical evidence to back up the claimed destruction activities, insisting that all of the relevant documents had been destroyed. Given the regime's meticulous record keeping in other areas, this statement did not ring true. The gaps in the evidence prevented UNSCOM from concluding definitively that the missing weapons had not been hidden rather than destroyed. These discrepancies were left to fester until, several years later, they became part of the rationale for the second U.S.-led war against Iraq.

CHAPTER SIXTEEN

WHISTLE-BLOWER

AFTER MUCH PERSONAL anguish and soul-searching, Vil Mirzayanov finally decided to go public with his concerns about the Soviet nerve agent development program. The precipitating event came in April 1991, when President Mikhail Gorbachev secretly awarded the Lenin Prize, the Soviet Union's highest honor, to GosNIIOKhT director Petrunin, General Kuntsevich, and General Igor Yevstavyev for their successful development and pilot-scale production of the Novichok agents. Mirzayanov was convinced that the Kremlin intended to conceal the existence of the Soviet binary program so that it would not have to be declared and eliminated under the future Chemical Weapons Convention. On October 10, 1991, he published an article titled "Inversion" in the Moscow newspaper *Kuranty* in which he exposed the duplicity of the Soviet military-chemical complex. Despite Gorbachev's claim in 1987 to have halted all manufacture of chemical weapons, Mirzayanov wrote, the Soviet Union was continuing in secret to develop a new class of nerve agents of extraordinary potency.

The article's publication in *Kuranty* was overshadowed by the tumultuous political events leading to the breakup of the Soviet Union. On December 8, 1991, the presidents of Russia, Ukraine, and Belarus signed a treaty creating the Commonwealth of Independent States. Most of the other former Soviet republics joined two weeks later, and on December 25, President Gorbachev resigned as president of the USSR and turned the powers of his office over to Boris Yeltsin, the leader of the Russian Federation. At midnight, the hammer-and-sickle flag was pulled down from the dome of the Kremlin and replaced with the Russian tricolor. Yeltsin had won Russia's first presidential election on June 12 and become world-famous on August 18 by standing defiantly atop an armored personnel carrier and challenging a hard-line coup against Gorbachev that had ultimately failed. Now, as the president of independent Russia, Yeltsin was responsible for the aging Soviet stockpile of chemical weapons, which were stored at seven depots on Russian soil.

Although few people read Mirzayanov's article in *Kuranty*, it came to the attention of the directors of GosNIIOKhT, who summarily fired him on January 6, 1992. With few prospects of finding another job, Vil tried to make ends meet by selling some of his possessions at the Moscow flea market. Several weeks later, many of his former colleagues also became unemployed when the GosNIIOKhT budget was slashed and roughly half of the scientific staff was laid off.

In mid-1992, Mirzayanov met Lev Fedorov, a professor of organic chemistry at the Vernadsky Institute of Geochemistry and Analytical Chemistry in Moscow. Although Fedorov had no ties to GosNIIOKhT and had never done classified research, he had a strong personal interest in the history of the Soviet chemical warfare program. The two men agreed to collaborate on an article for the weekly newspaper *Moskovskiye Novosti* (Moscow News), which was published on September 16, 1992, under the headline A POISONED POLICY. The article alleged that because of inadequate safety systems, GosNIIOKhT was venting toxic fumes into the Moscow air that threatened the health and safety of city residents. In the event of a major fire or explosion at the institute, Mirzayanov and Fedorov wrote, eight to ten kilograms of superlethal nerve agents might be released into the atmosphere, giving rise to a humanitarian disaster that could rival the 1986 nuclear accident at Chernobyl.

Mirzayanov also granted an interview to journalist Will Englund, the Moscow correspondent for the *Baltimore Sun*, who subsequently wrote two detailed articles on the Foliant program. The first, published on September 15, was titled "Ex-Soviet Scientist Says Gorbachev's Regime Created New Nerve Gas in '91." Four days after the second Englund article appeared on October 18, the Russian authorities moved into action. At seven in the morning of October 22, agents from the Russian Federal Security Service (FSB), the successor to the Soviet KGB, hammered on the door of Mirzayanov's two-room apartment in Moscow. While his wife and two young sons looked on in terror, the secret police arrested Vil and searched the apartment for classified documents. Mirzayanov was then taken to Lefortovo, the infamous prison for political dissidents in downtown Moscow. Although no sensitive materials had been found in his home and the newspaper articles had not revealed any technical details about the Foliant nerve agents, Mirzayanov was charged with divulging state secrets in violation of Article 75 of the Russian criminal code. The FSB also arrested and interrogated Federov, but he was released because he did not have access to secret information.

Mirzayanov was imprisoned at Lefortovo for eleven days without access to a lawyer. His first two days were spent in solitary confinement and the other nine sharing a cell with two other prisoners. Finally he was granted a hearing before a judge. Mirzayanov argued that because he posed no danger to society and had two young children, he should be released from prison and kept under house arrest. The judge agreed, on the condition that Vil remain in Moscow and report daily to FSB headquarters for interrogation. After a struggle, Mirzayanov was granted permission to retain counsel, but neither he nor his lawyer was allowed to review the secret law under which he was being charged or the prosecutor's list of counts. Paradoxically, the FSB gave Mirzayanov access to dozens of top secret Foliant documents to help him prepare his defense. The chemist suspected that the Russian government intended to make an example of him and that his chances of getting a fair trial were slim. Indeed, GosNIIOKhT deputy director Alexander Martinov vowed that Mirzayanov would be convicted and sent to prison for the rest of his life.

AT THE SAME TIME that the Mirzayanov drama was playing out in Moscow, the CWC negotiations at the Conference on Disarmament were reaching a critical stage. The discovery of Iraq's massive chemical arsenal in the aftermath of the Persian Gulf War had injected a new sense of urgency into the Geneva talks. Although Saddam Hussein had fortunately not resorted to his stockpile of chemical weapons, the possibility of their use against coalition forces had highlighted the fact that chemical proliferation posed a clear and present danger to international security. The CWC negotiators recognized that they had a narrow window of opportunity to conclude the treaty before political interest waned and consensus again became elusive. To make the best use of this momentum, the CWC Working Group set the end of 1992 as the deadline for completing its work and the participating states agreed to remain in continuous session until then. In August 1991, the French government also made an important symbolic gesture by formally canceling the ACACIA binary weapons program and committing to join the future treaty.

The final phase, or "endgame," of the CWC negotiations focused on working out detailed provisions for verification and on-site inspection, which accounted for much of the two-hundred-plus pages of the rolling text. To accelerate the process, the Australian government took the initiative of developing a "model treaty" in which all of the bracketed sections in the rolling text were replaced with compromise language that aimed to bridge the gaps among national positions on the outstanding issues. To seek American support, the Australian negotiating team flew to Washington and conducted secret talks for several days with senior U.S. officials. Although the model treaty differed from the U.S. positions on a number of important points, the Bush administration supported the Australian initiative in the hope that it would move the process forward. The Australians then made a tour of other major capitals in the fall of 1991.

On March 19, 1992, Australian Foreign Minister Gareth Evans presented the model treaty in Geneva. Although most delegations found something to criticize in the Australian draft, it marked a turning point in the negotiations. In May 1992, the chairman of the Ad Hoc Committee on Chemical Weapons, German Ambassador Adolph Ritter von Wagner, launched his own effort to seek out the delicate compromises needed to conclude the treaty. Drawing on the agreed-upon portions of the rolling text, elements of the Australian model treaty, "vision papers" prepared by informal working groups called "friends of the chair," and a great deal of backroom consultation, Wagner prepared his own "chairman's text" that offered compromise language on the major unresolved issues.

The introduction of the chairman's text on June 22, 1992, led to a phase of intensive negotiations at the ambassadorial level during which some key issues—such as the monitoring of chemical industry plants and the conduct of short-notice "challenge" inspections at facilities suspected of illicit activity—were resolved or brought close to resolution by the end of June. On July 23, the U.S. delegation accepted the chairman's text, but a number of other countries remained unsatisfied. After a major push to hammer out the details of the challenge-inspection procedure, Wagner prepared a second revised chairman's text that was introduced on August 7, 1992. This version finally won the general approval of all delegations. Even so, the timely conclusion of the CWC was possible only through the liberal use of "creative ambiguity," or language that could be interpreted in different ways, to paper over substantive differences among delegations. One contentious issue that was finessed in this manner was the continued existence of the Australia Group, which some developing countries opposed as discriminatory.

On September 3, 1992, the last day of the negotiating session, the Conference on Disarmament adopted the CWC and sent it on to the U.N. General Assembly in New York, which endorsed it by consensus. The treaty was then opened for signature at a formal ceremony in Paris on January 13–15, 1993, at which time 130 countries signed. In one of the last acts of the presidency of George H. W. Bush, Acting Secretary of State Lawrence Eagleburger signed the CWC for the United States. The timing of this event was significant because President Bush had worked hard to conclude the chemical weapons ban and considered it one of the major achievements of his administration.

At the CWC signing ceremony in Paris, the 130 initial signatories adopted a separate resolution establishing a Preparatory Commission, or PrepCom, to address twenty-three issues that had not been fully resolved during the Geneva talks and to negotiate detailed procedures for implementation where the treaty text was vague or did not provide sufficient guidance. Examples of such details included declaration formats, lists of inspection equipment, and other technical issues related to the conduct of on-site inspections. Another responsibility of the PrepCom was to establish a new international organization that would oversee the implementation of the CWC after it entered into force. Named the Organization for the Prohibition of Chemical Weapons (OPCW), this new entity would be headquartered in The Hague, the capital of the Netherlands, and would include a Technical Secretariat with an international staff of about five hundred people, including some two hundred inspectors.

The CWC was considered a major step forward in the field of arms control because it greatly extended the Geneva Protocol's prohibition on the use of chemical weapons in war by banning their development, production, stockpiling, and transfer. It also required member states to destroy within a decade their existing stockpiles of chemical arms and to eliminate all former chemical weapons production facilities or convert them irreversibly to peaceful purposes. Most ambitiously, the treaty included extensive verification provisions for monitoring chemical industry plants, so as to permit the peaceful applications of chemistry while preventing its exploitation for hostile purposes.

Despite the great promise of the CWC, the launch of the treaty was marred by the refusal of a number of important Arab states to sign. Egypt, Syria, Lebanon, Libya, and Iraq all argued that chemical weapons could be eliminated from the Middle East only in the context of a regional ban on all weapons of mass destruction, including Israel's undeclared nuclear arsenal. Iran, for its part, signed the CWC but denied possessing a stockpile of chemical arms or any current chemical weapons production facilities, despite U.S. government allegations to the contrary. One week after the CWC signing ceremony in Paris, President Bush's successor, William J. Clinton, took the oath of office. From then on, it would be his responsibility to persuade the U.S. Senate to approve the ratification of the CWC.

In Russia, meanwhile, the Mirzayanov case continued to spark controversy. On February 4, 1993, Vladimir Uglev, now fifty, granted an interview to the Russian magazine *Novoye Vremya* (*New Times*). Outraged by Mirzayanov's arrest, he had decided to corroborate the chemist's allegations and disclose his own involvement in the Foliant development program. Having discounted the security rationale for acquiring chemical weapons, he believed that the Russian military had no real concept for their use and viewed them simply as a vehicle for obtaining state prizes, perks, and research grants.

In the *New Times* interview, Uglev described the development of A-232 by Kirpichev and himself. He warned that unless the charges against Mirzayanov were dropped, he would disclose the chemical formulas of the Novichok agents, which could easily be manufactured by other countries once their molecular structures were known. Because Uglev was a people's deputy from Volsk and Shikhany township, he enjoyed immunity from prosecution. Infuriated by Uglev's remarks, his superiors at GITOS locked him out of the laboratory and asked the Volsk City Council to strip him of legal protections so that he could be put on trial for revealing state secrets.

The Russian government also tried to crack down on the journalists and periodicals involved in the Mirzayanov and Uglev exposés. On April 8, 1993,

FSB officials interrogated *Baltimore Sun* reporter Will Englund and tried unsuccessfully to intimidate him into testifying against Mirzayanov. Then in June, an FSB colonel from Saratov District arrived at the Moscow offices of *New Times* and *Moscow News,* which had published interviews with Uglev, and demanded all tapes, notes, or original documents pertaining to the scientist. Both newspapers provided copies of their published articles but courageously refused to hand over any source materials.

Despite the ongoing controversy, GosNIIOKhT continued to develop the Novichok binary agents. In the fall of 1993, Professor Georgi Drozd discovered a new formulation called Novichok-7, which had a volatility similar to that of Soman but was about ten times more potent. A few dozen tons of Novichok-7 were produced for experimental testing at Nukus and Shikhany. In addition, two more binary agents, Novichok-8 and Novichok-9, were in the development pipeline.

As Mirzayanov's case worked its way through the Russian legal system, he was featured repeatedly on national television and in the press. His plight also attracted the attention of scientific and human rights organizations in Germany, Britain, the Netherlands, Canada, Italy, Sweden, and the United States. Two human rights activists in Princeton, New Jersey, Gale M. Colby and Irene Goldman, decided to devote themselves to Mirzayanov's case and tirelessly lobbied journalists, opinion leaders, and members of Congress on his behalf. They persuaded two senior members of the Senate Foreign Relations Committee, Senators Bill Bradley (D.–New Jersey) and Jesse Helms (R.–North Carolina), to write letters of protest to President Yeltsin. Finally, acting under instructions, U.S. Ambassador Tom Pickering held a news conference in Moscow in which he defended Mirzayanov for "telling the truth about an activity which is contrary to treaty obligations."

Despite the growing international protests, the Russian Office of the Prosecutor General moved forward with Mirzayanov's closed trial, which began in Moscow City Court on January 24, 1994. If convicted of revealing state secrets, the chemist faced up to eight years in prison. Six weeks into the trial, however, the Yeltsin government realized that a conviction would confirm Mirzayanov's allegations, and on March 11, the acting prosecutor general dismissed the case for "lack of evidence." Although the People's Court ordered GosNIIOKhT to pay Mirzayanov 30,000 rubles in financial and emotional damages, the institute refused to pay and instead filed a counter-suit for 33 million rubles. Vil's lawyer urged him to continue the legal battle for compensation, but as a divorced father of two young sons, he decided to abandon the struggle.

On February 16, 1995, Mirzayanov traveled to Atlanta, Georgia, to speak at the annual meeting of the American Association for the Advancement of Science and receive the organization's Scientific Freedom and Responsibility Award. During the ceremony, he met Gale Colby, the Princeton woman who had campaigned for his release. After his return to Moscow, the two began a correspondence, and several months later, Mirzayanov, then sixty-three, moved to Princeton and married Colby. Following his arrival in the United States, the Russian chemist was debriefed extensively by the CIA and offered a research position at Edgewood. As a condition of his employment, he had to undergo a security background investigation and a polygraph exam. During the lie detector test, the CIA examiner asked him repeatedly, "Are you a spy?" Mirzayanov was outraged by this question. "Why do you ask me that?" he shouted. "You can judge for yourself whether or not my information is correct." Deeply insulted, he withdrew his application.

Over the next few years, Mirzayanov wrote his memoirs and continued to follow developments in Russia. Much as he had feared, the toxic contamination caused by nerve agent production at Volgograd and Novocheboksarsk had left a bitter legacy of chronic illness, birth defects, and environmental damage. In February 1995, the chairman of the Union of Khimprom Workers at Novocheboksarsk wrote an open letter to the international community that read in part, "Our health has been ruined. Many of us see that our work in V-gas production affected [the] health of our children as well. Ecology of our town has been undermined. . . . Our health steadily deteriorates [and] the nervous system (central and peripheral) collapses, as does the liver, the heart fails."

IN RESPONSE to Mirzayanov's revelations about the Novichok program, the Clinton administration conducted a behind-the-scenes dialogue with the Russian government about the accuracy of Moscow's declarations under the 1989 Wyoming MOU. The second phase of the data exchange had been delayed for several years by the collapse of the Soviet Union and the ensuing disarray in the Russian government. In January 1994, Presidents Bill Clinton and Boris Yeltsin had held a summit in Moscow at which they agreed to implement a scaled-down version of Phase II of the Wyoming MOU. The two sides had exchanged data in April and May 1994 and then conducted five "practice" inspections at declared government chemical weapons facilities in each country between August and December. Subsequently, both Washington and Moscow raised questions about the completeness of the data submitted by the other.

The United States had three areas of concern about the Russian declarations. First, the total size of the declared Russian stockpile, at 40,000 metric tons (80 percent nerve agents, 20 percent blister agents), appeared too low to be consistent with other evidence. Second, given the scale of the Soviet chemical weapons program, the Russians had listed very few development facilities. Under the terms of the Wyoming MOU, both countries were supposed to identify all buildings that devoted more than 50 percent of their manpower, floor space, or funding to chemical weapons development. Although the United States had declared more than a hundred such buildings, Russia had declared only one—despite reports in the Russian press that at least three clandestine chemical weapons development centers existed in the Moscow region alone.

Finally, the Russian government had provided no information in its Wyoming MOU declarations to clarify the allegations by Mirzayanov and Uglev about the development and production of the Novichok agents. During discussions with U.S. officials, the Russians did not dispute the facts that Mirzayanov had disclosed—only their interpretation. They admitted having conducted research on a new class of nerve agents but maintained that the Wyoming MOU and the BDA required declaring only stockpiled weapons, not small amounts of agent produced for development and testing purposes. Although some members of the U.S. intelligence community suspected that the Russians had manufactured significant quantities of the Novichok agents, the evidence was not clear-cut. Despite several rounds of discussions with Russian officials, the open questions were never resolved to Washington's satisfaction.

Although Mirzayanov no longer had direct contacts with GosNII-OKhT, he believed that secret work on the Novichok agents continued. He had learned, for example, that Pyotr Kirpichev, now in his early fifties, was working at a secret military research institute in Shikhany, about a mile and a half from GITOS. Mirzayanov worried that the Novichok agents posed a serious proliferation risk because their production could be concealed within commercial chemical plants, greatly complicating the verification of the CWC.

Meanwhile, Lieutenant General Anatoly Kuntsevich was facing some legal problems of his own. On April 7, 1994, President Yeltsin dismissed him from the post of senior adviser on chemical and biological arms control for "gross violation of his duties" and replaced him with his former deputy, Pavel Pavlovich Syutkin. It appeared that Kuntsevich had signed an agreement with the Syrian government in 1992 to create a Syrian Center for Ecological Protection and had supplied it with laboratory equipment and materials. In October 1995, the general was aboard a government aircraft preparing to depart Moscow on an official visit to Damascus when the flight was halted and FSB agents took him into custody. Kuntsevich was charged with having shipped 800 kilograms of V-agent precursors to Syria in 1993 and the attempted smuggling of an additional 5.5 tons in 1994. At the time, Kuntsevich was running as a right-wing candidate for the Russian State Duma (the lower house of Parliament) and claimed that the charges against him were politically motivated. Several months later, however, the charges were dropped under mysterious circumstances, and Kuntsevich retained his prestigious post of academician with the Institute of Chemical Physics of the Russian Academy of Sciences until his death in September 2002. Reportedly, he was on a plane en route from Moscow to Damascus when he suffered a fatal heart attack.

IN APRIL 1994, more than three years after the end of the Persian Gulf War, the UNSCOM Chemical Destruction Group overseeing the elimination of Iraq's stockpile of chemical weapons finally completed its work. Ron Manley led a small team of inspectors to the Muthanna State Establishment in June to certify that the facility was "clean" and handed the keys back to the Iraqi authorities. Despite this milestone, UNSCOM's work was far from over. From then on, the U.N. inspectors would monitor Iraq's chemical manufacturing plants with air-sampling systems, closed-circuit video cameras, and surprise on-site inspections to ensure that Saddam Hussein did not secretly reconstitute his chemical arsenal. UNSCOM would also continue to restrict imports of chemical precursors and dual-use production equipment, and attempt to clarify several remaining uncertainties and discrepancies, particularly with respect to the production of VX.

Beyond the specific case of Iraq, the U.S. government was increasingly concerned that chemical weapons would fall into the hands of so-called rogue states that sponsored terrorism and were hostile to the United States. Officials at the Departments of State and Defense distinguished between two types of proliferation: "horizontal," meaning the spread of basic chemical warfare capabilities to additional countries; and "vertical," meaning the acquisition of more advanced agents and delivery systems by states with established chemical warfare programs, such as Iran and Syria. There was also the phenomenon of "secondary" proliferation, in which a country with an established weapons program transferred relevant equipment, materials, and know-how to other states.

Governments had various motives for pursuing chemical arms, including the search for status and prestige, the need for a low-cost "force multiplier" to enhance the effectiveness of their conventional forces, and the desire to deter attack from a hostile neighbor or to balance an unconventional threat. General Mamdouh Hamed Ateya, the former head of Egypt's Chemical Warfare Directorate, argued in 1989 that Arab countries were justified in acquiring chemical weapons as a counterweight to Israel's undeclared nuclear arsenal. Another factor promoting proliferation was the worldwide diffusion of chemical production technology and know-how. For many poor nations, acquiring a domestic chemical industry was vital to their economic and social development, giving them the ability to manufacture agricultural chemicals and essential drugs. Yet any nation with a moderately advanced chemical industry was potentially capable of producing blister and nerve agents.

Equally troubling was the role of unscrupulous foreign suppliers in abetting the spread of chemical weapons. Despite the existence of the Australia Group, the huge volume and globalization of chemical trade made tracking shipments of dual-use chemicals extremely difficult. Proliferators could circumvent the Australia Group controls by ordering precursors from nonmember states or by using middlemen, front companies, transshipment points, falsified end-use certificates, and other forms of deception. As Julian Perry Robinson of the University of Sussex observed, "There are so many brokers, so many intermediaries, that it takes a skilled investigator to track these things down. A single trainload of chemicals can change hands six times on its way from the factory to the port, so all trace of its origin gets lost."

Finally, intelligence information about the smuggling of chemical weapons precursors was often unreliable. In July 1993, for example, the U.S. intelligence community received a tip that the Chinese cargo ship *Yin He* (Galaxy) was on its way to Iran with supplies of two chemical weapons precursors: thiodiglycol and thionyl chloride. U.S. warships tailed the *Yin He,* and American officials requested to search the vessel at one of its ports of call in the Persian Gulf. After resisting the U.S. demand for several weeks, China finally agreed under duress to allow the ship to be inspected in the Saudi port of Damman from August 24 to September 4, prior to its arrival in Iran. Although inspectors searched the *Yin He* from top to bottom, they failed to find the prohibited chemicals. Some observers theorized that the illicit cargo—if it had actually existed—had been dumped at sea. Another rumor was that the chemicals had been delayed in the Chinese rail system and had arrived in Shanghai after the ship had left. According to this view, Chinese officials had agreed to the inspection only when they knew that nothing would be found. In any event, the *Yin He* incident embarrassed the U.S. government and sharply increased tensions with Beijing, which accused Washington of acting "in an utterly indiscreet and irresponsible manner."

IN ADDITION TO the proliferation of chemical weapons to rogue states such as Saddam Hussein's Iraq, an equally frightening scenario was that they would fall into the hands of terrorist organizations. During the mid-1990s, the nightmare of chemical terrorism materialized in an unexpected place—Japan. The individual responsible for this development was equally improbable: a chubby, half-blind yoga instructor and cult leader known as Shoko Asahara, whose real name was Chizuo Matsumoto.

Matsumoto had been born in 1955 to a poor family of tatami mat-weavers on Japan's southern island of Kyushu. A case of infantile glaucoma had left him blind in his left eye and partially sighted in the right, exposing him to constant teasing and harassment from his childhood classmates. To escape this situation, Chizuo's parents enrolled him in a boarding school for the blind, where he took advantage of his partial eyesight and stocky build to bully and dominate his classmates. After completing high school, Matsumoto aspired to wealth and fame, but he failed the entrance exam to Tokyo University—the Harvard of Japan—leaving him angry and bitter. His luck finally improved when he met a young college student and married. With money from her parents, Chizuo opened an alternative health clinic in Tokyo that treated patients with acupuncture and yoga and sold quack herbal remedies at exorbitant prices. The clinic soon became a thriving business, yet despite his newfound success and wealth, he continued to search for spiritual fulfillment.

Matsumoto dabbled in geomancy (an ancient form of divination involving the "reading" of handfuls of soil scattered in the ground), Chinese fortune-telling, and meditation. In February 1984, inspired by the popularity of the "new religions" craze in Japan, the twenty-nine-year-old founded a yoga school called the Aum Association of Mountain Wizards. Matsumoto offered his followers "karmic cleaning," or absolution for past sins, and supernatural powers such as the ability to levitate and see through walls. These promises of psychic awakening attracted alienated Japanese young people seeking to fill the spiritual void in their lives, and the cult grew rapidly.

In July 1987, after a trip to the Himalayas during which he supposedly achieved enlightenment, Matsumoto renamed his yoga school Aum Shinrikyo, after the Sanskrit word *aum,* meaning the power of creation and destruction in the universe, and the Japanese word *shinrikyo,* or "teaching supreme truth." He cobbled together a belief system from an eclectic mix of sources, including Tibetan Buddhist meditation, worship of the Hindu god Shiva, the prophecies of the sixteenth-century French seer Nostradamus, the

Shoko Asahara, the leader of the Aum Shinrikyo doomsday cult, appears on Japanese television in 1995. The Japanese characters read: "Aum Shinrikyo Representative Shoko Asahara."

Christian Book of Revelations, and occult and pseudoscientific ideas. Matsumoto also shed his prosaic identity and assumed the charismatic persona of a guru, a transformation that involved adopting the name "Shoko Asahara," growing a thick black beard and shoulder-length hair, and dressing in the white robes of a holy man. Asahara claimed to be the reincarnation of Jesus Christ and the first enlightened being since the Buddha. He began to make millenarian prophecies in which he described an apocalyptic nuclear war between the United States and Japan in the year 2003 that would devastate Tokyo and other major Japanese cities, and that only members of Aum would survive.

By 1988, Aum Shinrikyo had about 3,000 members and had opened branch offices in major cities throughout Japan. The cult established its main headquarters about seventy miles from Tokyo in Kamikuishiki, a village of 1,700 people in the Mount Fuji foothills, a picturesque region of parks, golf courses, and dairy farms. After a few years of intense construction, the cult's walled compound, called the Mount Fuji Center, consisted of a motley array of wooden shacks, prefab buildings, warehouses, and ramshackle dormitories. Believers paid $2,000 to attend weeklong meditation seminars and training courses. Those who decided to become full-time "monks" or "nuns" lived in cult housing and had to donate their entire net worth to Aum, including real estate. In this way, the cult accumulated many valuable tracts of land at a time of skyrocketing property values.

In April 1989, Asahara and about two hundred of his followers visited the Tokyo Metropolitan Building to protest the city government's delay in recognizing Aum Shinrikyo as a religious organization. The cultists aggressively lobbied city officials and bombarded the vice governor with telephone calls at home. Four months later, the Tokyo government granted Aum official religious status, entitling it to special tax breaks and legal protections.

In addition to the money donated by members, the cult invested in real estate and launched a series of lucrative businesses: cut-rate computer manufacturing and retail, yoga centers, a restaurant chain, and sales of books, videos, and religious paraphernalia. The latter included samples of Asahara's blood, beard clippings, and even his dirty bathwater (known as "Miracle Pond"), which sold for $800 a quart. Aum also engaged in criminal activities such as extortion, kidnapping, insurance fraud, and the manufacture of illicit drugs and explosives, and formed links with Japanese organized-crime syndicates (the *yakuza*) to acquire firearms and sell drugs. Through these diverse ventures, Aum amassed a vast fortune worth several hundred million dollars and valuable properties in Japan and abroad, including a trading company in Taiwan, a tea plantation in Sri Lanka, and a sheep station in Australia.

Aum's organizational structure was rigidly hierarchical, with various levels of priests and laypersons reporting to Asahara, who was known as "His Holiness the Master." Senior cult officials had the title "Seidaishi" (High Master) and, one rank below, "Seigoshi" (Master). To maintain control over the rank and file, cult leaders employed a blend of indoctrination, discipline, coercion, and physical violence. Whereas Asahara lived a decadent, luxurious lifestyle, Aum devotees were required to cut all ties to family and friends, eat a meager diet of boiled vegetables, sleep as little as three hours a night in spartan dormitories, and devote all of their waking hours to manual labor and the study of Asahara's teachings. During Aum meditation services, cult members wore battery-powered caps containing electrodes that delivered light shocks to the scalp, supposedly synchronizing the wearer's brain waves with those of the guru. Individuals who deviated from the rules or attempted to leave the cult were punished with beatings, solitary confinement, and even death. These effective "mind-control" techniques gradually caused Aum monks and nuns to suspend their personal judgment and conscience and subordinate their personality to Asahara's will.

In order to attract more adherents, Asahara gave sermons in which he depicted the coming nuclear Armageddon not as something to be feared but as an opportunity to eradicate the evils of society and bring about a spiritual transformation. From the ruins of the postapocalyptic world, he preached, the members of Aum would create a new race of superhuman beings. To heighten his followers' sense of urgency, Asahara gradually moved up the predicted date of doomsday. "People who have acquired the power of God through the right kind of training will be the ones to create a new world after 1997," he said. In addition to ordering the harsh treatment of deviant members, Asahara began to direct violent attacks against external critics, rivals, and other perceived enemies. To this end he portrayed extreme acts, including murder, as "challenges" to be overcome in the process of spiritual training. Asahara also developed religious concepts that rationalized killing: *poa,* or sacrifice for the benefit of the victim's soul, and *vajirayana,* which justified eliminating anyone hostile to the Aum faith.

Asahara's dictatorial authority within the cult fed his megalomania and lust for power. Aspiring to high government office, he established a political wing called Shinrito, or Supreme Truth Party. In the February 1990 parliamentary elections, Asahara headed a slate of twenty-five candidates and ran for a seat from the fourth district of Tokyo. Much to his shock and dismay, none of the Aum candidates was elected and he himself received only 1,783 votes, fewer than the number of cult members living in his district. Because Asahara had never imagined that his election bid would fail, the defeat came as a crushing blow to his self-esteem. He also faced growing legal problems as Aum's rapid growth and aggressive tactics embroiled it in a series of bitter disputes with its neighbors and the civil authorities, including lawsuits against the cult for land fraud.

In the face of these setbacks, Asahara came to believe that the Japanese government had rigged the parliamentary election to deny him his rightful victory and that the lawsuits were conspiratorial acts of oppression by officials bent on persecuting and destroying him. Obsessed with these paranoid delusions, he abandoned the peaceful road to political power and began to plot the violent overthrow of the Japanese state, with the goal of establishing a theocratic regime under his unquestioned leadership. To prepare for the future takeover, Asahara organized the Aum leadership into a shadow government modeled closely after the Japanese executive branch, with twenty-two "ministries" led by top cultists. For example, the "Minister of Welfare" was Seichi Endo, thirty-four, who had studied biochemistry at Kyoto University and researched AIDS and cancer before joining Aum; the "Minister of Intelligence" was Yoshihiro Inoue, twenty-five, a former freelance writer; and the "Minister of Science and Technology" was Hideo Murai, thirty-six, who had studied astrophysics at Osaka University and worked at the Kobe Steel Company before becoming one of the founding members of Aum.

FOR ASAHARA, ACQUIRING powerful weapons became the path to achieving his political ambitions, which he kept secret from all but the most senior cult leaders. At first he planned to establish a militia of several hundred armed troops, who would stage an uprising against the Japanese government and kill all high-ranking officials. This scheme entailed the construction of a factory to mass-produce Kalashnikov assault rifles with computer-controlled machine tools, and efforts to recruit members from the Japan Self-Defense Forces. Asahara also sought to acquire biological, chemical, and even nuclear weapons by recruiting brilliant but alienated young scientists from leading Japanese universities. Developing unconventional arms was the primary task of Aum's Ministry of Science and Technology, whose 250 members made it the largest of the cult ministries.

In 1990, Seichi Endo launched an effort to produce germ weapons. He and a team of cultists built a microbiological laboratory inside a prefab building on the Mount Fuji compound. Endo ordered a vial containing a sample of anthrax bacteria from a commercial supplier and cultivated it into a liquid suspension of bacterial spores. On July 1, 1993, Aum members staged a biological attack with the intent of killing many thousands of people. They dispersed the liquid slurry of anthrax spores with an aerosol sprayer mounted on the roof of an eight-story Aum building in Kameido, an area of eastern Tokyo. During the dispersal, neighbors complained to the local environmental health authority about a foul-smelling mist emanating intermittently from the building's cooling tower. Much to Asahara's disappointment, however, the release of anthrax spores did not cause any illnesses or deaths. Endo, trained in virology rather than bacteriology, had inadvertently ordered the Sterne strain, a harmless form of the anthrax bacterium that is widely used as a veterinary vaccine.

Meanwhile, Aum was expanding its activities in Russia, where it had a large number of followers, a prime-time radio program, and several affiliated companies. When the cult had begun to operate in the Soviet Union in 1990, it had tried to build close ties with the ruling Communist Party. After the Soviet breakup in December 1991, Asahara moved to cultivate people close to Russian President Boris Yeltsin. Post-Soviet Russia was a "Wild East" of unbridled capitalism and corruption in which almost everything was for sale, and Aum operatives carried suitcases full of cash with which to bribe Russian officials. One such individual was Oleg Lobov, a top security adviser and longtime associate of Yeltsin from his hometown of Sverdlovsk. In return for payoffs of as much as $12 million to the Russia-Japan Foundation that Lobov chaired, he granted Aum access to senior officials in the Yeltsin government, arranged for the military training of cult members at Russian army bases, and had Aum removed from the list of organizations that the FSB kept under routine surveillance.

In early March 1992, Aum's public affairs officer, Fumihiro Joyu, organized a trip to Moscow for Asahara. Through Lobov's personal intervention, the cult leader met with Russian Vice President Alexander Rutskoi and

Speaker of the Supreme Soviet Ruslan Khasbulatov. Another purpose of the Russia trip was to further Aum's pursuit of unconventional weapons. Asahara ordered one of his trusted aides, the engineer Kiyohide Hayakawa, to stay behind in Russia and collect useful information. Lobov helped the cult scientist to acquire Soviet chemical weapons expertise. Reportedly, Hayakawa purchased the blueprints of a Sarin production facility for about $100,000.

In March 1993, Asahara met with Masami Tsuchiya, thirty, the head of the cult's Department of Chemistry. Tsuchiya had earned a master's degree in physical organic chemistry from Tsukuba University but had abandoned a promising scientific career to become a devout member of Aum. Sporting a crew cut and a goatee, he was an introvert whose sole obsession was the synthesis of interesting chemicals, including powerful explosives and illicit drugs such as LSD, methamphetamine, and mescaline.

Asahara ordered Tsuchiya to synthesize a variety of chemical warfare agents in order to decide which one to mass-produce. Over the next few weeks, the chemist worked intensively in his small personal laboratory in the Kushtigarba building, a windowless prefab at the Mount Fuji compound that had double walls to prevent toxic gases from escaping. To handle lethal chemicals safely, he used a homemade fume cabinet with long rubber gloves into which he inserted his hands.

By April, Tsuchiya had selected Sarin as the best agent because of its lethality, its relative ease of production, and the availability of raw materials. Asahara agreed that Sarin would be the ideal weapon to fulfill his apocalyptic prophecies and trigger the widespread chaos that the cult would need to take over the Japanese government. The cult leader was also intrigued by the fact that Sarin had been invented in Nazi Germany and was closely associated with his hero, Adolf Hitler.

CHAPTER SEVENTEEN

THE TOKYO SUBWAY

In June 1993, Aum construction crews finished work on a windowless, three-story building at the Mount Fuji Center called Satian 7. (All major structures on the compound were given the name "Satian," derived from the Sanskrit word for "truth," followed by a number.) From the outside, the building was a shabby warehouse surrounded by piles of dirt, rubble, and empty boxes. A cluster of large air ducts emerged from one wall and fed into a shack containing powerful ventilation fans. Despite its nondescript appearance, Satian 7 was said to be so holy that only monks who had reached a high level of spiritual enlightenment could enter. Security guards stood at the front door and limited access to cultists wearing a special badge.

Behind the heavy steel door of Satian 7 was a maze of narrow, dimly lit corridors that opened up into a shrine to Shiva, the cult's chief deity, consisting of an altar and a golden statue of the multiarmed Hindu god of destruction. In fact, the shrine was an elaborate facade. Hidden behind the statue was a two-story distillation column, and above the false ceiling were stainless-steel holding tanks for raw materials and chemical intermediates.

On the second floor of the building, behind a submarine-hatch door, was the heart of the Sarin manufacturing plant, which the cult had ordered through front companies at a total cost of about $10 million. Suspended on a steel scaffolding were five reactor vessels made of corrosion-resistant Hastalloy, along with heat exchangers, injectors, and pumps, interconnected with pipes and electrical wiring. A computerized process control system automatically regulated the flow of chemicals and the reaction temperatures. Next to the manufacturing area was a control room where

The main compound of the Aum Shinrikyo doomsday cult, located in the vicinity of Mount Fuji in Japan. The three-story building in the foreground is Satian 7, where Aum attempted to manufacture large quantities of Sarin. Ventilation pipes can be seen protruding from the side of the building at the lower left.

Aum members, wearing white robes and electronic meditation caps, monitored the production sequence on closed-circuit television screens.

On the third floor of Satian 7 were a dozen small offices for Aum chemists and engineers, divided by thin partitions and crammed with laboratory equipment, computers, and books on chemistry and weapons technology. Another room contained a bar stocked with bottles of plum, grape, and rice wine. Although Asahara prohibited ordinary members from consuming alcohol, cult leaders held secret drinking parties there.

Day and night, about a hundred Aum members labored in shifts to get the Sarin plant into full operation, but technical problems led to repeated delays. Leaks of toxic fumes from faulty welds caused nosebleeds, eye irritation, and severe fatigue. Meanwhile, the nearby residents of Kamikuishiki village began to smell foul odors reminiscent of burnt plastic. Their dairy cows stopped producing milk and the leaves on the trees bordering the Aum compound wilted and died. Frightened and angry, the villagers called the district police and demanded an investigation, but cult officials denied responsibility for the odors and threatened legal action. Intimidated by Aum's aggressive tactics and wary of its protected status as a religious organization, the police backed off.

To deliver the tons of Sarin that would be produced in Satian 7, Asahara planned to use a large helicopter to spray the nerve agent over key targets, such as the Japanese Parliament building. Aum's Russian affiliate company, Mahaposya, bribed senior Russian officials to purchase a MIL Mi-17 military helicopter, which was shipped to the Mount Fuji Center. In December 1993, Asahara sent two cult members to the United States to study for helicopter pilot licenses.

By mid-February 1994, Tsuchiya had synthesized a total of 44 pounds of Sarin in Satian 7 and his personal laboratory in the Kushtigarba prefab next door. Asahara demanded that Satian 7 be ready for mass production by April 25. In the meantime, Aum began using Sarin on an experimental basis as an assassination weapon. The cult's first target was Daisaku Ikeda, the honorary chairman of the Soka Gakkai, the largest and most popular of Japan's new religions. Deeply jealous of his rival's success, Asahara ordered three cultists to murder Ikeda while he gave an evening lecture. The attackers planned to use an industrial sprayer, mounted on a truck parked near the building where Ikeda was speaking, to vaporize two pounds of Sarin. But the device malfunctioned and sprayed the nerve agent on the Aum security chief, Tomomitsu Niimi. He collapsed in convulsions and was saved only by a prompt injection of antidote.

To verify that Sarin would be an effective weapon when delivered as a vapor, Aum scientists conducted field tests at a sheep station in Western Australia that the cult had purchased in April 1993. The 500,000-acre ranch was located near the remote town of Banjawarn, about 375 miles north of Perth. Tsuchiya set up a crude testing facility there, including a small laboratory equipped with computers, Bunsen burners, glassware, and chemicals. Wearing protective suits and gas masks, Aum scientists tethered twenty-nine sheep to wooden stakes and exposed them to lethal mists of Sarin. The animals collapsed, convulsed, and died within minutes, demonstrating the potency of the nerve agent. Several years later, Australian police officials investigating the cult discovered the bones of the dead sheep, scattered over the ground.

IN JUNE 1994, Aum leaders planned their first large-scale operation involving the use of Sarin. A lengthy trial against the cult had recently ended in Matsumoto, a tourist and industrial city at the foot of the Japanese Alps, about a hundred miles northwest of Tokyo. Several local landowners had filed suit against Aum in May 1992, alleging that the cult had fraudulently purchased land through a front company. Although the stated purpose of the land acquisition had been to construct a food-processing plant, the two-story building had turned out to be a new branch office of Aum. Angered by the deception, the original owner had sued to invalidate the sale. A decision on the case by the three-judge panel was scheduled for July 19, and Aum's legal adviser had predicted that the court would rule against the cult. To prevent this outcome, Asahara ordered the assassination of the judges. He was also eager to test the effectiveness of Sarin against a populated target.

On June 20, Asahara met at his private office in Satian 6 with Aum chief scientist Murai and three other high-ranking cult leaders to discuss the operation. The plotters decided to target the local office of the Nagano District Court. To prepare for the attack, Aum technicians under the direction of mechanical engineer Kazumi Watanabe repaired and improved the Sarin vaporizer. Installed in the back of a customized refrigerator truck, the new system consisted of a steel tank filled with Sarin that was bolted to the truck's loading platform, an electric heater powered by thirty car batteries that weighed about 1,000 pounds, and a fan that would blow the lethal vapor out through a small window in the side of the vehicle.

Because Matsumoto was experiencing cool temperatures and rain showers, which would reduce the effectiveness of Sarin, the plotters waited for a change in the weather. On June 27, the skies cleared and the temperature soared, and Asahara ordered the cult leaders to take advantage of the hot, dry conditions to stage their attack. That afternnoon, Murai and his team drove the modified refrigerator truck to Matsumoto, followed by a group of lower-ranking cultists in a rented minivan. They took back roads to avoid observation by the highway police and because the weight of the batteries prevented the truck from going more than thirty miles per hour. By the time the team arrived in Matsumoto, the work day was over and the court-house was closed. They therefore decided to attack the judges' dormitory, which was located in a residential neighborhood called Kaichi Heights on the north side of Matsumoto Castle, a city landmark dating from the six-teenth century.

Around 10:00 p.m., the refrigerator truck pulled into a supermarket parking lot in Kaichi Heights. While the occupants of the minivan kept lookout, the senior cultists prepared for the operation: the team doctor injected them with Sarin antidote, and they donned homemade gas masks consisting of a plastic bag connected by a tube to an oxygen cylinder. The two vehicles then drove to a vacant lot about sixty yards upwind of the judges' dormitory and parked behind a stand of trees. Surrounding the dor-mitory were private homes, town houses, and a midrise apartment building owned by the Meiji Insurance Company.

At about 10:40 p.m., with the truck's headlights turned off but the motor running, the cultists opened the side window, switched on the bank of car batteries, and opened a spigot that began to drip liquid Sarin onto the heating plate. Because the Sarin solution contained excess hydrochloric acid, the drops hitting the hot surface erupted into a dense white mist that filled the interior of the truck with acrid fumes. Struggling for air, one of the cultists began to panic. Soon, however, the fan began to blow the toxic fog out the window. After about twenty minutes, several liters of Sarin had been vaporized, forming a cloud that, hugging the ground, floated on a light breeze in the direction of the judges' dormitory. The Aum cultists then fled Matsumoto, forgetting in their haste to turn off the vaporizer and spreading the poison further.

After the attackers had left, the wind shifted and began to carry the Sarin cloud away from the judges' dormitory and toward the Meiji Insurance apartment building. Because the summer night was pleasantly warm, many of the residents had left their doors and windows open, allowing the lethal gas to enter freely. One man was smoking a cigarette on his balcony when he became aware of a pungent, irritating odor and then developed a throbbing headache, dizziness, shortness of breath, and a blurring and darkening of vision. A woman was taking a bath when her nose began to flow with watery mucus; she experienced severe stomach cramps before losing consciousness.

At 11:09 p.m., Yoshiyuki Kono, forty-four, a machinery salesman who lived in a house near the judges' dormitory, telephoned the Matsumoto Emergency Services to report that his wife had suddenly cried out in pain and then collapsed and begun to convulse. Five minutes later, a Fire Department ambulance arrived on the quiet residential street. When Kono came out to meet the emergency medical team, he looked pale and disoriented and said that his wife had fallen into a coma. The ambulance evacuated the stricken couple and their two daughters to Kyoritsu Hospital. During the trip, a paramedic performed CPR on Kono's wife, but she had already suffered massive brain damage from oxygen starvation.

Over the next few hours, paramedics discovered five residents of Keichi Heights dead in their homes and evacuated some three hundred others to local hospitals, where fifty-four were admitted. When blood tests indicated that the victims had extremely low levels of cholinesterase, doctors treated them symptomatically for organophosphate poisoning. Two more victims died in the hospital, bringing the total number of deaths to seven, but the cause of the incident remained unknown.

AT 5:35 THE NEXT MORNING, June 28, Matsumoto police detectives wearing protective clothing and portable air supplies investigated the crime scene. Under a grove of trees near the vacant lot they found several dead animals, including dogs, sparrows, a dove, and a large number of caterpillars, and the nearby leaves and shrubs were shriveled and discolored. Kono, the first victim to report the incident, became a prime suspect when the police discovered chemicals in his home and dead fish and crayfish in a small pond in his yard. The local police investigators, who lacked scientific or forensic training, theorized that the salesman had been preparing a homemade batch of herbicide for use in his garden when the reaction had gone out of control and generated a cloud of lethal gas.

Although the Matsumoto police pressured Kono to confess, he claimed that he was innocent and had purchased the chemicals for his photography hobby. During a monthlong stay in the hospital to recuperate from minor brain damage, he was hounded relentlessly by the Japanese news media. On July 3, however, analytical chemists at the Nagano Police Science Investigation Institute identified traces of a Sarin breakdown product in samples collected near the vacant lot, where eyewitnesses had seen a van releasing a whitish vapor. Because the chemicals found in Kono's house could not have reacted to produce Sarin, it was clear that he was innocent and that the real perpetrator remained at large.

ASAHARA WAS PLEASED with the success of the Matsumoto operation—seven dead, more than two hundred injured—and the fact that Aum had not been implicated. Although the three judges targeted by the cult had survived the attack, they were too ill to attend court, forcing an indefinite postponement of the ruling in the land-fraud case. To eliminate any incriminating evidence of Aum's involvement in the incident, the attack squad had dismantled the truck-mounted vaporizer and cleaned the truck and rental van with decontamination solution. Still, in commemoration of the Matsumoto attack, an anonymous cult member penned a song that appeared in a secret Aum manual on Sarin production. The lyrics ran:

> *It came from Nazi Germany,*
> *A dangerous little chemical weapon,*
> *Sarin, Sarin!*
> *If you inhale the mysterious vapor,*
> *You will fall with bloody vomit from your mouth.*
> *Sarin, Sarin, Sarin,*
> *The chemical weapon.*
> *Song of Sarin, the Brave.*

> *In the peaceful night of Matsumoto City,*
> *People can be killed, even with our own hands,*
> *Everywhere there are dead bodies.*
> *There! Inhale Sarin, Sarin,*
> *Prepare Sarin! Prepare Sarin!*
> *Immediately poison gas weapons will fill the place.*
> *Spray! Spray! Sarin, the brave Sarin.*

Determined to stockpile enough nerve agent for mass-casualty attacks in Tokyo and major U.S. cities, Asahara set a production target for Satian 7 of two tons of Sarin per day. His goal was to amass a total of seventy tons in roughly forty days of operation. Under Murai's supervision, Satian 7 began to manufacture DMMP, a key Sarin intermediate. Because of the highly corrosive nature of the reactions and the poor quality of the welding, foul-smelling chemicals leaked from pipe fittings, dripped onto the floor, and crystallized on surfaces. Plant workers jury-rigged repairs with buckets, plastic sheeting, and duct tape, all the while suffering from dim vision, nosebleeds, and muscle spasms caused by exposure to the toxic fumes.

On July 9, 1994, a serious malfunction at Satian 7 caused DMMP to overflow the reaction tank and leak into the soil outside the building, causing foul odors and patches of dead vegetation. When villagers complained to the district police, cult officials said they were producing agricultural chemicals and threatened to sue the local authorities for harassment. Four months later, on November 16, another leak occurred, but this time chemists from the National Police Agency took soil samples from the patches of dead grass. Chemical analysis at the National Research Institute of Police Science revealed the presence of methylphosphonic acid, the Sarin degradation product that had been detected after the Matsumoto incident in June. The police also discovered that an Aum front company had purchased large quantities of precursor chemicals used in Sarin production. This circumstantial evidence clearly implicated the cult in the Matsumoto attack. Yet Japanese law enforcement agencies did not pursue the investigation aggressively because they lacked strong leadership, were overly dependent on confessions, and hesitated to confront Aum because it was a religious organization and was belligerent and litigious.

In December 1994, cultists working at Satian 7 converted the DMMP into sixty liters of DF, the immediate precursor of Sarin, which was stored in large plastic tanks inside the factory. Nevertheless, the chronic leaks in the production line made it too hazardous to start combining DF with iso-propyl alcohol to make Sarin. Fumihiro Joyu, the chief of Aum's Moscow operation, attempted to recruit Russian chemical engineers who had experience with Sarin manufacture and might be able to fix the defective equipment. When that effort failed, Tsuchiya decided to shut down the production line. The planned aerial delivery system for Sarin was also unusable. Although a Russian military helicopter had been delivered to the

Mount Fuji compound in June, the cult engineers had been unable to get it in working order.

After the failure to mass-produce Sarin in Satian 7, Tsuchiya worked in his personal laboratory to synthesize 340 grams of VX, which cult members used for a series of attempted assassinations. The targets included a lawyer working on behalf of Aum victims, an old man sheltering a dissident member of the cult, the head of the Aum Shinrikyo Victims' Association, and the leader of a rival religious organization called the Institute for Research into Human Happiness. Only one of these attacks was successful. On December 12, 1994, Niimi and another cultist posing as joggers used a syringe connected to a plastic tube to sprinkle drops of VX on the neck of Takahito Hamaguchi, twenty-eight, a former Aum member living in Osaka who was threatening to investigate the cult for production of illicit drugs. Hamaguchi chased his assailants for a hundred yards before collapsing. Comatose, he was rushed to Osaka University Hospital and died several days later without regaining consciousness. The cause of his death remained a mystery to the Osaka police.

By EARLY 1995, Aum Shinrikyo had reached the height of its power and influence. The cult had twenty facilities scattered throughout Japan, branches in six countries, and a total membership of more than 40,000, of which about 10,000 were in Japan and 30,000 in Russia. Yet Asahara felt increasingly under siege. On January 1, 1995, Japan's largest newspaper, *Yomuri Shimbun,* reported that police scientists had detected Sarin degradation products in the soil near the Aum compound in Kamikuishiki village and suspected that the cult had been involved in the Matsumoto poisoning incident.

Alerted by the newspaper article that Aum was under police surveillance, Asahara feared an imminent raid on the Mount Fuji compound and told Murai to organize a cover-up by destroying the left-over stocks of Sarin in Tsuchiya's laboratory and converting the ground floor of Satian 7 into a worship hall. Cult members washed the interior of the building with neutralizing solution, removed contaminated topsoil from around its base, destroyed documents, emptied tanks of phosphorus trichloride and refilled them with kerosene, and dumped a large quantity of DMMP into a nearby well. Although Asahara ordered cult scientists to discard the remaining stock of DF, the physician Tomomasa Nakagawa did not fully implement this directive. Because he considered the chemical intermediate too precious to destroy, he secretly buried three pounds of it and later told Murai what he had done.

Once the cleanup was complete, Asahara sought to refute the *Yomuri Shimbun* allegations by inviting journalists to visit Satian 7. On January 4, several reporters were ushered into the dimly lit shrine, which was filled with clouds of incense and dominated by a fifteen-foot Styrofoam statue of the Buddha. Although the rest of the building was off-limits to visitors, the reporters were told that the second floor was used for meditation classes and the third floor for storage. At a press conference, an Aum attorney accused the Japanese government of attacking the cult with nerve gas, a claim elaborated in a videotape titled *Slaughtered Lambs.*

In early 1995, Asahara self-published a book called *Disaster Looms for the Land of the Rising Sun* in which he warned that Armageddon was imminent and that only those who followed his teachings would be saved. Over the next few months, the pressure on Aum continued to build. In early March, provoked by the brazen abduction and murder of a Tokyo notary public whose sister had fled the cult, the Tokyo Metropolitan Police Department planned simultaneous raids on the Mount Fuji Center and other Aum facilities around the country. Police officials, fearing attacks with nerve agents, asked the Japan Self-Defense Forces to supply them with three hundred gas masks and chemical protective suits. On March 16, 1995, two Japanese army sergeants who were members of Aum learned of the police request and tipped off the cult leadership.

On the evening of Saturday, March 18, Asahara held a crisis meeting with three of his top lieutenants—Murai, Niimi, and Inoue—at a cult-owned restaurant in Tokyo. Asahara said that it would be necessary to prevent the impending police raid by causing a major diversion. Murai thought for a while and then suggested spraying Sarin on the Tokyo subway. "That would cause panic," Asahara said approvingly, and he ordered Murai to take charge of the operation.

EACH WORKDAY, more than five million commuters in the Tokyo metropolitan area ride thirteen subway lines covering 293 kilometers of underground and surface tracks. The system is clean, safe, efficient, and almost unfailingly punctual. During rush hour, the trains are so densely packed that white-gloved attendants push riders into the cars so that the doors can close. To ensure that a Sarin attack on the Tokyo subway would have maximum impact, the Aum plotters decided that the primary target should be the Kasumigaseki transfer station in the heart of Tokyo. Because Kasumigaseki was close to the Tokyo Metropolitan Police Department, the National Police Agency, the Ministry of Foreign Affairs, the Ministry of Finance, and other Japanese government agencies, a Sarin attack at the height of the morning rush hour would cause enormous disruption.

Murai planned the simultaneous release of Sarin in five subway trains traveling on three different lines—the Hibiya, Marunouchi, and Chiyoda—that would all converge on Kasumigaseki station between 8:00 and 8:10 a.m. Conducting the operation would be five two-man teams of Aum cultists, each consisting of a subway rider and a getaway car driver. As the subway riders, Murai selected four members of his Ministry of Science and Technology, along with Dr. Ikuo Hayashi, Aum's sadistic Minister of Health. Although some of the chosen cultists were terrified by the assignment, they were too cowed to resist. Expressing even the slightest doubt about Asahara's directives was considered a sign of shallow faith and inadequate religious training.

Because Asahara had ordered the subway attack for early Monday morning, the cult scientists had only twenty-four hours to produce an adequate supply of Sarin and devise a suitable delivery system. At Murai's direction, Endo and Tsuchiya decided to synthesize the nerve agent from the leftover supply of DF that Nakagawa had buried. Murai told Nakagawa to recover the stash and bring it to Endo's personal laboratory in a windowless prefab, which was equipped with a crude fume hood made of wood and glass that vented to the outdoors.

Under Tsuchiya's supervision, Endo and Nakagawa, wearing homemade oxygen masks as an additional precaution, worked through the night of March 18 to produce about seven liters of Sarin under the fume hood. They performed the synthesis in three-necked flasks containing a mixture of DF, hexane solvent, and a catalyst. Controlling the reaction temperature, the Aum chemists slowly dripped isopropyl alcohol into the DF solution to produce Sarin. Because there was no time to distill the final product, it was only about 30 percent pure. The cultists also added a small amount of acetonitrile, a volatile solvent, to the Sarin solution to help jump-start its evaporation.

Concerned that the nerve agent was too dilute to inflict the required number of casualties, Endo went to Asahara's quarters in Satian 6 to seek his blessing. The guru shrugged and said that given the need to disrupt the imminent police raid, dilute Sarin was better than nothing. Endo conveyed Asahara's message to the others. Then, using the glove box, the cultists injected the Sarin solution into empty plastic bags made of tough nylon-polyethylene, each of which held about twenty ounces. There was enough liquid to fill eleven bags, which were then heat-sealed. To prevent leakage, each bag of Sarin was sealed inside another, slightly larger plastic bag. Murai told Endo to fill five similar bags with water so that he could train the cultists who would carry out the subway attack.

On the evening of March 19, Inoue met with the five two-man teams at an Aum hideout in Tokyo and assigned them their targets. At about 10:00 p.m., the subway riders left the meeting and traveled to their respective stations, where they boarded trains and identified the spots where they would release the Sarin and the escape routes. After midnight, the group met for a late dinner at a Thai restaurant. At 1:30 a.m., Murai ordered the team members to drive to the Mount Fuji Center and pick up the Sarin-filled bags. He then telephoned Inoue and told him to go to an all-night convenience store and purchase seven vinyl umbrellas.

Shortly after 3:00 a.m., the operatives met on the ground floor of Satian 7. A cultist had sharpened the umbrella tips with an electric lathe, and the subway riders practiced using them to puncture the water-filled bags. To make the Sarin-filled bags appear less conspicuous, the cultists decided to wrap them in morning newspapers before boarding the trains. Murai handed out antidote tablets containing PAM, which he told the subway riders to take two hours before the operation began. At 6:00 a.m., having stayed up all night, the five operatives swallowed their pills and left the hideout with their drivers, heading for downtown Tokyo.

SHORTLY BEFORE 8:00 a.m., the five Aum operatives boarded their respective trains and prepared to carry out the attack. Jammed into one of the packed cars was Kenichi Hirose, thirty, the cult's Vice Minister of Science and Technology. He had graduated at the top of his university class in applied physics and turned down an offer from a major electronics firm to join Aum. Though aware that he might die in carrying out the attack, he did not feel overwhelmed by fear. He wanted to avoid death if at all possible and continue his spiritual training, but his chief desire was to implement Asahara's orders as efficiently as possible.

Hirose was about to remove the Sarin-filled bags from his knapsack and place them on the floor when he met the eyes of a pretty young woman standing directly in front of him. Realizing that he was about to kill a living human being, he felt queasy and was unable to proceed. His heart pounding, he got off the car at the next station. After meditating for a few minutes to calm his thoughts, he boarded the next train, telling himself that he was engaged in a "deliverance mission." As he dropped the bags of Sarin on the floor and pierced them with the sharpened umbrella tip, he silently recited a mantra. Visualizing Asahara's face, he prayed that his sins would not adversely affect the transmigration of the victims' souls and that they would be joined in eternal bliss with the Holy Master.

After puncturing the bags of Sarin, each operative got off the subway train at the next stop and met his driver at a prearranged location. During the drive back, the cultists stopped on the bank of a river in the Tokyo suburbs, where they burned their contaminated clothes and threw the sharpened umbrellas into the water. On arriving at the hideout, some of the subway riders felt ill from low-level Sarin exposure and were given injections of atropine and PAM. The cultists then met with Asahara, who led them in a celebratory mantra.

Eight of the eleven plastic bags had been punctured successfully during the operation, releasing a total of 159 ounces of dilute Sarin. The nerve agent seeped through the layers of newspaper surrounding each bag and formed a puddle on the floor of the subway car. As the liquid slowly evaporated, the invisible fumes filled the lower half of the car, drifting out onto the station platform each time the doors opened. Some of the passengers aboard the affected trains smelled a noxious odor and felt intense eye irritation and darkened vision; others had trouble breathing or experienced muscle weakness.

THE FIRST NEWS of the disaster broke at 8:09 a.m., when the Tokyo Metropolitan Police Department received a garbled report of an explosion in the subway system, with numerous casualties suffering burns and carbon-monoxide poisoning. Minutes later, ambulances from several major Tokyo hospitals converged on the fifteen affected underground stations, along with fire trucks, rescue squads, and police emergency vehicles. As disoriented commuters began streaming out of the subway exits into the sunlight, dozens of firefighters, policemen, and paramedics raced down the steps in the opposite direction. Lacking gas masks and other protective gear, the responders were soon overcome themselves by the toxic fumes.

Meanwhile, scenes of horror were playing out underground. At Kamiya-cho station next to Kasumigaseki, stricken people lay sprawled on the platform, vomiting uncontrollably, foaming at the mouth, bleeding from the nose, or wracked by convulsions, their arms and legs thrashing. Other victims leaned against the station walls and benches, gasping for breath, unable to see clearly, and terrified that they were dying. Because of the confusion, subway trains continued to stop at the affected stations, spreading the contamination further. Outside the subway exits, the sidewalks were covered with casualties. The growing chaos was accompanied by the throbbing hum of TV helicopters hovering overhead, transmitting live images of the terrifying scene to the nation and the world.

The Sarin attack was the largest disaster to hit Tokyo since World War II. In addition to the hundreds of victims taken to hospitals by 131 ambulances, thousands more arrived in minibuses provided by the fire department, in taxis and private cars, and on foot. More than eighty hospitals and clinics in Tokyo were inundated with a total of 3,227 victims, of whom 493 were admitted. Because of the general lack of preparedness, paramedics did not intubate seriously injured patients or insert intravenous lines until after they had reached the hospital. Not knowing the cause of the poisoning, emergency physicians began treating the victims symptomatically.

About two hours after the attack, the Tokyo Metropolitan Fire Department misidentified the toxic chemical as acetonitrile, a solvent that the cultists had added to the Sarin solution to accelerate its evaporation. Around 11:00 a.m., the National Research Institute of Police Science finally identified the agent as Sarin but failed to inform the hospitals; doctors treating the victims learned the identity of the poison only by watching television.

In addition to ten immediate deaths, two victims died a few weeks later from the complications of irreversible brain damage caused by oxygen starvation, bringing the total number of fatalities to twelve. The nonfatal casualties were divided into three categories. Seventeen patients were in critical condition with generalized convulsions, cardiac or respiratory arrest, or other symptoms requiring intensive care; some of these individuals suffered permanent neurological damage that left them in a vegetative state or severely disabled. Thirty-seven patients were in serious condition, complaining of shortness of breath, vomiting, severe headache, muscular twitching, or gastrointestinal problems. Finally, nearly 1,000 victims had mild symptoms, such as impaired vision, runny nose, and headache; they were treated with antidotes and recovered fully. The remaining casualties, most of whom had self-reported to hospitals, were "worried well": they had not been exposed to Sarin poisoning but instead were suffering from severe anxiety and psychosomatic symptoms.

The nerve agent release also produced some unexpected ripple effects. Because of the failure to decontaminate victims before evacuating them to hospitals, the off-gassing of Sarin from their clothes caused the secondary exposure of 135 ambulance workers and 110 medical staff, thirty-three of whom required hospitalization. The Sarin attack also terrorized Tokyo for months, causing some residents to develop posttraumatic stress syndrome and deterring people from riding the subways. For the first time in the postwar era, the vast metropolitan area felt like a city under siege.

Despite the grave consequences of the subway attack, the damage could have been far worse. The low purity of the Sarin solution, the crude method of delivery, and the excellent ventilation in the subway system all helped to reduce the number of fatalities. Had the Sarin been 80 percent pure rather than 30 percent and delivered in a more efficient manner—both of which might well have been the case if the cult had had more time to prepare the attack—the dead might well have numbered in the thousands.

Despite the small volume of Sarin released, a major cleanup effort was required to decontaminate the subway system and the victims' belongings. On March 21, the day after the attack, Mitsumasa Furuya, the president of a Tokyo waste disposal company, was called to Metropolitan Hiroo General Hospital, which had treated scores of casualties. In the laundry room in the basement of the hospital, he was shown about two hundred plastic bags filled with five tons of Sarin-tainted clothes, possessions, and uniforms of hospital staff. To assess the level of contamination, Furuya stuck one end of a plastic tube into a bag of clothes and the other end into a sealed cage of laboratory mice. Two hours later, the mice had developed pinpoint pupils and begun to spasm. Furuya moved a large steel vat into the parking lot next to the hospital laundry room and filled it with a concentrated solution of caustic soda. Donning rubber gloves, plastic goggles, and an industrial gas mask, he set to work, soaking the contents of each bag. It took him half a day to finish the job. Because the caustic soda ruined the clothes, they could not be returned to their owners.

AFTER THE SUBWAY ATTACK, the Aum cultists fled throughout Japan and attempted to destroy documents and other incriminating evidence. On March 22, two days after the incident, some 2,500 members of the Tokyo Metropolitan Police Department launched the largest law enforcement operation in the nation's history. Wearing gas masks, they conducted simultaneous raids on twenty-five Aum facilities and arrested more than 400 cult members. Because the police lacked field detectors for nerve agents, they went to pet markets and purchased dozens of caged canaries, which they carried with them as crude warning devices. A search of the Mount Fuji Center turned up huge quantities of precursor chemicals used in Sarin production, including fifty tons of phosphorus trichloride and ten tons of sodium fluoride.

Asahara and the other top cult leaders, most of whom had gone underground, were now the focus of intense attention from the news media. On the evening of April 23, 1995, Hideo Murai was about to give a press conference in Tokyo when a small-time gangster stabbed him in the stomach with a seven-inch butcher's knife. The murder was carried live on prime-time television and the killer was arrested on the spot. Although he was almost certainly a contract assassin, it was not clear who had hired him. One theory was that Japanese organized crime had wanted Murai dead to prevent him from exposing Aum's links with the *yakuza* in the trafficking of illicit drugs. But it was also possible that Asahara had ordered Murai's murder to prevent him from saying too much to the police.

Meanwhile, the cult leader remained at large. During a search of Satian 6 on May 16, 1995, a police investigator noticed that the positions of the ceiling panels had shifted from a previous visit. On inspecting the ceiling, he discovered a secret crawl space where Asahara was sitting cross-legged in the darkness, and placed him under arrest. That same day, forty-one other Aum leaders were taken into custody on suspicion of accessory to murder.

To track down the remaining suspects, the police distributed 1.6 million posters and flyers. Over the next few months, 192 Aum members were indicted for serious crimes.

The attack on the Tokyo subway caused police and fire departments in New York, London, and other major cities to worry that they lacked the equipment and training to handle the deliberate release of a nerve agent. In October 1995, the U.S. Senate Governmental Affairs Committee held hearings on the threat of chemical and biological terrorism and heard a detailed staff report on Aum Shinrikyo. A CIA counterterrorism official testified that Aum had not been on the "radar screen" of the U.S. intelligence community because it was a religious cult rather than a traditional politically motivated terrorist organization.

The Tokyo subway attack also drew attention to the fate of the Chemical Weapons Convention, which had been submitted to the U.S. Senate on November 23, 1993, but continued to languish in the Foreign Relations Committee. Although the United States had signed the CWC in January 1993, it would become a full party only if the Senate gave its consent to ratification by a two-thirds majority vote. Because the CWC had not been a high priority for incoming president Bill Clinton, the administration had passed up the opportunity to seek a ratification vote in 1993 or 1994, when the Democrats held a majority in the Senate. This delay proved to have been a serious miscalculation when, during the midterm congressional elections of November 1994, the Republican Party unexpectedly won control of the Senate.

One consequence of the Republican victory was that Senator Jesse Helms became chairman of the Senate Foreign Relations Committee, which has jurisdiction over treaties. Helms and other conservative senators opposed the CWC on the grounds that it was "unverifiable" and would impose excessive burdens on the U.S. chemical industry. Supporters of the treaty countered that it was "effectively verifiable" and that the chemical industry's leading trade organization, the Chemical Manufacturers Association, not only endorsed the CWC but was actively lobbying for its approval. This attitude was in marked contrast to the spoiler role that the U.S. chemical industry had played in 1925, when it had helped to block Senate ratification of the Geneva Protocol.

One reason the Chemical Manufacturers Association supported the CWC was that its representatives had worked closely with U.S. government negotiators for several years to ensure that the treaty provisions related to inspections of commercial plants and the protection of trade secrets were acceptable to the chemical industry. The association also worried that if the United States did not ratify the CWC, American companies would be subjected to restrictions on trade in certain treaty-controlled chemicals. Finally, the U.S. chemical industry wanted to improve its public image by demonstrating that it no longer manufactured chemical weapons and strongly favored their elimination. Despite industry's support for the CWC, however, Senator Helms exploited his power as chairman of the Foreign Relations Committee to block a ratification vote in the Senate for the next two years.

CHAPTER EIGHTEEN

THE EMERGING THREAT

In 1996, five years after the end of the Persian Gulf War, an UNSCOM inspection team led by Dutch chemical weapons expert Cees Wolterbeek began to investigate the ruins of the administration building at the Muthanna State Establishment. The three-story concrete structure had received a direct hit during the coalition bombing campaign and collapsed with the floors sandwiched together. Although Iraqi workers had attempted to clean out the building in the spring of 1991, before the U.N. weapons inspectors arrived, the presence of unexploded bombs had made the ruins too dangerous to disturb. Wolterbeek suspected that secret documents remained inside filing cabinets and safes buried deep in the rubble, a potential gold mine of information about the history of the Iraqi chemical weapons program.

From February 24 through March 12, 1996, the U.N. weapons inspectors supervised a team of Iraqi workers who excavated six sections of the collapsed administration building, using large cranes and special drilling and cutting equipment flown in from abroad. After several days of work, the structure was finally safe enough for the inspectors to enter. Amid the debris of the shattered offices, Wolterbeek and his team discovered more than five thousand pages of classified Iraqi documents, including memos, organization charts, official letters, records, computer disks, and scientific papers. One particularly useful item was a telephone directory listing the scientists and senior managers who had worked in the Iraqi chemical weapons program. The U.N. inspectors also recovered a safe from the office of the former director of Muthanna and forced it open. Inside were several technical reports dealing with VX production, including the purchase of raw materials and the development of manufacturing processes. This new evidence suggested that prior to the 1991 Persian Gulf War, Iraq had produced as much as fifty to a hundred tons of VX, all of which remained unaccounted for.

Another legacy of the Gulf War took more than five years to come to light. Immediately after the war, a U.S. engineering battalion had blown up Iraqi munitions bunkers at the Khamisiyah ammunition depot in southeastern Iraq. On June 21, 1996, the U.S. Department of Defense disclosed new information obtained by UNSCOM that one of the destroyed bunkers and an open pit had contained roughly eight tons of artillery rockets filled with a mixture of Sarin and Cyclosarin. The explosions had sent up huge plumes of tainted smoke and dust that had drifted downwind. Initially, the Pentagon estimated that only about four hundred American soldiers had been "presumed exposed" to low levels of nerve agents. Subsequently, however, computer modeling of the incident suggested that the prevailing winds could have carried the Sarin/Cyclosarin plume in a southerly direction as far as 300 miles from the blast site, dropping toxic fallout on as many as 100,000 U.S. troops deployed in southern Iraq, Kuwait, and northern Saudi Arabia.

In the wake of this stunning revelation, an intense controversy developed over the possibility that low-level exposures to nerve agents from the Khamisiyah incident might be responsible for some of the chronic health problems reported by roughly 100,000 of the 696,000 U.S. troops who had fought in the Gulf War, including fatigue, muscle and joint pain, memory loss, and severe headaches. Shortly after the bunker demolitions at Khamisiyah, chemical agent alarms had gone off, but the concentration of nerve agent in the air had been too low to produce acute signs and symptoms of exposure such as pinpoint pupils or shortness of breath. Factors complicating the effort to establish a possible causal link between nerve agent exposure and Gulf War illnesses were the puzzling variety of symptoms reported by sick veterans, the paucity of reliable epidemiological data on chemical exposures during the war, and the lack of scientific evidence that low-level doses of nerve agents could cause delayed or chronic health problems in humans. Some researchers hypothesized that "synergistic" exposures to nerve agents, organophosphate pesticides, pyridostigmine bromide (PB), and other chemicals that inhibit cholinesterase might produce such effects, however.

Given these uncertainties, it seemed unlikely that the medical debate over the possible role of nerve agent exposures in the cause of Gulf War illnesses would ever be resolved conclusively. Nevertheless, an intrepid biomedical researcher named Robert W. Haley, M.D., at the University of Texas Southwestern Medical Center, ventured into this political minefield. He did a systematic study of sick Gulf War veterans and found that a high percentage of those suffering from severe cognitive symptoms (such as mental confusion, memory loss, and dizziness) reported likely exposures to nerve agents in Iraq. In addition, MRI scans of these individuals indicated that they had suffered physical damage to specific areas of the brain that were functionally related to their mental deficits. Finally, Haley found that sick Gulf War veterans tended to have low levels of an enzyme called type Q paraoxonase, which helps to break down nerve agents, making such individuals more susceptible to harm from low-level exposures.

These findings clearly refuted the claims that sick Gulf War veterans were either malingerers or were suffering from the psychosomatic effects of "stress." Because the Veterans Administration had refused for many years to fund Haley's research, he had been supported initially by private grants from the eccentric Texas billionaire Ross Perot, who was a strong believer in the reality of Gulf War illnesses despite the intense skepticism of the Pentagon, the VA, and a series of blue-ribbon committees.

IN THE SPRING OF 1997, the United States faced a moment of truth with respect to its participation in the Chemical Weapons Convention. Although the Republican leaders of the Senate had finally scheduled a vote on CWC ratification in September 1996, the Clinton administration had withdrawn the treaty after calculating that it would not pass by the required two-thirds majority. According to a provision written into the CWC, it would automatically go into effect 180 days (approximately six months) after it had been ratified by sixty-five states. When Hungary became the sixty-fifth state to ratify the treaty on October 31, 1996, the clock began ticking down to the target date of April 29, 1997. If the United States did not ratify the CWC by that deadline, it would not become an original state party and hence would be deprived of an influential role in shaping the future treaty organization. Yet because of the uncompromising attitude of Senate Foreign Relations Committee chairman Helms, the odds of making the April deadline seemed increasingly slim.

Belatedly recognizing the high stakes involved, the Clinton administration launched a full-court press to get the treaty approved. In order to win the support of conservative Republican senators who were skeptics on arms control, it was clear that the administration would have to address their concerns. To this end, Robert Bell, the senior director for arms control on the National Security Council staff, began to negotiate with Senator Helms's foreign policy aides.

Helms and other conservative Republicans demanded the inclusion of thirty-three conditions in the ratifying legislation for the CWC. One of the more troublesome conditions was that samples collected at chemical plants in the United States could not be removed from U.S. territory for analysis in foreign laboratories because of the potential loss of industrial trade secrets. The Republican skeptics also wanted the future implementing legislation for the CWC to contain a provision authorizing the president to deny a request for an on-site inspection if it could "cause a threat to U.S. national security interests." Clinton administration officials protested that other countries would respond by enacting similar exemptions, weakening the ability of the CWC inspectorate to verify that member states were complying with their treaty obligations.

Despite these concerns, Senator Helms refused to budge. Because he effectively held the fate of the CWC hostage, the Clinton administration had little choice but to accept twenty-eight of the Republican conditions in the ratifying legislation. Helms took further advantage of his political leverage to force the elimination of the U.S. Arms Control and Disarmament Agency (ACDA), a small but dedicated bureaucracy that President John F. Kennedy had created in 1961 to negotiate and implement arms control treaties. According to the Helms plan, ACDA's functions would be folded into the Department of State, eliminating a separate institutional voice for arms control with the U.S. government.

On April 23, 1997, the CWC finally came up for a ratification vote on the Senate floor. After a heated debate that ran late into the night, the Senate approved U.S. participation in the treaty—with the Republican conditions—by a margin of 74 to 26, seven more votes than the required two-thirds majority. On April 25, President Clinton signed the U.S. instrument of ratification, which was immediately couriered to the treaty office at United Nations headquarters in New York. As a result, the United States met the ratification deadline by a whisker, becoming an original party to the CWC only days before the treaty entered into force on April 29, 1997.

Washington was now committed to forswear any future development or production of chemical weapons and to destroy its existing stockpile over ten years, with a deadline of April 29, 2007. Although Congress had ordered the elimination of obsolete unitary chemical munitions back in 1982, this effort was proceeding slowly because of the difficulty and expense of destroying blister and nerve agents in a safe and environmentally responsible manner. Moreover, because of public opposition to transporting chemical weapons across state lines, they would all have to be destroyed at the U.S. Army depots where they were already stored. JACADS, on Johnston Island, was the first of nine planned chemical weapons destruction facilities. The other eight would be built at Army storage depots scattered across the continental United States: Umatilla, Oregon; Tooele, Utah; Pueblo, Col-

The chemical weapons destruction facility on Johnston Island, known as the Johnston Atoll Chemical Agent Disposal System (JACADS), operated from the spring of 1990 through November 2000. Its purpose was to destroy obsolete U.S. chemical weapons transferred to the island from Okinawa and West Germany.

U.S. chemical weapons are stored in earth-covered concrete igloos like the one shown at the Umatilla Chemical Depot in Umatilla, Oregon. Under the Chemical Weapons Convention, which went into force in 1997, all 31,000 tons of chemical weapons in the U.S. stockpile must be destroyed.

orado; Newport, Indiana; Aberdeen, Maryland; Lexington, Kentucky; Pine Bluff, Arkansas; and Anniston, Alabama. At some of the storage sites, local opposition to the use of high-temperature incinerators slowed the pace of weapons elimination by forcing the Army to develop alternative destruction technologies, such as chemical neutralization.

By the end of 1997, several other important countries had ratified the CWC, including China, Russia, Iran, India, Pakistan, and all of the members of the European Union. Russia, which had inherited the vast Soviet chemical weapons stockpile, now faced the daunting task of destroying, within a decade, some 40,000 metric tons of blister and nerve agents stored at seven far-flung depots on its territory. In contrast to the United States, Moscow opted for a two-step destruction process involving chemical neutralization followed by "bitumenization," or mixing the neutralized waste product with asphalt to form giant blocks, which would then be buried in landfills. Because the BDA had never entered into force, the planned U.S.-Russian bilateral verification mechanism did not exist. As a result, all on-site inspections of U.S. and Russian chemical weapons storage and destruction facilities would have to be performed by the international CWC inspectorate.

As of early 1998, several countries of chemical weapons concern remained outside the CWC, including Egypt, Iraq, Israel, Libya, Syria, and North Korea. Although Saddam Hussein refused to join the new treaty, Iraq continued to be subjected to highly intrusive monitoring by UNSCOM. In April and May 1998, for example, U.N. weapons inspectors dug up fragments of twenty "special" Al-Hussein warheads, which Iraq had unilaterally destroyed after the 1991 Gulf War and buried in two pits near the town of Nibai. UNSCOM took wipe samples from the warhead fragments and sent them for analysis at three laboratories in the United States, France, and Switzerland. The results of the analyses were inconsistent. Although the U.S. laboratory identified traces of VX degradation products and a VX stabilizer, the French and Swiss labs were unable to confirm this finding. In an effort to clarify the contradictory results, UNSCOM convened a Technical Evaluation Meeting at its New York headquarters on October 22–23, 1998, attended by twenty-one chemical weapons experts from seven countries. After much discussion and debate, they agreed unanimously that the analyses indicating the presence of VX on the warhead fragments had been "conclusive and valid," raising new questions about Iraq's earlier denials that it had loaded VX into missile warheads.

In December 1998, after a series of escalating confrontations with Saddam Hussein over access to presidential palaces and other sensitive sites in Iraq, UNSCOM withdrew its inspectors at the request of the United States and Britain. The two countries then carried out a bombing campaign against Iraq to punish Baghdad for its defiance of U.N. Security Council resolutions and failure to cooperate fully with the inspections process. After this military action, which was dubbed Operation Desert Fox, a furious Saddam refused to let the U.N. weapons inspectors back into the country.

ISRAEL'S DECISION not to ratify the CWC was also troubling. Unlike the frontline Arab states, which had boycotted the signing ceremony in Paris in January 1993, Israel had been among the first group of countries to sign the treaty, and Israeli experts and diplomats had participated actively in the four years of PrepCom deliberations in The Hague. During the brief period of optimism that followed the Oslo peace process and the famous handshake between Yitzhak Rabin and Yasser Arafat on the White House lawn on September 13, 1993, the Israeli government had planned to submit the CWC to the Knesset, or parliament, for ratification. Treaty supporters argued that ratification would improve Israel's international position, demonstrate its willingness to pursue arms control, and enable the United States and others to pressure the holdout Arab states to join, benefiting Israeli diplomatic and strategic interests. Conversely, a failure to ratify the CWC would subject Israel's chemical industry to international trade restrictions on certain dual-use chemicals covered by the treaty, imposing potentially significant economic costs.

By 1997, however, the Oslo peace process had failed to live up to its promise and Israel's Arab neighbors were continuing to stockpile nerve agents and missile delivery systems at a frightening pace. Some Israeli analysts argued that chemical weapons were not an essential element of the nation's deterrence equation and could be abandoned, but others countered that doing away with the chemical option would force Israel to rely on nuclear weapons to deter a Syrian or Iranian chemical attack, dangerously lowering the nuclear threshold. Israeli opponents of the CWC also expressed concern that other countries might request "frivolous or abusive" challenge inspections at the Dimona reactor complex in the Negev Desert, where the Israeli nuclear weapons program was based. Because of these drawbacks, Israel announced that it would not move to ratify the CWC until after all of its Arab neighbors had joined. For their part, Egypt, Iraq, Libya, Syria, and Lebanon continued to boycott the CWC until Israel acceded to the nuclear Non-Proliferation Treaty. Since Israel had no intention of giving up its nuclear deterrent, the stalemate continued.

In 1998, an investigation into the mysterious crash of an El Al cargo plane in Amsterdam six years earlier focused attention on Israel's activities in the chemical weapons field. Although Israel accused its Arab neighbors of stockpiling chemical arms, it had never acknowledged possessing a chemical arsenal of its own. Some reports suggested, however, that in 1955 the Jewish state had launched an urgent effort to develop chemical weapons as a stop-gap deterrent until it acquired nuclear arms. Other evidence indicated that France had provided technical assistance, such as the fact that Israeli scientists had visited the French chemical weapons testing site in Algeria in 1960.

Unconfirmed reports also suggested that in 1969, two years after the Six-Day War, Israel had expanded its production of chemical weapons to counter Egypt's growing arsenal, and that Iraq's use of nerve agents during the Iran-Iraq War of the 1980s had caused Israel to upgrade its capabilities further.

The facility linked most closely to the Israeli chemical warfare program was the Israel Institute for Biological Research (IIBR), a top secret government laboratory near the town of Nes Ziona, about twelve miles southeast of Tel Aviv. IIBR had been founded in 1952 as an offshoot of the Weizmann Institute of Science and consisted initially of a single building hidden in an orange grove. In the early 1960s, the institute's Organic Chemistry Department began to conduct research on organophosphorus nerve agents, and a paper published in 1963 described several steps in the synthesis of V agents. Over the years, IIBR gradually expanded into a cluster of low buildings on several acres of land, with a staff of about 300 chemists, biologists, and other scientists. The institute was subordinated directly to the Israeli prime minister's office and remained so secret that it was not shown on local or aerial survey maps. Protection of IIBR and its information was the responsibility of the Bureau of Security of the Defense Establishment, known by its Hebrew acronym MALMAB. The institute compound was ringed with a six-foot-high concrete wall topped with intrusion-detecting sensors, and government security vehicles continually patrolled the perimeter road outside the fence.

A tragic accident offered a rare glimpse through the dense veil of secrecy surrounding IIBR. On October 4, 1992, El Al cargo flight 1862 to Tel Aviv, originating at John F. Kennedy International Airport in New York City, departed from Schiphol Airport on the outskirts of Amsterdam. Approximately ten minutes after takeoff, the Boeing 747 developed serious mechanical problems and began to lose altitude. Schiphol air-traffic control repeatedly urged the Israeli pilot to attempt an emergency landing in nearby Ijsselmeer Lake, but he ignored these instructions and instead turned back toward the airport. Falling short of the runway, the cargo plane plowed into a cluster of high-rise apartment buildings in the suburb of Bijlmermeer and burst into flames. The crash and ensuing fire destroyed or damaged hundreds of apartments and killed all three crew members on board and forty-three people on the ground.

From the outset, the El Al disaster was wrapped in mystery and intrigue.

Israeli and Dutch officials claimed that the flight had been transporting "perfumes and gift articles" from the United States to Israel via Amsterdam. Yet for several hours after the accident, local residents saw men in white hazardous-materials suits and respirators picking through the smoldering crash site and removing certain items. Over the next few years, more than a thousand emergency workers and residents of Bijlmermeer began to suffer from a variety of chronic medical and psychological conditions that they attributed to exposure to unknown toxic materials from the crash. Finally, the Dutch Parliament launched a formal inquiry into the incident and the allegations of a cover-up.

On September 30, 1998, six years after the crash of El Al flight 1862, the Rotterdam daily newspaper *NRC Handelsblad* published a lengthy report on its own investigation. According to freight documents obtained by the paper, the cargo on the ill-fated flight had been destined for IIBR in Nes Ziona and had included three precursor chemicals used in the production of Sarin: ten plastic drums, each containing 18.9 liters of DMMP (dimethyl methylphosphonate), as well as smaller quantities of isopropyl alcohol and

Rescue workers pick through the wreckage of an El Al 747 cargo plane that crashed into an apartment complex near Schiphol Airport outside Amsterdam in October 1992. An investigation by the leading Dutch newspaper NRC Handelsblad *uncovered the contents of the plane's cargo: 189 liters of a Sarin precursor for an Israeli research institute.*

hydrogen fluoride. The 189 liters of DMMP, enough for the production of 270 kilograms of Sarin, had been manufactured by Solkatronic Chemicals of Allentown, Pennsylvania. Although DMMP was a dual-use chemical that was subject to strict export controls, the U.S. Department of Commerce had granted Solkatronic a license to export it to Israel. Several weeks after the accident, a company called Shalom Chemicals of Nes Ziona had ordered an identical shipment from Solkatronic, which again received an export license from the U.S. government. According to the Israeli newspaper *Ha'aretz,* no one had heard of Shalom Chemicals, which was believed to be an IIBR front company.

In response to the *NRC Handelsblad* exposé, David Bar-Ilan, a media adviser to Israeli Prime Minister Benjamin Netanyahu, adamantly denied that the El Al flight had carried precursor chemicals for Sarin. Twelve hours later, however, he was contradicted by an El Al spokesman, who admitted that the flight had carried DMMP and the other chemicals but claimed that the shipment had been intended for strictly defensive purposes, such as the testing of gas masks and protective filters. "We fly sugar, which can be used for cake," he explained. "But that doesn't mean we're flying a cake."

Israel's preoccupation with chemical defense was not surprising given the fact that several of its neighbors had advanced chemical warfare programs. Syria, for example, had SS-21 missiles armed with nerve agent warheads that could reach Tel Aviv and other cities on Israel's densely populated coastal strip. Ze'ev Schiff, a leading Israeli journalist, warned in 1999 that the growing chemical arsenals of Syria and Iran "could seriously alter the regional balance of power." Nevertheless, it remained an open question whether Israel's chemical weapons program was strictly defensive or included offensive R&D and perhaps an active stockpile.

DESPITE THE GREAT normative and legal significance of the CWC, it was partially obsolete even before it entered into force. A product of the Cold War period during which it had been negotiated, the treaty was based on the assumption that the primary threat of chemical weapons came from large military stockpiles of traditional agents produced by nation-states in large, highly visible facilities. Yet the end of the Cold War in 1991, followed by the Sarin attack on the Tokyo subway in 1995, called this traditional paradigm into question. Whereas the West no longer saw the former Soviet chemical arsenal as a military threat, the acquisition of chemical weapons by subnational actors, such as insurgents and terrorists, was becoming a major concern—one that a traditional arms control treaty among sovereign states was not fully capable of addressing.

The fear of chemical terrorism became more acute with the rise of Islamic extremism, in particular the Al-Qaeda network founded by terrorist mastermind Osama bin Laden. Born in 1957 into a large and wealthy family in Saudi Arabia, bin Laden grew up under comfortable circumstances, but as a young man he began to identify with pan-Arab and Islamist ideology. Within days after the Soviet invasion of Afghanistan in December 1979, he joined the Afghani resistance and rose to prominence by financing the recruitment, training, and transportation of Arab volunteers to fight alongside the local *mujahidin.* In 1988, bin Laden founded a network of Islamist recruits called Al-Qaeda ("The Base") and organized paramilitary training camps in Afghanistan and Pakistan. After the Soviets withdrew their forces from Afghanistan in 1989, bin Laden returned to Jeddah, Saudi Arabia, to work in his family's construction company while continuing to support militant Islamist groups throughout the region.

At the end of 1990 Saudi Arabia expelled bin Laden, and he and his associates relocated to Khartoum, the capital of Sudan. That country was attractive because it was geographically close to the Arab world and had a radical Islamist government, the National Islamic Front (NIF), which had taken power in a bloodless coup in 1989. The leader of the NIF, Hasan al-Turabi, and bin Laden quickly established a close working relationship. Over the next several months, the Saudi financier embarked on numerous business ventures with wealthy members of the NIF, including a construction company, an import-export firm, a bank, and a financial operation called Taba Investments. Bin Laden also bankrolled civil infrastructure development projects on behalf of the regime, such as an airport and a road linking Khartoum with Port Sudan, and he supported the development of an indigenous armaments industry under the Sudanese Military Industrial Corporation.

During and after the 1991 Persian Gulf War, bin Laden viewed the permanent stationing of U.S. military forces in Saudi Arabia as a grave threat to conservative Islam and the sanctity of the Muslim holy places. He therefore turned against the United States, his former ally in fighting the Soviets in Afghanistan, and committed himself to driving American military bases and cultural influences out of the Middle East. Bin Laden also continued to expand his network of militant veterans of the Afghanistan War, financing the travel of more than three hundred of them to Sudan in May 1993. During this period, U.S. intelligence agencies learned that Iraqi chemical weapons experts were visiting Khartoum, raising concern that Iraq might transfer chemical weapons production technologies to Sudan so as to hide them from the UNSCOM inspectors. Other intelligence suggested that the Sudanese government was seeking chemical weapons for its brutal war against non-Muslim rebels in southern Sudan. It was also rumored that bin Laden had requested al-Turabi's help in obtaining nerve agents for terrorist attacks against U.S. military personnel in Saudi Arabia. After bin Laden moved to Afghanistan in 1996, he remained on good terms with al-Turabi, who allowed Al-Qaeda to use Sudan as a safe haven.

One of the projects partially funded by the Sudanese Military Industrial Corporation with suspected financial involvement by bin Laden was the Al-Shifa Pharmaceutical Factory, a joint venture by a Sudanese engineer and a Saudi Arabian shipper. From 1992 to 1996, the sprawling chemical plant rose in a mixed residential-industrial district of northern Khartoum. The production equipment was imported from the United States, Sweden, Italy, Switzerland, Germany, India, and Thailand. When the Al-Shifa Factory began operation in 1996, it employed over 300 workers and supplied more than half of Sudan's pharmaceutical needs, including antibiotics, pain relievers, drugs for malaria and tuberculosis, and veterinary medicines. The U.S. intelligence community, however, received reports suggesting that the Al-Shifa plant was secretly linked to the manufacture of nerve agents. In 1997, a CIA informant stated that three sites in Khartoum, including Al-Shifa, were involved in chemical weapons production. The U.S. National Security Agency, which specializes in electronic eavesdropping, also intercepted telephone conversations between the general manager of Al-Shifa and a senior scientist in Baghdad: Dr. Emad Husayn Abdullah Ani, the "father" of the Iraqi VX program.

In December 1997, the CIA obtained what appeared to be a highly incriminating piece of evidence. Agency officials reasoned that if the Al-Shifa Factory was producing nerve agents or related precursors, trace amounts of these chemicals would escape into the air or the liquid runoff from the plant and be deposited on the ground nearby. Accordingly, the CIA sent a trained Egyptian operative to take a soil sample about sixty feet from the main entrance of Al-Shifa. This sample was flown to the United

States, where it was split into three parts and sent to private contractor laboratories for analysis. The CIA later claimed that all three analyses had detected an organophosphate compound called EMPTA, which was known to be an intermediate in the Iraqi process for manufacturing VX.

Although EMPTA has peaceful applications in the production of fungicides, nothing suggested that any facility in Sudan was using it commercially for that purpose. Based on the intelligence findings, Clinton administration officials suspected that the Sudanese government had hired Iraqi chemical weapons scientists to manufacture nerve agents in Khartoum. According to this hypothesis, EMPTA had either been produced or stored at the Al-Shifa factory and then moved to another location for conversion into VX and loading into artillery shells.

Meanwhile, Al-Qaeda was preparing a devastating attack against U.S. interests in East Africa. On August 7, 1998, suicide terrorists from the bin Laden network carried out near-simultaneous bombings of the U.S. embassies in Nairobi, Kenya, and Dar es Salaam, Tanzania. In both cities, trucks packed with conventional explosives rammed into the embassy building and exploded, killing a total of 224 people, including twelve Americans, and injuring about 5,000 others. Stunned Clinton administration officials debated how to respond. On August 10, three days after the embassy bombings, National Security Adviser Samuel "Sandy" Berger convened a high-level meeting in the White House Situation Room to discuss options for military retaliation. Held under conditions of extreme secrecy, the meeting was limited to six top officials known as the "Small Group." Sitting around a mahogany conference table in the Situation Room, the six—President Clinton, Berger, Secretary of State Madeleine Albright, Secretary of Defense William S. Cohen, JCS Chairman General Henry H. Shelton, and Director of Central Intelligence George J. Tenet—reviewed the evidence obtained by the CIA Counterterrorism Center, which had attributed the two embassy bombings to Al-Qaeda.

President Clinton and other senior officials agreed with a recommendation by General Shelton to rule out U.S. military operations involving ground troops or piloted aircraft. Instead, the retaliatory strike would be carried out with unmanned Tomahawk cruise missiles. Because Al-Qaeda had targeted U.S. embassies in two countries, Berger wanted to retaliate against two separate locations. He suggested bombing the terror network's six main training camps in Afghanistan in the hope of killing bin Laden, and a related site in a second country. General Shelton presented a list of possible targets in Sudan, and the Al-Shifa Factory was high on the list.

Because the evidence linking Al-Shifa to bin Laden and VX production was circumstantial, President Clinton faced a difficult decision. Should he order an attack against an ostensibly civilian pharmaceutical plant in a country with which the United States was not at war? Several members of the Small Group argued that the threat of chemical terrorism justified preemptive military action against Al-Shifa, even though the case was less than airtight. If the Islamic regime in Sudan was manufacturing VX at the facility, it might be prepared to share the deadly agent with Al-Qaeda for use in terrorist attacks against the United States. As Berger pointed out, "What if we do not hit [Al-Shifa] and then, after an attack, nerve gas is released in the New York City subway? What will we say then?"

After further discussion, the members of the Small Group agreed unanimously to target Al-Shifa and a second facility in Khartoum that had been linked to chemical weapons production, and President Clinton signed off on the decision. Less than twenty-four hours before the operation was to begin, however, Clinton had second thoughts and decided to drop the second facility from the target list. Not only were its links to Al-Qaeda fairly tenuous, but it was located in a densely populated area where a missile strike might cause high civilian casualties.

On August 20, less than two weeks after the U.S. embassy bombings in East Africa, President Clinton gave the final order to proceed with the retaliatory strike against Al-Qaeda, code-named Operation Infinite Reach. U.S. Navy warships in the Arabian Sea and the Red Sea moved into position and fired a total of seventy-five Tomahawk cruise missiles, each worth about $750,000, at the designated targets in Afghanistan and Sudan. Thirteen cruise missiles targeted on the Al-Shifa Factory were launched in rapid succession from vertical tubes beneath the decks of two destroyers in the Red Sea. The missiles roared into the sky on plumes of flame and smoke. Once aloft, their turbofan engines and sophisticated guidance systems switched on, carrying them on programmed flight paths toward the target at 400 miles per hour.

At approximately 7:30 p.m. local time, the whine of low-flying jet engines filled the air over northeast Khartoum. Suddenly a swarm of winged cruise missiles appeared, skimming low over the industrial zone and homing in on the Al-Shifa Factory. Screaming out of the sky in a blur of metal, the missiles exploded into a series of brilliant fireballs that shook the ground like an earthquake. The blasts ignited an intense blaze that gutted the factory, leaving it a smoking ruins. Although the timing of the strike had been chosen to minimize "collateral damage," a security guard was killed and ten other Sudanese civilians were injured, five of them seriously.

After Operation Infinite Reach had been carried out, President Clinton gave a televised address in which he claimed that the Al-Shifa Factory had been linked to bin Laden and involved in the production of VX nerve agent, one of the deadliest poisons ever invented. "I ordered our armed forces to strike at terrorist-related facilities in Afghanistan and Sudan because of the imminent threat they presented to our national security," the president explained. "Our target was terror. Our mission was clear: to strike at the network of radical groups affiliated with and funded by Osama bin Laden, perhaps the preeminent organizer and financier of international terrorism in the world today."

As it happened, the cruise missile attack on the terrorist training camps in Afghanistan missed the primary target because bin Laden had left the

The ruins of the Al-Shifa Pharmaceutical Factory in Khartoum, Sudan, after a U.S. cruise-missile strike on August 20, 1998. President Bill Clinton's administration alleged that the Sudanese government had plotted with Osama bin Laden to produce a VX precursor at the civilian plant.

area a few hours earlier. Critics also began to second-guess the strike on the Al-Shifa Factory, casting doubt on the U.S. government's evidence for VX-related production at the site. Journalists who visited the ruins of the plant reported that it had not been under heavy security prior to the attack, had manufactured urgently needed medicines for the civilian population, and was a simple formulation and packing facility that lacked the specialized equipment needed to produce a complex chemical such as EMPTA. The CIA was also forced to admit that it had been unaware that the original owners had sold the Al-Shifa plant for $32 million in March 1998 to Salah Idris, a Sudanese-born businessman and adviser to Saudi Arabia's largest bank. Some Republican critics accused President Clinton of attacking an innocent country in order to distract attention from the Monica Lewinsky scandal. This allegation was called the " 'Wag the Dog' scenario" because it resembled the plot of a popular movie by that name.

Three years later, new evidence emerged supporting the Clinton administration's claim that Al-Qaeda had sought to produce nerve agents in Sudan, although not necessarily at the Al-Shifa Factory. During the 2001 trial in New York City of several suspects in the 1998 U.S. embassy bombings, a former Al-Qaeda operative named Jamal Ahmed al-Fadl was the chief prosecution witness. Under questioning by Assistant U.S. Attorney Patrick J. Fitzgerald, he testified that his job had been to shuttle back and forth between Afghanistan and Sudan to supervise Al-Qaeda's involvement in the Sudanese chemical weapons program. Al-Fadl said that in late 1993 or early 1994, he had accompanied Mamdouh Mahmud Salim, a top bin Laden lieutenant, on a visit to a large compound owned by the Sudan National Security Agency in the Hilat Koko district of northern Khartoum. According to al-Fadl, Salim had told him that Al-Qaeda was planning to help the Sudanese ruling party, the National Islamic Front, manufacture chemical weapons at the compound in Hilat Koko.

Because the Al-Shifa Factory was located a few miles away from Hilat Koko, al-Fadl's testimony suggested that the United States might have hit the wrong target. But Richard A. Clarke, the White House coordinator for counterterrorism, testified before a closed session of the House and Senate Intelligence Committees on June 11, 2004, that he continued to believe that the Al-Shifa Factory had been involved in VX production. According to Clarke's heavily censored testimony, Al-Shifa "was the type of plant that you would use to make the precursor [and it] appeared there was no other plant in the area where you would make that precursor." He also noted that Al-Shifa had been targeted because it could be destroyed with little collateral damage.

THE HORROR and devastation of the 1998 embassy bombings in East Africa were dwarfed by the Al-Qaeda attacks against the World Trade Center and the Pentagon on September 11, 2001. One week later, the first of two sets of letters contaminated with anthrax bacterial spores was sent through the U.S. mail. Targeting an odd assortment of media and political figures, the tainted letters ultimately killed five people, infected seventeen others, and frightened millions of Americans. Although the perpetrator of the anthrax attacks remained unknown, there were no obvious links to Al-Qaeda; the only individual that the FBI identified as a "person of interest" in the case was a former government scientist who had worked in the U.S. Army's biodefense lab at Fort Detrick, Maryland. Nevertheless, President George W. Bush, who had taken office eight months before, began to emphasize the threat that Al-Qaeda terrorists might use chemical or biological weapons against American cities. In his January 2002 State of the Union address, Bush made the case for invading Iraq because Saddam Hussein had retained stockpiles of unconventional arms and could provide them to terrorists. "I will not wait on events while dangers gather," Bush declared. "I will not stand by as peril draws closer and closer. The United States of America will not permit the world's most dangerous regimes to threaten us with the world's most destructive weapons."

In August 2002, the Cable News Network (CNN) broadcast a series of Al-Qaeda training videotapes obtained in Afghanistan, including disturbing images of crude experiments in which dogs were killed by exposure to a lethal gas of unknown composition. Although the CNN tapes indicated that Al-Qaeda was interested in acquiring chemical weapons, they also suggested that the group's level of technical sophistication was rudimentary. The Bush administration, however, continued to promote the unlikely scenario that a country hostile to the United States would arm terrorists with sophisticated chemical or biological weapons for use against American targets, despite the clear risks of retaliation and loss of control. To counter this hypothetical threat, the administration developed a doctrine of "preemptive" war that called for toppling regimes that sponsored terrorism and possessed unconventional weapons before they could stage an attack. In September 2002, the White House released a new edition of the U.S. National Security Strategy stating that the United States would "act against such emerging threats before they are fully formed" and would "not hesitate to act alone, if necessary, to exercise our right of self defense by acting preemptively."

As the Bush administration began to prepare for war with Iraq, Secretary of State Colin Powell persuaded the president to seek greater international legitimacy for military action by obtaining the political endorsement of the United Nations Security Council. On September 12, 2002, President Bush addressed the U.N. General Assembly and laid out a bill of indictment against Iraq for retaining prohibited weapons and defying the will of the international community. Almost as an afterthought, Bush said that the U.S. government was prepared to work with the Security Council to address the "common challenge in Iraq," but he stressed that U.S. military action would be "unavoidable" if Saddam Hussein did not cooperate fully. Four days later, U.N. Secretary-General Kofi Annan announced that he had received a letter from the Iraqi government stating that it would permit the return of U.N. weapons inspectors "without conditions."

In October, the U.S. intelligence community hastily prepared a secret National Intelligence Estimate (NIE) on Iraq's weapons of mass destruction (WMD). One of the key findings was that Iraq had rebuilt its chemical weapons program after the departure of U.N. weapons inspectors in December 1998 and currently possessed large quantities of Sarin, Cyclosarin, and VX, as well as chemical bombs, artillery shells, and rockets. The NIE stated, "Although we have little specific information on Iraq's CW stockpile, Saddam has probably stocked at least 100 metric tons (MT) and possibly as much as 500 MT of CW agents—much of it added in the last year." A supplementary CIA memo warned that Iraq might possess "dusty" agents: mustard or Sarin that had been adsorbed onto a fine, talcum-like powder to facilitate aerosolization, creating an extreme respiratory threat.

On November 8, the U.N. Security Council unanimously adopted Resolution 1441, which provided for unfettered weapons inspections in Iraq by international experts from the U.N. Monitoring, Verification and Inspection Commission (UNMOVIC), a new weapons inspectorate that had been established in December 1999 as a successor to UNSCOM. The resolution stated that any building in Iraq was potentially subject to inspection, including Saddam's presidential palaces, and authorized serious consequences if Iraq failed to cooperate. Even when confronted by a united Security Council, Baghdad's compliance appeared to be grudging. In early December, Iraq submitted a new declaration of its prohibited WMD programs that appeared to be a compilation of earlier versions and contained no new information. The Iraqi declaration denied the existence of any current stocks of chemical or biological weapons and failed to answer questions about unaccounted-for stockpiles and materials.

In late November 2002, teams of UNMOVIC inspectors began searching hundreds of suspect sites throughout Iraq. Over the next three and a half months, Iraq made no attempt to impede the inspections, even at presidential palaces, and the U.N. experts found no hidden caches of chemical or biological weapons. Much of the specific intelligence information supplied by the United States turned out to be inaccurate, resulting in numerous "wild WMD chases." Yet Bush administration officials repeatedly criticized the competence of the U.N. inspectors and insisted that Iraq was continuing to hide large stocks of chemical and biological arms.

On February 5, 2003, Secretary Powell gave a dramatic speech to the U.N. Security Council in which he described the various elements of Iraq's illicit arsenal and laid out declassified intelligence data, including satellite imagery and communications intercepts, suggesting that the Iraqi government was actively deceiving the U.N. inspectors. With respect to the Iraqi chemical weapons program, Powell declared, "Our conservative estimate is that Iraq today has a stockpile of between 100 and 500 tons of chemical weapons agent. That is enough to fill 16,000 battlefield rockets. Even the low end of 100 tons of agent would enable Saddam Hussein to cause mass casualties across more than 100 square miles of territory, an area nearly five times the size of Manhattan." Although Secretary Powell enjoyed the most international respect and credibility of any U.S. government official, his speech failed to sway the votes of nine of the fifteen countries on the Security Council. The dissenting states believed that the U.N. weapons inspectors should be given more time to finish their work and that the United States was moving recklessly toward war without having exhausted the diplomatic options.

President Bush was not to be dissuaded, however. After failing to obtain a second Security Council resolution authorizing military action, he ordered the invasion of Iraq on March 17, 2003, without the imprimatur of the international community. As American and British units fought their way toward Baghdad, they discovered several caches of Iraqi chemical protective suits and syringes containing nerve agent antidotes. Pentagon officials warned that Saddam might unleash his chemical weapons as soon as the invading forces crossed a "red line" surrounding the capital. Although the coalition troops braced themselves for chemical attacks, they never encountered as much as a whiff of mustard or nerve agent. On April 9, 2003, U.S. soldiers took control of central Baghdad and toppled the famous statue of Saddam in Firdos Square.

Immediately after the invasion, a U.S. military

United Nations Secretary-General Kofi Annan (bottom center) and French Foreign Minister Dominique de Villepin (bottom left) listen as U.S. Secretary of State Colin Powell addresses the U.N. Security Council on February 5, 2003. Chief U.N. weapons inspector Hans Blix is second from right in the second row. Secretary Powell described intelligence evidence—later shown to be incorrect—that Iraq possessed stocks of chemical arms.

intelligence team called the 75th Exploitation Task Force claimed to have discovered several suspect caches of chemical weapons, but all of them turned out to be false alarms. Indeed, much to the embarrassment of the CIA, no trace of Iraq's supposedly large chemical stockpile was found. As for the chemical suits and antidote stocks that coalition forces had encountered on the road to Baghdad, Iraqi sources explained that these defensive preparations had been inspired by fears that Israeli troops would join the U.S.-British invasion and use nerve agents against the Iraqi army.

During the summer and fall of 2003, investigations by members of the Iraq Survey Group (ISG), a U.S.-led fact-finding mission reporting to the Pentagon and the CIA and numbering some 1,200 personnel, concluded that the Iraqi chemical weapons production complex had not been rebuilt after 1998. In an interim report issued in October 2003, ISG director David Kay wrote, "Multiple sources with varied access and reliability have told ISG that Iraq did not have a large, ongoing, centrally controlled CW [chemical weapons] program after 1991. . . . Iraq's large-scale capability to develop, produce, and fill new CW munitions was reduced—if not entirely destroyed—during Operations Desert Storm and Desert Fox, thirteen years of UN sanctions and UN inspections." These preliminary findings raised serious doubts about the prewar intelligence assessment, which had apparently been based on outdated or false information.

On October 6, 2004, Charles A. Duelfer, who had replaced David Kay as the head of the ISG, released a 918-page final report summarizing the results of the eighteen-month investigation into Iraq's unconventional weapons programs prior to the March 2003 war. The Duelfer report concluded that Iraq had destroyed its undeclared chemical stockpile in 1991 and had not resumed production thereafter. Although a network of clandestine laboratories operating under the Iraqi Intelligence Service had conducted research and testing on various toxic chemicals and poisons for assassination purposes, this effort did not meet the definition of a militarily significant capability. Moreover, the ISG concluded that the 1998 U.S. finding of VX degradation products on Iraqi missile warheads had been a false-positive, possibly caused by the cross-contamination of samples.

At the same time, the Duelfer report judged that Saddam had planned to relaunch chemical weapons production after the U.N. sanctions were lifted. In the late 1990s, Iraq had reorganized its chemical industry "to conserve the knowledge base needed to restart a CW program, conduct a modest amount of dual-use research, and partially recover from the decline of its production capability caused by the effects of the Gulf war and UN-sponsored destruction and sanctions." By employing existing chemical plants and importing key precursor chemicals, Iraq would have been able to produce mustard agent within a period of months and nerve agent within two years. Nevertheless, the ISG did not find any chemical production units that had been configured to produce key precursors or warfare agents, nor was there evidence of "explicit guidance from Saddam" on how to reconstitute the program. The ISG report also noted that, ironically, the war's chaotic aftermath may have contributed to the proliferation of chemical weapons. Not only were many Iraqi weapons scientists unaccounted for, but dual-use chemical plants had been "systematically" looted.

In response to the ISG report, former U.N. weapons inspectors argued that international sanctions and inspections had successfully kept Iraq's illicit arms programs in check from 1991 until the U.S.-led invasion in 2003, and that ongoing monitoring and verification would have contained the Iraqi threat indefinitely without the need for war. Supporters of the invasion countered that international sanctions had been crumbling, Saddam's motivations had remained unchanged, and Iraq's embryonic unconventional weapons programs had posed a significant long-term threat.

With respect to Iraqi intentions, Duelfer concluded that Saddam Hussein had an almost mystical faith in the power of chemical weapons, which he believed had preserved his rule through repeated military crises. The former Iraqi leader was convinced that Iraq had been "saved" by the use of mustard and nerve agents to neutralize Iranian human-wave offensives during the Iran-Iraq War, and that chemical weapons had also been effective in suppressing the Kurdish uprising in 1988. For this reason, Saddam had deliberately created ambiguity about whether he possessed chemical weapons so as to deter Iran from attacking and to intimidate his domestic enemies. UNMOVIC Executive Chairman Hans Blix likened Saddam's behavior to that of someone who does not own a dog but tries to discourage thieves by posting a BEWARE THE DOG sign on his door.

At the same time that Saddam bluffed possessing chemical weapons as a deterrent, he sought to persuade the U.N. Security Council that he had disarmed so that the crippling economic sanctions would be lifted. The Duelfer report found that Iraqi regime was unable to resolve the contradiction inherent in its pursuit of these competing objectives, and that the resulting mixed messages had confused the U.S. and British intelligence services. "Ultimately," the Duelfer report concluded dryly, "foreign perceptions of these tensions contributed to the destruction of the regime."

IN LATE 2004, the veil of secrecy that had surrounded the tragic case of Ronald Maddison for more than fifty years was finally lifted. Ronald's sister, Lillias Craig, had been determined to discover the truth about her brother's untimely death at Porton Down in May 1953 and had fought a stubborn battle with the British authorities to reopen the investigation. In 2003, Chief Justice Lord Woolf had finally quashed the original verdict and ordered a new hearing.

On November 15, 2004, the inquest jury made headlines in the British newspapers with its finding that Maddison had been "unlawfully killed" and that the cause of death had been "the application of a nerve agent in a non-therapeutic experiment." The reopened inquest also provided new information about the circumstances of Ronald's death. After 1953 it had been learned that fat content is a critical factor influencing the ability of Sarin to penetrate the skin. According to the postmortem report, Maddison had been quite thin and his skin fat had been "practically absent," an unusual characteristic that had probably contributed to the fatal outcome of the experiment.

After the second inquest, Professor Sir Ian Kennedy, a leading expert in medical ethics, argued that the human trials with Sarin at Porton Down had been "beyond the bounds of what was ethically permissible, despite the imperative of the Cold War." The Wiltshire police made clear that no criminal charges would be brought against the Porton scientists implicated in Maddison's death. Nevertheless, the jury decision was considered likely to result in compensation claims in the millions of pounds from 550 former British servicemen who had undergone nerve agent experiments at Porton Down and had since developed chronic health problems.

TOWARD ABOLITION

ON THE MORNING of January 5, 2005, a small group of people stood in the rain at Pine Bluff Arsenal in Arkansas, watching as bulldozers pulled down two scrubber towers that had been part of the pollution control system for the difluor (DF) production plant. The towers were among the last structures remaining from the former Integrated Binary Production Facility, which had manufactured chemical components for the M687 Sarin projectile and the Bigeye VX bomb.

Its engine straining, one bulldozer tugged on a cable attached to the top of the metal tower, while a second bulldozer pushed at its base. After about fifteen minutes, the tower toppled and crashed to the ground, and the bulldozers repeated the same operation for the second tower. The demolition process was carefully recorded on videotape for later review by verification officials at the Organization for the Prohibition of Chemical Weapons (OPCW) in The Hague.

Construction of the binary manufacturing complex at Pine Bluff Arsenal had begun in 1982 but had stopped abruptly eight years later, after the United States and the former Soviet Union signed the 1990 Bilateral Destruction Agreement ending all further production of chemical arms. Work at the Pine Bluff complex never resumed. With the entry into force of the Chemical Weapons Convention (CWC) in April 1997, the United States was required to destroy all of its former chemical weapons production facilities or to request permission from the OPCW to convert them to peaceful purposes. For example, the Marquardt Company plant in Van

Nuys, California, which had produced canisters and OPA for the M687 projectile, was converted into a sound stage.

Dismantling of the binary production complex at Pine Bluff Arsenal had started in October 2003. Most of the old equipment was sold off as scrap metal, but "specialized" items that had been tagged by the OPCW, such as corrosion-resistant pipes and reactors, had to remain on site until the inspectors could verify their destruction. The only building left standing, the Fill and Close Facility for the Multiple Launch Rocket System (MLRS) warhead, was converted into a neutralization plant to dispose of the remaining stocks of binary precursors stored at Pine Bluff. These stocks included some 56,000 DF-filled canisters for the M687 projectile, six 55-gallon drums of DF, and 291 drums of QL for the Bigeye bomb. It would take about six months to neutralize these chemicals in the former Fill and Close Facility, which would operate twenty-four hours a day, seven days a week, until the job was done. In addition, Pine Bluff housed 3,850 tons of unitary nerve and blister agents, which would be burned in a special incinerator over a five-year period.

WITH THE DESTRUCTION of chemical weapons stockpiles and former production facilities in the United States, Russia, and other countries, the world is making slow but steady progress toward chemical disarmament. Nevertheless, the implementation of the CWC is at a crossroads that could lead either to the ultimate abolition of chemical weapons or to a weakening of the disarmament regime and further proliferation.

On the positive side of the ledger, the great majority of the world's nations have signed and ratified the CWC, and six countries (Albania, India, Libya, South Korea, Russia, and the United States) have declared stockpiles of chemical weapons and begun to destroy them. Libya's decision in December 2003 to renounce its chemical warfare capability was a major breakthrough. After becoming a member of the CWC on February 5, 2004, the Libyan government declared 23 metric tons of mustard agent and 1,300 tons of nerve agent precursors, which would subsequently be destroyed under the watchful eyes of international inspectors. Libya also obtained permission from the OPCW to convert its former chemical weapons production facility at Rabta into a pharmaceutical plant to produce AIDS drugs and other urgently needed medications.

Despite these successes, however, several countries in conflict-plagued regions such as the Middle East and northeast Asia continue to remain outside the CWC and to possess chemical arms. In January 2004, Syrian President Bashar al-Assad hinted in a newspaper interview that his country's chemical arsenal provided an affordable way to counterbalance Israel's nuclear capability. "It is natural for us to look for means to defend ourselves," he said. "It is not difficult to get most of these weapons anywhere in the world and they can be obtained at any time." Iran, a party to the CWC, denies possessing any chemical weapons, but the U.S. government alleges that Tehran retains clandestine stocks and production facilities. North Korea began developing chemical weapons in the 1960s, initially with technical assistance from the Soviet Union and then from China. According to U.S. intelligence estimates, North Korea has roughly 5,000 tons of blister and nerve agents, most of which have been loaded into artillery shells and missile warheads deployed near the Demilitarized Zone. In the event of another war on the Korean Peninsula, these weapons would pose a grave threat to the population of nearby Seoul.

Of course, the existence of holdout states such as Syria and North Korea does not invalidate the CWC, any more than the continued existence of street crime makes domestic laws irrelevant. Even without universal adherence, the CWC can slow the spread of chemical weapons by isolating the small number of nonparticipating countries, limiting their access to precursor chemicals, and exposing them to international political and economic pressure if they continue their illicit programs. As defense analyst Brad Roberts has observed, "Norms matter in international politics—not because they constrain the choices of the most malevolent of men but because they create the basis for consensus about responses to actions inconsistent with those norms."

DESPITE THE growing interest in chemical weapons on the part of terrorists, the technical hurdles to the acquisition of nerve agents remain significant. An annex to the 2004 Iraq Survey Group report disclosed that an Iraqi insurgent group known as the "Al-Abud network" had attempted for six months to produce chemical warfare agents, without success. The group recruited a Baghdad chemist and acquired chemicals from looted state companies and other sources, but it could not obtain the ingredients for Tabun

The headquarters in The Hague, the Netherlands, of the Organization for the Prohibition of Chemical Weapons (OPCW). This international body oversees the implementation of the Chemical Weapons Convention, a treaty banning chemical weapons that entered into force in 1997.

and also failed at subsequent attempts to manufacture mustard agent. Nevertheless, according to the ISG report, "The most alarming aspect of the Al-Abud network is how quickly and effectively the group was able to mobilize key resources and tap relevant expertise to develop a program for weaponizing CW agents."

The George W. Bush administration's major contribution to combating the threats of chemical weapons proliferation and terrorism has been the Proliferation Security Initiative (PSI), which President Bush announced in a speech in Kraków, Poland, on May 31, 2003. Under this agreement, a coalition of like-minded countries (initially eleven, now more than sixty) have agreed to conduct joint operations to interdict and seize illicit shipments by sea, air, or land of unconventional weapons, delivery systems, or related materials to and from "states and non-state actors of proliferation concern." PSI is organized around a set of "interdiction principles" that call upon the participating states to take specific actions permitted under inter-

OPCW inspectors count chemical artillery shells during an inspection in Russia.

national law, such as stopping and searching ships and denying air-transit rights.

Given that transfers of actual chemical weapons are unlikely, the effectiveness of PSI in interdicting clandestine shipments of chemical precursors and production equipment remains to be seen. The sheer volume of chemical trade and the reluctance of states to share sensitive intelligence will probably make it difficult for states to identify and track suspicious cargoes in a timely manner. Moreover, because a PSI strategy requires near-universal participation to be fully effective, the limited number of fully active members represents a serious weakness. For these reasons, a significant proportion of illicit traffickers and terrorist organizations are likely to evade any interdiction strategy.

ALTHOUGH INFORMAL arrangements such as the Australia Group and PSI can make a useful contribution to combating chemical proliferation, they cannot do the job on their own. It is also essential to reinforce the international norm embodied in the Chemical Weapons Convention, which remains a key instrument for pursuing the ultimate abolition of gas warfare. To achieve this goal, the CWC requires effective verification and enforcement measures, which still leave much to be desired. One negative trend is that member states have weakened some of the treaty's more intrusive inspection procedures, creating loopholes for potential cheaters. For example, the OPCW inspectorate has a powerful but portable instrument called a gas chromatograph–mass spectrometer (GC/MS). This device, which fits into a large trunk, can identify an unknown compound by breaking it into fragments that are sorted by molecular weight, generating a spectrum with distinctive peaks and valleys that a computer matches against a "library" of known spectra in an electronic database. Because of concerns by chemical manufacturers over the potential compromise of industrial trade secrets, CWC member states have constrained the use of GC/MS during routine inspections of chemical plants. In addition, the limited sampling and analysis that does occur (mainly with on-site equipment) must employ "blinding" software, which merely reports the presence or absence of known chemical warfare agents and precursors listed in an annex to the CWC. As a result, a determined cheater could develop and produce an unlisted compound (such as one of the Novichok agents) in a bid to avoid detection.

Despite these clear gaps in the verification regime, the parties to the CWC have been reluctant to increase the number of chemical agents and precursors subject to on-site sampling and analysis. For example, although the U.S. and British governments know the chemical structures of the Novichok agents and their binary precursors, they have decided not to add these compounds to the CWC's list of declarable chemicals because of concern that the information could be exploited by proliferators or terrorists. A possible solution to this dilemma would be to increase the number of chemical spectra stored in the GC/MS analytical database, making it possible to detect a larger range of toxic compounds, without formally expanding the list of declarable chemicals.

Closely related to verification of the CWC is the problem of enforcement. Historically, the international community has been unable or unwilling to punish violators of arms control treaties. Compliance has relied largely on moral suasion, yet determined cheaters may not be deterred without a credible threat of economic or military sanctions. Indeed, Iraq's flagrant violations of the Geneva Protocol during the 1980s demonstrate the need for enforcement to make disarmament treaties more effective.

Although the CWC does not constrain terrorists directly, the fewer the number of states that continue to possess or to pursue chemical arms, the harder it will be for terrorists to follow suit—either by stealing actual weapons or by obtaining the equipment, materials, and know-how needed to produce them.

A Russian military officer with a Scud missile warhead of the type that Soviet forces had filled with nerve agents, during a display on June 8, 2001, at the chemical weapons depot near the remote Russian town of Shchuch'ye. The Russian government invited diplomats from the United States, the European Union, and Canada to Shchuch'ye to appeal for financial support for Russia's long-delayed effort to destroy its vast chemical weapons stockpile.

ANOTHER WEAKNESS OF CWC implementation has been the failure of Russia and the United States, the world's two largest possessors of chemical weapons, to destroy their respective stockpiles according to the timetable set out in the treaty. The former Soviet stockpile, stored at seven depots on Russian territory, consists of about 40,000 metric tons of nerve agents (Sarin, Soman, and R-33) and blister agents (mustard and lewisite). Destroying this toxic legacy of the Cold War in a safe and responsible manner entails huge financial, political, and environmental challenges.

Because the total cost of destroying the Soviet chemical weapons stockpile has been estimated at more than $8 billion, Russia relies heavily on financial assistance from the United States and other countries. Many bureaucratic hurdles and technical problems have slowed the destruction effort, making it impossible to meet the 2007 deadline in the CWC for eliminating the entire stockpile. As permitted by the treaty, Moscow has requested a five-year extension until 2012, but even the new target date may not be realistic. In the meantime, inadequate physical protection and accounting at some of the Russian depots have rendered the chemical weapons they contain vulnerable to theft or diversion. A depot near the Russian town of Shchuch'ye, for example, consists of fourteen barn-like buildings containing 1.9 million nerve-agent-filled artillery shells, stacked on wooden racks like bottles in a wine cellar. The shells are small enough that a few could be smuggled out in a suitcase.

The United States has also encountered serious difficulties in destroying its aging stockpile of chemical weapons. A modest milestone occurred in November 2000, when the high-temperature incinerator on Johnston Island in the Pacific finished burning the munitions transferred from Okinawa and West Germany, accounting for 6.6 percent of the U.S. stockpile. The remaining weapons are stored at eight depots scattered across the continental United States and are slated for elimination at dedicated destruction facilities built at each site. Since 1985, however, technical, managerial, and political problems have slowed the pace of the "chemical demilitarization" effort and markedly increased its cost. When the program began, the projected price tag for destroying the entire stockpile was roughly $1.8 billion, but by 2004 it had ballooned to $26.8 billion. One reason for the delays and cost overruns has been opposition to the use of high-

U.S. Army workers prepare a pallet of thirty M55 rockets to be transported to the chemical weapons incinerator at Tooele Army Depot in Utah. In recent years, the U.S. effort to destroy its stockpile of chemical weapons has fallen behind schedule.

temperature incineration from communities near the U.S. Army depots where the weapons are stored. Public pressure has derailed plans to build chemical-weapons incinerators at the depots in Indiana, Kentucky, Maryland, and Colorado, and forced the adoption of alternative technologies such as chemical or biological neutralization. Yet despite legitimate concern over the potential health hazards of incineration, an even greater risk lies in not destroying the weapons, which are plagued by leaks and provide an attractive target for terrorists.

A further obstacle in the path of chemical disarmament is the effort by the United States and Russia to develop a new generation of "nonlethal" weapons, including powerful incapacitating agents that act on the nervous system. The CWC explicitly bans the combat use of incapacitants and tear gas because any release of toxic chemicals on the battlefield could easily escalate to the employment of lethal agents. Yet Russia and the United States both claim the right to employ incapacitating agents for counterterrorism operations under an exemption in the CWC for "domestic law enforcement."

On October 23, 2002, Chechen rebels stormed the Dubrovka Theater in downtown Moscow during an evening performance and took about 900 people hostage, threatening to set off bombs unless their demands were met. Russian Special Forces surrounded the theater and a standoff ensued. During the three-day siege, the militants released about 200 people, mostly women, children, and Muslims, but the Russian authorities refused to negotiate. The standoff ended on October 26 when the Special Forces, suspecting that the rebels had begun killing hostages, pumped a narcotic gas (related to the anesthetic fenantyl) into the building through the air-conditioning system. The agent subdued many of the militants, who were either executed at point-blank range by the government commandos or killed in the ensuing shootout. But 129 of the hostages also died, all but two of them from the effects of the gas. Weakened by fatigue and hunger, they succumbed to an overdose of the narcotic, which suppressed the breathing center of the brain. Contributing to the debacle was the Russian authorities' refusal to identify the agent, preventing paramedics from administering an antidote in time to save many lives.

The Moscow theater incident suggests that the "lethality" or "nonlethality" of a given chemical agent is not an intrinsic characteristic but is a function of a way it is used, including the concentration, the means of delivery, and the targeted population. Moreover, the use of chemical incapacitants in paramilitary operations is dangerous because it blurs the line between law enforcement and warfare, creating a "slippery slope" that makes the battlefield use of chemical weapons more likely. It is therefore essential to close this legal loophole in the CWC by defining the law enforcement exemption more narrowly.

A final challenge to the chemical disarmament regime is the rapid pace of technological innovation in the chemical and pharmaceutical industries, which could spawn new chemical warfare threats. For example, "combinatorial chemistry," which involves the automated synthesis of thousands of related compounds followed by their rapid screening for desired physiological effects, is a powerful tool of drug discovery. Yet this technique could easily be misused to identify incapacitating or lethal chemical warfare agents.

Similarly, progress in understanding the functional biochemistry of the brain could lead to the development of improved drugs for treating mental illness, but it could also spawn a new generation of chemical warfare agents that induce sleep, fear, paralysis, or rapid death. More compact and efficient chemical manufacturing technologies, such as "microreactors," could also make it easier to conceal an illicit chemical weapons production facility.

THE HISTORY of the discovery, proliferation, and control of chemical weapons offers grounds for hope as well as concern. It is clear that the taboo against poison warfare has a source deep in the human psyche, giving rise to an international behavioral norm of great antiquity and wide cross-cultural character. During World War I, the advent of industrial synthetic chemistry, which made the large-scale production and use of poisons feasible and cheap, and the pressure of "military necessity" to escape the bloody stalemate of trench warfare, combined to erode the existing legal and moral restraints on chemical warfare. In the 1930s, the accidental discovery of the nerve agents in Germany and their production and stockpiling by the Nazi regime further undercut the norm, but fears of Allied retaliation (based in part on incorrect intelligence assessments) ultimately discouraged Hitler from using his "secret weapon." The postwar competition among the victorious Allies for the secrets of the German nerve agents culminated in the chemical arms race of the Cold War, a "war of nerves" in which each superpower tried to deter attack by the other while pressing for political and strategic advantage. Beginning in the late 1960s, nerve agents spread to the developing world, leading to their alleged use in the Yemen civil war and their known use in the Iran-Iraq War. The trickle-down process reached its logical conclusion in the 1990s, when nerve agents became an instrument of terror by a doomsday cult in Japan.

Over the same period, however, there were some positive countervailing trends. Because of the deeply rooted nature of the poison taboo, the norm gradually became reestablished as the twentieth century advanced. Chemical weapons were progressively delegitimated under international law, beginning in 1925 with the Geneva Protocol's limited ban on use in war and culminating in 1993 with the CWC's sweeping prohibitions on the development, production, stockpiling, transfer, and use of chemical arms, to which the great majority of the world's states now adhere. Toxic weapons that a few decades ago were being mass-produced and stockpiled in the thousands of tons by the major powers are now considered "beyond the pale," and billions of dollars are being spent on their destruction. Only a handful of states persist in acquiring or retaining chemical arms, and none of them admit possession.

Although the CWC is binding only on countries that join voluntarily, it can be of value in slowing the spread of toxic weapons to rogue states and terrorist organizations, such as Al-Qaeda, by strengthening the restrictions on trade in precursor chemicals and increasing the vigilance of the international chemical industry about the proliferation threat. Since the 9/11 terrorist attacks, the OPCW has encouraged member states to pass national legislation making the provisions of the CWC binding on their citizens and corporations, and imposing penal sanctions for violations. The international behavioral norm embodied in the CWC remains fragile, however, and scientific or technological advances (such the development of new incapacitating agents) could once again undermine the taboo against poison warfare. In view of the continuing threat of chemical weapons proliferation and use, strengthening the legal and moral barriers against the use of chemistry for hostile purposes will be vital for human well-being and survival in the twenty-first century.

GLOSSARY

ACACIA French binary weapons program

ACDA Arms Control and Disarmament Agency

AEC Atomic Energy Commission

AG Aktiengesellschaft (German word for "corporation"); Australia Group

ALSOS Mission to investigate Nazi Germany's unconventional weapons programs

BDA Bilateral Destruction Agreement (U.S.-Soviet)

BDO Battle-Dress Overgarment

BIOS British Intelligence Objectives Subcommittee

BMEWS Ballistic Missile Early Warning System

BWC Biological Weapons Convention

CAM Chemical Agent Monitor

CBIC CW/BW Intelligence Committee

CBR Chemical-Biological-Radiological

CCD Conference of the Committee on Disarmament (U.N.)

CCTC Chemical Corps Technical Committee (U.S. Army)

CD Conference on Disarmament (U.N.)

CDC Communicable Disease Center (Centers for Disease Control and Prevention)

CDEE Chemical Defence Experimental Establishment (Porton Down, U.K.)

CDG Chemical Destruction Group (UNSCOM)

CDTF Chemical Defense Training Facility

CERS Center for Scientific Studies and Research (Syria)

CHASE "Cut holes and sink 'em"

CIA Central Intelligence Agency

CIOS Combined Intelligence Objectives Subcommittee

CML C Chemical Corps (U.S. Army)

CNN Cable News Network

CPSU Communist Party of the Soviet Union

Glossary

CSDIC	Combined Staff Defence Intelligence Committee (U.K.)
CW	Chemical warfare
CWC	Chemical Weapons Convention
CWS	Chemical Warfare Service (U.S. Army)
DAP	Destruction Advisory Panel (UNSCOM)
DC	Dichlor (methylphosphonic dichloride)
DCI	Director of Central Intelligence
DF	Difluor (methylphosphonic difluoride)
DFP	Diisopropyl fluorophosphate
DMHP	Dimethyl hydrogen phosphite
DMMP	Dimethyl methylphosphonate
DPC	Defense Planning Committee (NATO)
DSB	Defense Science Board
EFI	Early U.S. Army code name for Ethylsarin
EMPTA	VX precursor chemical
EUCOM	European Command (U.S.)
FBI	Federal Bureau of Investigation
FIAT	Field Information Agency, Technical
FRG	Federal Republic of Germany
FROG	Free Rocket over Ground (Soviet)
FSB	Federal Security Service (Russia)
FY	Fiscal year
G-2	U.S. Army Intelligence
GA	U.S. Army code name for Tabun
GB	U.S. Army code name for Sarin
GD	U.S. Army code name for Soman
GE	U.S. Army code for Ethylsarin
GF	U.S. Army code for Cyclosarin
GH	U.S. Army code for Isopentylsarin
GITOS	State Institute for Technology of Organic Synthesis (Shikhany, Russia)
GosNIIOKhT	State Scientific Research Institute of Organic Chemistry and Technology (Moscow, Russia)
HAZMAT	Hazardous materials
HF	Hydrogen fluoride
ICI	Imperial Chemical Industries (U.K.)
IDF	Israel Defense Forces
IG	Interessengemeinschaft (German conglomerate)
IIBR	Israel Institute for Biological Research
IVA	Intermediate-volatility agent
JCS	Joint Chiefs of Staff
JIOA	Joint Intelligence Objectives Agency
Le-100	Early German code name for Tabun (Le = Leverkusen)
M20	Canister containing DF for Sarin binary projectile
M21	Canister containing OPA for Sarin binary projectile

MCE   Early U.S. Army code name for Tabun
MFI   Early U.S. Army code name for Sarin
MIT   Massachusetts Institute of Technology
MLRS   Multiple-launch rocket system
MOPP   Mission-Oriented Protective Posture
MOU   Memorandum of Understanding
MRI   Magnetic Resonance Imaging
NAC   North Atlantic Council (policy-making body of NATO)
NATO   North Atlantic Treaty Organization
NBC   National Broadcasting Company, Nuclear Biological Chemical
NDRC   National Defense Research Committee
NEPA   National Environmental Policy Act
NIE   National Intelligence Estimate
NII-42   Scientific Research Institute No. 42
NORAD   North American Air Defense Command
NPT   [Nuclear] Non-Proliferation Treaty
NSA   National Security Agency
NSC   National Security Council
OPA   Binary component (alcohol, stabilizer, catalyst) for M687 artillery shell
OPCW   Organization for the Prohibition of Chemical Weapons
PAM   Pyridine aldoxime methiodide
PB   Pyridostigmine bromide
PSI   Proliferation Security Initiative
QL   Immediate precursor of VX
R-33   Soviet version of VX
R-35   Soviet code name for Sarin
R-55   Soviet code name for Soman
RAF   Royal Air Force
R&D   Research and development
SBCCOM   Soldier and Biological Chemical Command (U.S. Army)
SEPP   State Enterprise for Pesticide Production (Iraq)
SHAD   Shipboard Hazard and Defense
SHAEF   Supreme Headquarters Allied Expeditionary Force
Site A   Dichlor plant at Muscle Shoals, Alabama
Site B   Sarin plant at Rocky Mountain Arsenal, Colorado
SSO   Special Security Organization (Iraq)
SW   Code name of intermediate in VX production
TAB   Antidote consisting of TMB4, atropine, and benactyzine
T-Force   Technical Intelligence Unit (U.S. Army)
Trilon-83   German code name for Tabun
Trilon-46   German code name for Sarin
TS   Technical Secretariat (of OPCW)
TTCP   Tripartite Technical Cooperation Program

Glossary

TVA	Tennessee Valley Authority
U.N.	United Nations
UNMOVIC	United Nations Monitoring, Verification and Inspection Commission
UNSCOM	United Nations Special Commission on Iraq
USIB	United States Intelligence Board
V-1	Vergeltungswaffe 1 (German buzz bomb)
V-2	Vergeltungswaffe 2 (German ballistic missile)
WMD	Weapons of mass destruction
WO	War Office (U.K.)

NOTES

Note: Books cited in short form in the Notes can be found in the Bibliography.

ABBREVIATIONS OF SOURCES

FBIS Foreign Broadcast Information Service

FOIA Document declassified and released under the U.S. Freedom of Information Act

MHI Military History Institute, U.S. Army War College, Carlisle Barracks, Pa.

NARA National Archives and Records Administration, College Park, Md.

NSA National Security Archive, George Washington University, Washington, D.C.

PRO Public Record Office, Kew, U.K.

RG Record Group (U.S. National Archives)

SPRU CBW Archive, Science Policy Research Unit, University of Sussex, U.K.

PROLOGUE: LIVE-AGENT TRAINING

4 *Osama bin Laden quote:* Jamal Ismail, "I Am Not Afraid of Death" [interview
with Osama bin Laden], *Newsweek,* January 11, 1999, p. 37.

5 *Chemical Defense Training Facility at Fort Leonard Wood:* Staff Sergeant Kath-
leen T. Rhem, American Forces Press Service, "We've Got the Nerve," August 1,
2000; "Battling an Unseen Enemy," *Retired Officer Magazine,* March 2001, pp.
60–66; Steve Goldstein, "I Volunteered for the Front Lines of Chemical Warfare," *Inquirer Magazine* (*Philadelphia Inquirer*), January 14, 2001; James Dao,
"Pentagon's Worry: Iraqi Chemical Arms," *New York Times,* May 19, 2002.

5 *Center for Domestic Preparedness, Anniston, Alabama:* See www.cdptraining.com.

CHAPTER ONE: THE CHEMISTRY OF WAR

9 *Early battles of World War I:* Tuchman, *The Guns of August,* p. 438.

9 *Trench warfare in World War I:* Fussell, *The Great War and Modern Memory,* p. 40.

9–10 *Proposed chlorine shell in American Civil War:* Jeffrey K. Smart, "History Notes: Chemical & Biological Warfare Research & Development During the Civil War," *CBIAC Newsletter,* vol. 5, no. 2 (Spring 2004), pp. 3, 11.

10 *Constraints on the use of poisons in antiquity:* John Ellis van Courtland Moon, "Controlling Chemical and Biological Weapons Through World War II," in R. D. Burns, ed., *Encyclopedia of Arms Control and Disarmament,* vol. 2 (New York: Charles Scribner's Sons, 1993), pp. 567–574.

10 *Pledge during the Middle Ages:* Tony Freemantle, "Toxic Warfare Bloomed on a Belgian Field," *Washington Times,* January 5, 1998.

10 *Lieber code of conduct in the Civil War:* Friedman, *The Law of War,* pp. 158–186.

10 *1874 Brussels Declaration:* Organization for the Prohibition of Chemical Weapons, "Fact Sheet 1: The Chemical Weapons Convention and the OPCW—How They Came About," p. 1, www.opcw.org/docs/fs1.pdf.

11 *Hague gas projectile declaration:* Richard M. Price, "A Genealogy of the Chemical Weapons Taboo," *International Organization,* vol. 49 (Winter 1995), p. 83.

11 *Use of irritant gases in 1914:* Jeffrey K. Smart, "History of Chemical and Biological Warfare: An American Perspective," in Sidell, Takafuji, and Franz, eds., *Medical Aspects of Chemical and Biological Warfare,* p. 14.

11 *Fritz Haber:* "Fritz Haber: Chemist and Patriot," www.woodrow.org/teachers/chemistry/institutes/1992/Haber.htm.

12 *Shortage of artillery shells; von Falkenhayn:* Price, *The Chemical Weapons Taboo,* pp. 47–51.

12 *Haber conversation with Otto Hahn:* Hahn, *Mein Leben,* p. 117.

13 *General von Deimling quote:* Berthold von Deimling, *Aus der alten in die neue Zeit* (Berlin, 1930), p. 201, cited in Brauch and Müller, *Chemische Kriegführung—Chemische Abrüstung,* p. 84.

13 *Shipment and emplacement of steel cylinders:* Smart, "History of Chemical and Biological Warfare," p. 14.

14 *Chlorine attack at Ypres:* Captain Edward F. Fitzgerald, "Gas!," *Armed Forces Chemical Journal,* vol. 14, no. 6 (November–December 1960), p. 16; Haber, *The Poisonous Cloud,* p. 34; Harris and Paxman, *A Higher Form of Killing,* pp. 2–6.

15 *Anthony Hossack diary entry:* Firstworldwar.com, "Memoirs & Diaries: The First Gas Attack," www.firstworldwar.com/diaries/firstgasattack.htm.

15–16 *Casualties from chlorine attack:* Ulrich Trumpener, "The Road to Ypres: The Beginnings of Gas Warfare in World War I," *Journal of Modern History,* vol. 47, no. 3 (1975), pp. 460–480.

16 *Suicide of Clara Haber:* Dan Charles, "The Tragedy of Fritz Haber," National Public Radio, July 11, 2002.

17 *British use of chlorine gas at Loos:* Smart, "History of Chemical and Biological Warfare," p. 14.

17 *Development of chemical defenses:* Ibid., pp. 15–16.

17 *"Dulcet et Decorum Est":* Owen, *Collected Poems of Wilfred Owen*, p. 55.

18 *American Expeditionary Force:* Charles E. Heller, "Chemical Warfare in World War I: The American Experience, 1917–1918," *Leavenworth Papers*, no. 10 (Fort Leavenworth, Kans.: Combat Studies Institute, 1984), p. 17.

18 *Amos Fries:* "Amos Alfred Fries, Brigadier General, United States Army," Arlington National Cemetery Web site, www.arlingtoncemetery.net/aafries.htm.

18 *Haber introduction of mustard:* Terry M. Weekly, "Proliferation of Chemical Warfare: Challenge to Traditional Restraints," *Parameters,* vol. 19 (December 1988), pp. 52–53.

19 *Effects of mustard agent:* David Zeman, "Duty, Honor, Betrayal: How U.S. Turned Its Back on Poisoned WWII Vets," *Detroit Free Press,* November 10, 2004.

19 *Gunpowder Reservation:* Chemical Weapons Working Group, "Background Info on CW Stockpile Site in Aberdeen, Maryland," available online at www.cwwg. org/maryland.html.

19–20 *Hitler injured by mustard:* Bullock, *Hitler: A Study in Tyranny,* pp. 24–25.

20 *Major combatants used 124,000 tons of 39 different agents:* Organization for the Prohibition of Chemical Weapons, "Fact Sheet 1," www.opcw.org/docs/fs1.pdf.

20 *Impact of chemical warfare on American troops:* Peter Grier, "US Was One of History's First Victims of Gas Warfare," *Christian Science Monitor,* December 14, 1988, p. B4; Rothschild, *Tomorrow's Weapons*, p. 3.

20 *Fries quote:* Fries and West, *Chemical Warfare,* pp. 435–439.

21 *Treaty of Versailles, Washington Disarmament Conference:* U.S. Arms Control and Disarmament Agency, *Arms Control and Disarmament Agreements,* p. 10.

21 *1925 Geneva Protocol:* Bernauer, *The Chemistry of Regime Formation,* p. 18; Charles C. Floweree, "The Politics of Arms Control Treaties: A Case Study," *Columbia Journal of International Affairs,* vol. 37 (1984), p. 271; Jean Pascal Zanders, "The CWC in the Context of the 1925 Geneva Debates," *The Nonproliferation Review,* vol. 3, no. 3 (1996), pp. 38–45.

22 *German-Soviet collaboration at Tomka:* Brauch and Müller, eds., *Chemische Kriegführung—Chemische Abrüstung,* pp. 32–38.

23 *Haber's exile and death:* "Fritz Haber: Chemist and Patriot."

CHAPTER TWO: IG FARBEN

24 *IG Farben:* CIOS, "Chemical Warfare: I.G. Farbenindustrie A.G., Frankfurt-Main," File No. XXX-19, Item No. 30, CIOS Target No. 8/59a, August 20, 1945 [NARA, RG 319]; L. Wilson Greene, "Military Government Control of the German Chemical Industry," *Chemical Corps Journal,* October 1947, pp. 42–45.

25 *Gerhard Schrader biography:* Groehler, *Der lautlose Tod,* p. 356; Pfingsten, *Dr. Gerhard Schrader,* pp. 6–19.

26–28 *Schrader invention of Tabun (Le-100):* BIOS, "The Development of New

Insecticides and Chemical Warfare Agents, by Gerhard Schrader" (Secret), Final Report no. 714, item no. 8, undated [NARA, RG 319]; Gerhard Schrader, "Nervengas" (letter), *Der Spiegel*, no. 13, March 23, 1970, pp. 18, 21.

29 *Dr. Gross's assessment of toxicity:* BIOS, "The Development of New Insecticides and Chemical Warfare Agents," pp. 25–27.

29 *Italian use of mustard in Abyssinian War:* Aram Mattioli, "Entgrenzte Kriegsgewalt: Der italienische Giftgaseinsatz in Abessinien 1935–1936," *Vierteljahrshefte für Zeitgeschichte,* heft 3 (Juli 2003), pp. 311–227.

30 *Organization of Army Ordnance Office:* CIOS, "War Gas Production and Miscellaneous Chemical Warfare Information, Anorgana G.m.b.H., Germany" (Secret), file no. XXVII-34, item no. 8, July 25, 1945, Appendix 2, "Organization of the Heereswaffenamt and Montan Industriewerke G.m.b.H" [NARA, RG 319].

30 *Visit by Sicherer and Wirth to Elberfeld:* Groehler, *Der lautlose Tod,* p. 327.

30 *Army Gas Protection Laboratory:* Gebhard Schultz, "Militärisches Sperrgebiet: Die Zitadelle Spandau im Nationalsozialismus," *Berlinische Monatsschrift,* vol. 7, no. 2 (2001), pp. 51–59.

31 *Visit by Schrader to Spandau Citadel, May 1937:* BIOS, "The Development of New Insecticides and Chemical Warfare Agents," p. 27; Groehler, *Der lautlose Tod,* p. 327.

31 *Systematic search for new war gases:* Military Intelligence Division, Great Britain, "Translation of a Top Secret German Document Dealing with the Sequence of Development of the German Agent Trilon 83 (Tabun)" (Secret), January 24, 1946, p. 3 [FOIA].

32 *IG Farben welcomed the Army's decision to take charge:* Ibid.

32 *Screening process for new war gases:* Gellermann, *Der Krieg, der nicht stattfand,* pp. 80–81.

32 *Raubkammer Chemical Weapons Testing Site:* CIOS, "Chemical Warfare Installations in the Munsterlager Area, Including Raubkammer," Report no. XXXI-86, item no. 8, compiled April 23–June 3, 1945 [NARA, RG 319]; Major General Alden H. Waitt, "Why Germany Didn't Try Gas," *Saturday Evening Post,* vol. 218, no. 36 (March 9, 1946), pp. 137–138.

34 *Ochsner memo (October 27, 1937):* Hermann Ochsner, "Stellungnahme von Oberstleutnant Ochsner zu einer Anfrage der 8. (technischen) Abteilung des Generalstabs des Heeres über neuzeitliche Kampfstoffe und Kampfstoffverwendung, vom 27. Oktober 1937," Dokument 35, in Brauch and Müller, eds., *Chemische Kriegführung—Chemische Abrüstung,* pp. 148–152.

35 *Schrader not allowed to accept Army contract:* BIOS, "Examination of Various German Scientists" (Secret), Final Report no. 44, item no. 8, BIOS Trip no. 1103, Target no. C8/134, 1945 [SPRU].

35 *Karl Krauch report:* Köhler, *Und heute die ganze Welt,* pp. 277–278; Rolf-Dieter Müller, "World Power Status Through the Use of Poison Gas? German Preparations for Chemical Warfare, 1919–1945," in Diest, ed., *The German Military in the Age of Total War,* pp. 186–187.

36 *Vz Tower:* Gellermann, *Der Krieg, der nicht stattfand,* p. 78.

36 *Field trials with Tabun at Raubkammer:* BIOS, "Interrogation of German Air Ministry (OKL) Technical Personnel, Luftwaffe Lager, near Kiel" (Secret), Final Report no. 9, item no. 28, compiled between July 17 and August 2, 1945 [SPRU]; Military Intelligence Division, Great Britain, "Translation of a Top Secret German Document Dealing with the Sequence of Development of the German Agent Trilon 83 (Tabun)" (Secret), January 24, 1946, pp. 11–14.

36 *Injuries during Tabun development work:* Military Intelligence Division, Great Britain, "Translation of a Top Secret German Document Dealing with the Sequence of Development of the German Agent Trilon 83 (Tabun)" (Secret), January 24, 1946, p. 19 [FOIA].

37 *Atropine and scopolamine:* "Atropine," "Scopolamine," *The Columbia Encyclopedia,* 6th ed. (New York: Columbia University Press, 2001) [online].

37–38 *Development of Sarin:* BIOS, "The Development of New Insecticides and Chemical Warfare Agents," pp. 40–42; BIOS, "Examination of Various German Scientists," Final Report no. 44, item no. 8, BIOS Trip no. 1103, Target no. C8/134, September 1945 [NARA, RG 319], "Vertragsnotiz des Heerenwaffenamtes für Hitler betreffend Trilon 46 vom 2. Dezember 1941," Dokument 44, in Brauch and Müller, eds., *Chemische Kriegführung—Chemische Abrüstung,* pp. 174–175.

38 *Tabun pilot plant "Vorwerk Heidkrug":* Military Intelligence Division, Great Britain, "Translation of a Top Secret German Document Dealing with the Sequence of Development of the German Agent Trilon 83 (Tabun)" (Secret), January 24, 1946, p. 17 [FOIA].

38 *Ochsner memo (June 28, 1939):* Hermann Ochsner, "Denkschrift von Oberst Ochsner 'Grundsätzliche Gedanken über den Einsatz von Kampfstoffen im Krieg' für die 1. Abteilung des Generalstabs des Heeres vom 28. Juli 1939," Dokument 39, in Brauch and Müller, eds., *Chemische Kriegführung—Chemische Abrüstung,* pp. 162–165.

38 *General von Brauchitsch's decision to procure Tabun:* Groehler, *Der lautlose Tod,* p. 185.

39 *Meeting of Army and IG Farben officials:* Ibid., pp. 185–186.

39 *Otto Ambros biography:* Heine, *Verstand & Schicksal,* p. 172–173; Anonymous, "Zum Tode eines Giftgasenthusiasten," *Pfalz-Forum,* no. 3 (October–December 1999), pp. 25–29.

40 *Founding of Anorgana:* Groehler, *Der lautlose Tod,* pp. 186–187.

40 *Origin of the name "Sarin":* BIOS, "Examination of Various German Scientists" (Secret), Final Report no. 44, item no. 8, BIOS Trip no. 1103, Target no. C8/134 [SPRU].

40 *Schrader's criticism of Sarin pilot plant development:* British Foreign Office, "Gerhardt [*sic*] Schrader," August 30, 1945 [PRO, FO 1031/105].

41 *Schrader's suspicions about human experimentation:* BIOS, "Interrogation of German CW Medical Personnel" (Secret), Final Report no. 138, item no. 8, August–September 1945 [SPRU].

41 *German conquest of France:* Shirer, *The Rise and Fall of the Third Reich,* p. 738.

41 *Churchill speech:* Sir Winston Churchill, "Their Finest Hour," speech to the House of Commons, June 18, 1940.

CHAPTER THREE: PERVERTED SCIENCE

42 *Founding of Luranil:* Bernd Appler, "The Production of Chemical Warfare Agents by the Third Reich, 1933–45," in Stock and Lohs, *The Challenge of Old Chemical Munitions and Toxic Armament Wastes,* pp. 77–102.

42 *Construction of Dyhernfurth plant:* Ibid., pp. 78–103; CIOS, "Chemical Warfare: I.G. Farbenindustrie A.G., Frankfurt-Main," file no. XXX-19, item no. 30, 1945 [NARA, RG 319]; " 'Die Pest ist denkbar unzuverlässig': Die B + C Rüstung des Dritten Reiches," *Der Spiegel,* no. 52 (December 22, 1969), pp. 98–99.

42 *Meeting of IG Farben staff in Ludwigshafen in August 1941:* CIOS, "Chemical Warfare: I.G. Farbenindustrie, A.G.," p. 8.

44 *Creation of the Speer ministry:* Simon, *German Research in World War II,* pp. 81, 87.

44 *Special Committee C:* FIAT, "Report on Chemical Warfare Based on the Interrogation and Written Reports of Jürgen E. von Klenck, Also Comments by Speer and Dr. E. Mohrhardt" (Secret), December 6, 1945, pp. 19–20.

44 *German memorandum on Trilon 83:* Military Intelligence Division, Great Britain, Military Attaché Report, "Translation of a Top Secret German Document Dealing with the Sequence of Development of the German Agent Trilon 83 (Tabun)," (Secret), A.11000, January 24, 1946, p. 22 [original German document dated February 14, 1942] [FOIA].

44–45 *Zyklon B:* "Zyklon B," Jewish Virtual Library, www.us-israel.org/jsource/Holocaust.Zyklon.htm (accessed January 30, 2004); Richard J. Green, "The Chemistry of Auschwitz," www.holocaust-history.org/auschwitz/chemistry (accessed January 30, 2004).

45–46 *Start of Tabun production at Dyhernfurth:* BIOS, "Interrogation of German CW Personnel at Heidelberg and Frankfurt" (Secret), Final Report no. 41, item no. 8, undated [SPRU].

46–47 *Tabun manufacturing process:* Military Intelligence Division, Great Britain, Military Attaché Report, "Translation of a Top Secret German Document Dealing with the Sequence of Development of the German Agent Trilon 83 (Tabun)"; CIOS, "Chemical Warfare: I.G. Farbenindustrie A.G.," file no. XXX-19, pp. 13–14; Capitaine Collomp, "Une arme secrète allemande: Les Trilons," *Forces Aériennes Françaises,* no. 37, October 1949, pp. 68–69.

47 *Two forms of Tabun:* Ibid., p. 76.

48 *High-fat diet:* CIOS, "Chemical Warfare: I.G. Farbenindustrie A.G.," p. 9.

48 *Fatal accidents at Dyhernfurth:* CIOS, "Chemical Warfare Installations in the Munsterlager Area" (Secret), file no. XXXI-86, item no. 8, compiled April 23–June 3, 1945 [SPRU].

48–49 *Security measures at Dyhernfurth:* CIOS, "Chemical Warfare: I.G. Farbenindustrie A.G., Frankfurt/Main," file no. XXX-19, item no. 30, August 20, 1945 [NARA, RG 319], pp. 2–3.

49 *Cover names for German nerve agents:* Capitaine Collomp, "Une arme secrète allemande: Les Trilons," *Forces Aériennes Françaises,* no. 37, October 1949, pp. 65–82; Groehler, *Der lautlose Tod,* pp. 184–185.

49 *U.S. intelligence report of July 1942:* Lieutenant Colonel Ralph W. Hufford, Intelligence Division, Chemical Warfare Service, Message to C Intel Div, July 9, 1942, Subject: "New German Poison Gas" [FOIA].

50 *Tabun filling line at Dyhernfurth:* Groehler, *Der lautlose Tod,* pp. 284–285.

50 *Initial Tabun production at Dyhernfurth:* CIOS, "War Gas Production and Miscellaneous Chemical Warfare Information, Anorgana GmbH, Germany" (Secret), Report no. XXVII-34, item no. 8, July 25, 1945 [NARA, RG 319]; SHAEF, Office of Assistant Chief of Staff, G-2, "Poison Gas: Abstract from Notes on Interrogations at Frankfurt, 21 April–4 May 1945" (Top Secret) [PRO, FO 1031/86]; Groehler, *Der lautlose Tod,* p. 238.

50–51 *Work camp at Dyhernfurth:* Groehler, *Der lautlose Tod,* p. 284.

51 *Human experimentation:* Ibid., p. 286.

51–52 *Richard Kuhn:* Nobel e-Museum, "Richard Kuhn—Biography," available online at www.nobel.se/chemistry/laureates/1938/kuhn-bio.html; Goudsmit, *ALSOS,* p. 80

52 *Dale and Loewi discoveries:* "The Nobel Prize in Physiology or Medicine 1936," www.nobel.se/medicine/laureates/1936/press.html.

53–54 *Pharmacological effects of nerve agents:* Dheraj Khurana and S. Prabhakar, "Organophosphorus Intoxication," *Archives of Neurology,* vol. 57 (April 2000), pp. 600–602.

54 *Industrial manufacture of Sarin:* CIOS, "Chemical Warfare: I.G. Farbenindustrie, A.G.," file no. XXX-19, pp. 23–24.

54–55 *Sarin plant at Falkenhagen (Seewerk):* FIAT, "Report on Chemical Warfare Based on the Interrogation and Written Reports of Jürgen E. von Klenck, Also Comments by Speer and Dr. E. Mohrhardt" (Secret), December 6, 1945, pp. 7–8 [SPRU]; Groehler, *Der lautlose Tod,* p. 328.

55 *German army officer captured in Tunisia:* CSDIC, "Report of German Chemical Weapons Research Based on Interview with an Unidentified Prisoner of War," March 7, 1943 [PRO WO 193.723 47353].

55–56 *Battle of Stalingrad:* Shirer, *The Rise and Fall of the Third Reich,* pp. 914–919, 925–932.

56 *Hitler's eastern headquarters at Wolf's Lair:* Ibid., p. 849.

56–58 *Meeting on May 15, 1943 at Wolf's Lair:* Otto Ambros, "Auszug aus der Denkschrift von Otto Ambros über die Lage auf dem Kampfstoffgebiet vom 20. März 1944," Dokument 49, in Brauch and Müller, eds., *Chemische Kriegführung—Chemische Abrüstung,* pp. 182–184; Groehler, *Der lautlose Tod,* pp. 252–256; Richard Halloran, "Gas Developed by Nazis, Who Weighed Using It," *New York Times,* August 19, 1970, p. 6.

58 *Arbusov and the Kazan school of organophosphorus chemistry:* Academician B. A.

Arbuzov, "The Chemistry of Organophosphorus Compounds and Their Application," *Vestnik Akademii Nauk SSSR* (Moscow), no. 3 (1960), pp. 103–105,
translated in CIA, *Scientific Intelligence Digest,* OSI-SD/61-26 (Secret), December 26, 1961 [NARA]; A. N. Pudovik and I. M. Aladzheva, "Chemistry of
Organophosphorus Compounds in the Kazan School of Organophosphorus
Chemists," *Uspekhi Khimii* (Advances in Chemistry, Moscow), vol. 36, no. 9
(1967), pp. 1499–1532, translated in U.S. Department of Commerce, Joint Publications Research Service, JPRS 44,591, March 6, 1968 [SPRU]; Joint Chiefs of
Staff, Intelligence Group, Report to the Joint Intelligence Committee, "Intelligence on Soviet Capabilities for Chemical and Bacteriological Warfare," JIC
156/12, January 27, 1949.

58–59 *Saunders research on DFP at University of Cambridge:* Saunders, *Some Aspects of
the Chemistry and Toxic Action of Organic Compounds Containing Phosphorus
and Fluorine,* pp. 42–86.

59 *Saunders experimented on himself with DFP:* Toy, *Phosphorus Chemistry in Everyday Living,* p. 198.

59 *DFP mainly regarded as a harassing agent:* Clarke, *The Silent Weapons,* p. 45.

59 *Division 9 of National Defense Research Committee:* Jeffrey K. Smart, Command
Historian, U.S. Army Chemical and Biological Defense Command, Aberdeen
Proving Ground, Md., "Oral History Interview: Elmer H. Engquist, February
26, 1993," transcript, p. 1; Johnston, *A Bridge Not Attacked,* p. 218.

59 *Test of DFP on U.S. soldiers:* Roger McCoy, "Ohioans Volunteered for Secret
Chemical Warfare Tests During World War II," *The Columbus Dispatch,* February 14, 2005.

60 *President Roosevelt's warning:* ACDA, *Arms Control and Disarmament Agreements,* pp. 10–11.

60 *Ambros recommendation to increase production capacity:* Rolf-Dieter Müller,
"World Power Status through the Use of Poison Gas? German Preparations for
Chemical Warfare, 1919–1945," in Diest, ed., *The German Military in the Age of
Total War,* p. 197.

61 *Level of Tabun production:* CIOS, "War Gas Production and Miscellaneous
Chemical Warfare Information, Anorgana G.m.b.H., Germany" (Secret), file
no. XXVII-34, item no. 8, July 25, 1945, Appendix 1, "Production of War Gases
in Germany" [NARA, RG 319].

61 *Establishment of a second labor camp at Dyhernfurth:* Groehler, *Der lautlose Tod,*
p. 286.

62 *Dr. Karl Brandt:* FIAT, "Report on Chemical Warfare Based on Interrogation
and Written Reports of Jürgen E. von Klenck" (Secret), December 6, 1945, pp.
20, 29.

62 *People's gas masks:* BIOS, "Interrogation of Certain German Personalities Connected with Chemical Warfare," Final Report no. 542, item no. 8, October
1946, p. 23 [NARA, RG 319].

62–63 *Kuhn discovery of Soman:* Ibid., pp. 4–7; BIOS, "Interrogation of Professor
Ferdinand Flury and Dr. Wolfgang Wirth on the Toxicology of Chemical War-

fare Agents," Final Report no. 782, item no. 8, BIOS Trip no. 1610 (undated), pp. 12–13 [NARA, RG 319].

CHAPTER FOUR: TWILIGHT OF THE GODS

64 *Invasion of Normandy (Operation Overlord):* Hastings, *Overlord,* p. 34.

64–65 *Gen. Omar Bradley quotes:* Bradley, *A Soldier's Story,* p. 279.

65 *Postwar report by Ochsner:* Hermann Ochsner, "U.S. Army Chemical Warfare Project: Bericht über Produktion von K-Stoffen, Raumexplosionen und Raumbränden, P-004a," Dokument 69, in Brauch and Müller, eds., *Chemische Kriegführung—Chemische Abrüstung,* pp. 217–236; Ochsner, *History of German Chemical Warfare in World War Two,* Part I: *The Military Aspect,* p. 19.

66 *German decision not to use nerve agents in V weapons:* BIOS, "Interrogation of Gen. Lt. Ochsner" (Secret), Final Report no. 187, item no. 8, December 5–6, 1945 [SPRU].

66 *July 20, 1944, plot against Hitler:* Shirer, *The Rise and Fall of the Third Reich,* pp. 1048–1054.

67 *Ley plan for the use of Tabun:* Groehler, *Der lautlose Tod,* p. 311.

67 *Speer conversation with Ley:* Speer, *Inside the Third Reich,* p. 413.

67–68 *Conflict over N-Stoff at Falkenhagen:* FIAT, "Report on Chemical Warfare Based on the Interrogation and Written Reports of Jürgen E. von Klenck, Also Comments by Speer and Dr. E. Mohrhardt" (Secret), December 6, 1945, pp. 8–11 [SPRU].

68 *Speer testimony at Nuremberg:* Speer testimony on June 21, 1946, International Military Tribunal, *Trial of the Major War Criminals before the International Military Tribunal, Nuremberg,* pp. 527–530.

68 *Speer message to Keitel:* Groehler, *Der lautlose Tod,* p. 311.

68–69 *Schieber meeting on November 2, 1944:* "Aktenvermerk über die K-Stoff-Besprechung beim Amtschef des Rüstungslieferungsamtes, Staatsrat Dr. Walther Schieber, am 2. November 1944," Dokument 54, in Brauch and Müller, eds., *Chemische Kriegführung—Chemische Abrüstung,* pp. 192–195.

69 *Destruction of chemical weapons documents:* Harris and Paxman, *A Higher Form of Killing,* p. 138.

69 *Volkssturm:* Hunt, *On Hitler's Mountain,* p. 188.

70 *Evacuation of Dyhernfurth:* Groehler, *Der lautlose Tod,* pp. 9–10.

70–71 *Murder of labor camp inmates:* Ibid., pp. 10, 284.

71 *German raid on Dyhernfurth:* Ibid., pp. 10–11; Gellermann, *Der Krieg, der nicht stattfand,* pp. 175–176; Duffy, *Red Storm on the Reich,* pp. 128–132.

72 *Hitler order of February 1945:* Groehler, *Der lautlose Tod,* p. 11.

73 *Goebbels response to the firebombing of Dresden:* Ibid., p. 314; Gellermann, *Der Krieg, der nicht stattfand,* p. 178.

73 *General Guderian statement:* Shirer, *The Rise and Fall of the Third Reich,* pp. 1097–1098; Trevor-Roper, *The Last Days of Hitler,* p. 123.

73 *Evacuation of Falkenhagen:* FIAT, "Report on Chemical Warfare Based on the

Interrogation and Written Reports of Jürgen E. von Klenck; Also Comments by Speer and Dr. E. Mohrhardt" (Secret), December 6, 1945, pp. 11, 14.

74 *Klenck's burial of documents in Gendorf:* FIAT, "Report on Chemical Warfare Based on the Interrogation and Written Reports of Jürgen E. von Klenck, Also Comments by Speer and Dr. E. Mohrhardt" (Secret), December 6, 1945, pp. 25, 34–35.

74 *"Scorched earth" policy:* Shirer, *The Rise and Fall of the Third Reich,* pp. 1102–1104.

75–76 *Speer plot to assassinate Hitler:* Speer, *Inside the Third Reich,* pp. 430–431; Trevor-Roper, *The Last Days of Hitler,* pp. 122–125.

76 *Hitler's fear of being captured alive:* O'Donnell, *The Bunker: The History of the Reich Chancellery Group,* p. 110fn.

76 *Speer quote about Hitler:* Speer, *Inside the Third Reich,* p. 431.

76 *Hitler said "If the war is lost . . .":* Shirer, *The Rise and Fall of the Third Reich,* p. 1104.

77 *Keitel order to evacuate chemical munitions:* "Befehl des Chefs des Oberkommandos der Wehrmacht über die Evakuierung von Kampfstoff-Munition vom 30. März 1945," Dokument 57, in Brauch and Müller, eds., *Chemische Kriegführung—Chemische Abrüstung,* pp. 197–198.

78–79 *Evacuation of Tabun-filled munitions, loading onto barges:* Groehler, *Der lautlose Tod,* pp. 12–14.

79 *U.S. air raid on German town of Lossa:* Groehler, *Der lautlose Tod,* pp. 6–8; Gellermann, *Der Krieg, der nicht stattfand,* p. 181.

80 *Captain Hemmen tour of chemical weapons depots:* Groehler, *Der lautlose Tod,* pp. 14–15; Gellermann, *Der Krieg, der nicht stattfand,* p. 182.

81 *Hitler's birthday (April 20, 1945):* Shirer, *The Rise and Fall of the Third Reich,* pp. 1107–1112.

81 *Evacuation of nerve-agent weapons to Lake Chiemsee:* Gwynne Roberts, "Hitler's Deadly Secrets," *Sunday Times* (London), February 22, 1981, p. 14.

81–82 *U.S. shelling of German barges:* Major General Alden H. Waitt, "Why Germany Didn't Try Gas," *Saturday Evening Post,* vol. 218, no. 36 (March 9, 1946), p. 138.

82 *Hitler's suicide:* Shirer, *The Rise and Fall of the Third Reich,* p. 1133.

82 *"Wotan" quote:* Ibid., p. 1100.

82 *Hitler was living in a fantasy world:* John Ellis van Courtland Moon, personal communication, February 29, 2004.

CHAPTER FIVE: A FIGHT FOR THE SPOILS

83 *T Forces:* Colonel Harry A. Kuhn, "German Technical Information," *Chemical Corps Journal,* January 1947, pp. 12–14.

83 *Mobile microfilm teams:* U.S. Army Military History Research Collection, Senior Officers Debriefing Program, "Conversations between Lieutenant General Andrew J. Boyle and Lieutenant Colonel Frank Walton," vol. 1 (Carlisle Barracks, Pa.: U.S. Army Military History Institute, 1971), p. 13.

83 *Establishment of CIOS:* Hunt, *Secret Agenda,* p. 7.

84 *Lieutenant Colonel Paul R. Tarr:* Major Bernard Tannor, "Cml C Intelligence in European Theater," *Chemical Corps Journal,* vol. 1, no. 3 (1947), p. 40; Bower, *The Paperclip Conspiracy,* pp. 94–95.

84–85 *British inspection of Raubkammer Testing Site:* Groehler, *Der lautlose Tod,* pp. 323–324; Major General Alden H. Waitt, "Why Germany Didn't Try Gas," *Saturday Evening Post,* vol. 218, no. 36 (March 9, 1946), p. 138.

85 *British capture of German Tabun munitions:* Wiseman, *Special Weapons and Types of Warfare,* vol. 1: *Gas Warfare,* p. 150.

85–86 *Analysis of "three-green-ring" bomb by Porton chemists:* Chief Superintendent, Chemical Defence Experimental Station, Porton Down, "German 250 Kg. bomb—3 Green Rings," Ptn/1371 (V.4256A), May 4, 1945; Lieutenant Colonel P. R. Tarr, Chief, Intelligence Division, Chemical Warfare Service, European Theater of Operations, Report No. 3816, "Porton Report on German 250 Kg Bomb—Three Green Rings," May 23, 1945 [FOIA].

86 *D. J. C. Wiseman quote:* Wiseman, *Special Weapons and Types of Warfare,* Vol. 1: *Gas Warfare,* p. 152.

86–87 *Interrogation of German chemical weapons experts:* BIOS, "Interrogation of German CW Personnel at Heidelberg and Frankfurt" (Secret), Final Report no. 41, item no. 8, undated [NARA, RG 319]; BIOS, "Interrogation of Gen. Lt. Ochsner" (Secret), Final Report no. 187, item no. 8, December 5–6, 1945; BIOS, "Interrogation of German CW Medical Personnel" (Secret), Final Report no. 138, item no. 8, August–September 1945; BIOS, "Examination of Various German Scientists" (Secret), Final Report no. 44, item no. 8, undated [NARA].

87 *Interrogation of Schrader:* CIOS, "A New Group of War Gases" (Secret), file no. XXIII-7, item no. 8, April 23, 1945 [SPRU]; Foreign Office, "Gerhardt [*sic*] Schrader," August 30, 1945 [PRO, FO 1031/105].

88 *Ambros in Gendorf:* DuBois, *The Devil's Chemists,* pp. 5–7; Hunt, *Secret Agenda,* p. 7.

88 *Tarr interrogation of Ambros:* FIAT, "Detention of Dr. Ambros," FIAT E 254-82 (AMBROS) [PRO, FO 1031.86]; Bower, *The Paperclip Conspiracy,* p. 96.

88–89 *ALSOS interview with Kuhn:* Goudsmit, *ALSOS,* pp. 80–83; Groehler, *Der lautlose Tod,* p. 329.

89 *Increases in CWS budget:* Harris and Paxman, *A Higher Form of Killing,* p. 116–117.

89 *CWS Development Laboratory at MIT:* Jeffrey K. Smart, Command Historian, U.S. Army Chemical and Biological Defense Command, Aberdeen Proving Ground, Md., "Oral History Interview: Elmer H. Engquist, February 26, 1993," transcript, p. 1 [FOIA].

90 *Coombs and Sauer analyses of Tabun:* Captain Robert D. Coombs and First Lieutenant Charles W. Sauer, "Investigations on MCE and MFI" (Secret), Chemical Warfare Service Development Laboratory, Massachusetts Institute of Technology, September 27, 1945 [FOIA].

90 *German Tabun-filled bombs shipped to United States:* Major General William N. Porter, Chief, Chemical Warfare Service, Memorandum to Commanding General, Army Service Forces, Subject: "Enemy Agent Munitions, LE-100 (Taboon)" (Secret), SPCWM 471 ASF, May 29, 1945 [FOIA].

90 *No. 1 Porton Group:* Carter, *Chemical and Biological Defence at Porton Down,* pp. 55–57.

91 *Large buried cache of microfilmed documents:* Groehler, *Der lautlose Tod,* p. 322.

91 *Soviet capture of Dyhernfurth and documents:* JCS, Joint Intelligence Group, Report, "Intelligence on Soviet Capabilities for Chemical and Bacteriological Warfare," Annex B, JIC 156/12, 1952, pp. 15–16 [NARA, RG 218].

91 *Volfkovich and Kargin trips to Germany:* Krause and Mallory, *Chemical Weapons in Soviet Military Doctrine,* pp. 114–115.

92 *Ocean dumping of German chemical weapons:* E. J. Hogendoorn, "A Chemical Weapons Atlas," *Bulletin of the Atomic Scientists,* vol. 53, no. 5 (September–October 1997), p. 39.

92 *British confiscation of Tabun bombs:* U.K., Chiefs of Staff Committee, "Note on COS(45) 400(0), Disposal of German Chemical Warfare Stocks," June 19, 1945 [PRO, WO 193/712]; UK, Combined Chiefs of Staff, "Disposal of German Chemical Warfare Stocks: Memorandum by the Representatives of the British Chiefs of Staff" (Secret), C.O.S. 883/2, August 20, 1945 [NARA].

92 *Contingency for possible use against Japan:* G. B. Carter and Graham S. Pearson, "Past British Chemical Warfare Capabilities," *RUSI Journal,* February 1996, p. 62.

92–93 *Transfer of Tabun bombs to Britain (Operation Dismal):* Sloan, *A Tale of Tabun,* pp. 38–43.

93 *Creation of FIAT:* Kuhn, "German Technical Information," p. 12.

93 *"One gets the idea that if the IG had not been fortunate enough to stumble . . .":* BIOS, "Interrogation of Professor Ferdinand Flury and Dr. Wolfgang Wirth on the Toxicology of Chemical Warfare Agents" (Secret), Final Report no. 782, item no. 8, undated, p. 8 [NARA, RG 319].

94 *None of the physiologists admitted involvement:* BIOS, "CW Investigation (BIOS Trip 1703)," September 5, 1945 [PRO, IFO 1031/105].

94 *"It does seem a matter for serious doubt . . .":* BIOS, "Interrogation of Professor Ferdinand Flury and Dr. Wolfgang Wirth on the Toxicology of Chemical Warfare Agents" (Secret), Final Report no. 782, item no. 8, undated, pp. 7–8 [NARA, RG 319].

94 *Klenk interrogation:* FIAT, "Report on Chemical Warfare Based on the Interrogation and Written Reports of Jürgen E. von Klenck, Also Comments by Speer and Dr. E. Mohrhardt" (Secret), December 6, 1945.

94 *FIAT interrogation of Ter Meer:* Sloan, *A Tale of Tabun,* p. 34.

94 *Human experimentation by Wimmer and Hirt:* Bower, *The Paperclip Conspiracy,* p. 96.

95 *Ambros transfer and disappearance, Tarr telegram:* FIAT, "Detention of Dr.

Ambros," FIAT E 254082 (AMBROS), undated [PRO FO 1031/86]; Bower, *The Paperclip Conspiracy,* pp. 96–97.

96 *Hirschkind meeting with Ambros:* Letter to Otto Ambros from W. Hirschkind, Dow Chemical Company, Western Division, Walnut Creek, Calif., dated July 21, 1967, available online at www.carr.lib.md.us/~stevenba/dowambro.html.

96 *Ambros protected by French government:* FIAT, "Detention of Dr. Ambros"; Bower, *The Paperclip Conspiracy,* p. 97.

96–97 *Schrader internment at Dustbin:* Major E. Tilley, Memorandum to Lieutenant Colonel P. M. Wilson, Subject: "Schrader," April 9, 1946 [PRO FO 1031/105 50146].

98 *IG Farben trial:* "The Farben Case, Military Tribunal VI, Case 6," in Nuremberg Military Tribunal, *Trials of War Criminals Before the Nürnberg Military Tribunals Under Control Council Law No. 10,* vol. 12, Nürnberg, October 1946–April 1949 (Washington, D.C.: U.S. Government Printing Office, 1953).

99 *"He is wily . . .":* FIAT, "Detention of Dr. Ambros."

99 *Ambros defense:* Anonymous, "Zum Tode eines Giftgasenthusiasten," *Pfalz-Forum,* no. 3 (October–December 1999), pp. 25–29.

99 *Peter Hayes quote:* Hayes, *Industry and Ideology,* p. 367.

99 *Statistics of IG Farben trial:* Plumpe, *Die I.G. Farbenindustrie A.G.,* pp. 746–759.

100 *Ambros sentencing:* Groehler, *Der lautlose Tod,* p. 331.

100 *DuBois book:* DuBois, *The Devil's Chemists.*

101 *"the Russians know that we possess this gas":* Air Staff, Royal Air Force, "Note on the Potential Value of Nerve Agents as C.W. Agents," September 6, 1947 [PRO, AIR 20/8730].

101 *Planned modification of Tabun bombs:* Chiefs of Staff Committee, Chemical Warfare Sub-Committee, "Chemical Warfare Reserve Policy, Note by the Air Ministry," CW(52) 9, July 30, 1952, p. 2 [PRO DEFE 41/157].

101 *Sea dumping of Tabun bombs (Operation Sandcastle):* J. O. C. Livesay, Memorandum, "Disposal of German Nerve Gas Stored at Llandwrog" (Secret), February 12, 1954 [PRO, AIR 20/8734]; Inspector-General, Royal Air Force, "Report on Visit to No. 31 Maintenance Unit—Llandwrog" [Disposal of German Nerve Gas Bombs Under Operation Sandcastle], Report No. 538 (Confidential), May 3, 1956 [PRO, AIR 20/8730]; Sloan, *A Tale of Tabun,* pp. 56–82.

102 *Pentagon strategists drew up plans* Kevin Sullivan, "U.S. Planned Chemical Attack on Japan," *Guardian* (London), August 6, 1991.

CHAPTER SIX: RESEARCH AND DEVELOPMENT

103 *New construction at Edgewood Arsenal:* U.S. Army Edgewood Arsenal, "An Introduction to Edgewood Arsenal" (brochure), September 1973 [SPRU].

104 *CWS assessment of Tabun:* War Department, The Adjutant General's Office, Washington, D.C., Memorandum, Subject: "New German Gas, MCE, Preliminary Report," June 19, 1945 [FOIA]; Jess W. Thomas, *Protection Afforded by Gas*

Mask Canisters Against Dimethylamino Cyano Ethoxy Phosphine Oxide (*MCE*), Technical Division Memorandum Report (Secret), CWS, Technical Command, Edgewood Arsenal, Control No. 5004-1127, September 12, 1945 [FOIA].

104 *Development of production process for Tabun:* Chemical Warfare Service, "Item No. 1445, Subject: CWS Project Program for Fiscal Year 1946" (Confidential), June 15, 1945 [FOIA].

104 *Static testing of Tabun-filled munitions at Edgewood:* U.S. Army Chemical Corps, Memorandum to the Chairman, Chemical Corps Technical Committee, Subject: "Chemical Corps Project Program for 1948" (Secret), June 16, 1947 [FOIA]; U.S. Army Chemical Corps, Research and Development Project Card (New Projects), "Munitions for G-Series Filling" (Secret), Project No. B10.5, 4-04-15-05, March 18, 1948 [FOIA].

105 *Adoption of U.S. nomenclature for G Agents:* Colonel J. H. Rothschild, Chief, Technical Division, CWS, "Symbols for German C.W.S. Agents," October 16, 1945 [FOIA]; Major James E. McHugh, Executive, Training Division, CWS, "Symbols for German C.W.S. Agents, Comment No. 2," October 17, 1945 [FOIA]; Colonel J. H. Rothschild, "Symbols for German C.W.S. Agents, Comment No. 3," October 23, 1945 [FOIA]; Colonel J. H. Rothschild, Memorandum to the Commanding General, A.S.F, Subject: "Adoption of Symbols for Chemical Warfare Agents," December 28, 1945 [FOIA].

106 *Establishment of U.S. Army Chemical Corps:* Russell, *War and Nature,* p. 177.

106 *Project A1.13-2.1:* Memorandum to the Chairman, Chemical Warfare Technical Committee, Subject: "Establishment of Project A1.13-2.1, GB Plant, Process Development" (Secret), CWTC Meeting no. 1, March 28, 1946 [FOIA].

106–07 *Soviet dismantlement of Dyhernfurth production line:* Krause and Mallory, *Chemical Weapons in Soviet Military Doctrine,* pp. 117–118.

107 *Origins of Beketovka plant:* Sonia Ben Ouagrham, "Conversion of Russian Chemical Weapons Production Facilities: Conflicts with the CWC," *The Nonproliferation Review,* vol. 7, no. 2 (Summer 2000), p. 51.

107 *Captured German scientists:* U.S. Department of the Army, Office of the Assistant Chief of Staff for Intelligence, Washington, D.C., Intelligence Staff Study, "Soviet Research and Development Capabilities for New Toxic Agents" (Secret), Project no. A-1735, July 28, 1958, pp. 6–7 [MHI].

107 *Soborovsky:* Author's interview with Vil Mirzayanov.

107 *Tabun mentioned in* Military Chemical Textbook: U.S. Department of the Army, Office of the Assistant Chief of Staff, G-2,, Intelligence Staff Study, "Chemical and Biological Capabilities in the Soviet Union and Satellites" (Secret), Project no. 7082, January 16, 1953, p. 5 [MHI].

108 *General Waitt quote:* Major General Alden H. Waitt, "Why Germany Didn't Try Gas," *Saturday Evening Post,* vol. 218, no. 36 (March 9, 1946), p. 138.

108 *Chemical weapons experimental station at Suffield, Canada:* Eggleston, *Scientists at War,* pp. 103–109; Hersh, *Chemical and Biological Warfare,* pp. 293–294; Paul A. D'Agostino and Cam A. Boulet, "Celebrating 60 Years of CB Research at

Defense Research Establishment Suffield (1941–2001)," *The ASA Newsletter,* vol. 01-4, issue no. 85 (2001).

108 *Tripartite Agreement:* G. B. Carter and Graham S. Pearson, "Past British Chemical Warfare Capabilities," *RUSI Journal,* February 1996, p. 63.

109 *Royal Air Force report of September 6, 1947:* Air Staff, Royal Air Force, "Note on the Potential Value of Nerve Gases as C.W. Agents," September 6, 1947 [PRO, AIR 20/8730].

109 *Research at Porton Down:* Chemical Defence Experimental Establishment, Porton Down, "Provisional Appreciation of the C.W. Value of Nerve Gases" (Restricted), Porton Memorandum, no. 32, December 10, 1946 [PRO WO 189/252]; Porton Memorandum no. 15, Addendum II, "Classified List of Compounds Examined Physiologically Since 1945, Part IV—Nerve Gases—(G Agents)" (Secret) [PRO WO 189/233]; Porton Memorandum no. 34, "Appreciation of the Potential CW Value of Nerve Gases Based on Information Available Up to 30.06.49," serial no. 149, June 30, 1949 (Restricted) [PRO WO 189/254]; Gradon Carter and Brian Balmer, "Chemical and Biological Warfare and Defence, 1945–90," in Bud and Gummet, *Cold War, Hot Science,* pp. 296–297.

109 *Edgewood report on human testing with Tabun:* L. Wilson Greene, "Psychochemical Warfare: A New Concept of War," Army Chemical Center, August 1949, cited in Hunt, *Secret Agenda,* p. 162.

110-11 *Pharmacology of nerve agents (muscarinic and nicotinic effects):* Frederick R. Sidell, "Nerve Agents," in Sidell, Takafuji, and Franz, eds., *Medical Aspects of Chemical and Biological Warfare,* pp. 129–179.

111 *Syrettes containing atropine:* Jeffrey K. Smart, "History of Chemical and Biological Warfare: An American Perspective," in Sidell, Takafuji, and Franz, eds., *Medical Aspects of Chemical and Biological Warfare,* p. 54.

111 *U.S. development of Sarin pilot plant:* U.S. Army Chemical Corps, Chemical Corps R&E Periodic Progress Report, "GB Plant, Process Development" (Secret), Project No. A1.13-2.1, 4-92-03-02, October 1, 1947 [FOIA]; U.S. Army Chemical Corps, R&E Periodic Project Progress Report, "Agents of the G-Series" (Secret), Project No. A1.13, 4-08-03-05, October 1, 1947 [FOIA].

112-15 *Colonel Loucks recruitment of Walther Schieber:* U.S. Army Military History Institute, Senior Officer Military History Program, "Major General Charles E. Loucks, USA, Retired, interviewed by Mr. Morris C. Johnson, 1984," Project 84-8, transcript [MHI].

115 *Loucks meetings with Schieber and Falkenhagen scientists:* Papers of General Charles E. Loucks, *Desk Diary 1948* [MHI].

116 *Project Paperclip and German scientists:* Oral History, "Conversations Between Lieutenant General Andrew J. Boyle and Lieutenant Colonel Frank Walton," vol. 1, 1971, U.S. Army Military History Research Collection, Senior Officers Debriefing Program [MHI].

116 *Operation Matchbox:* David Williams and Tom Rawstorne, "British Nerve Gas Deaths 'Had Nazi Scientists,' " *Daily Mail* (London), October 25, 1999, p. 17.

116–17 *Fritz Hoffmann:* Hunt, *Secret Agenda,* pp. 160–161; Russell, *War and Nature,* p. 178.

117 *Australian involvement in Operation Matchbox:* Roger Maynard, "Australia 'Hired Nazi Scientists After War,' " *The Times* (London), August 17, 1999.

117 *German chemists in United States:* Hunt, *Secret Agenda,* pp. 159–162; author's interview with William C. Dee.

117 *Schieber work for U.S. European Command (EUCOM):* Hunt, *Secret Agenda,* p. 169.

117 *French chemical weapons program after World War II:* Lepick, *Les armes chimiques,* pp. 97–98.

118 *French Experimental Station at Beni Ounif:* Vincent Jauvert, "Quand la France testait des armes chimiques en Algérie," *Le Nouvel Observateur,* October 23–29, 1997, pp. 10–22; Meyer, *L'arme chimique,* pp. 154–161.

118–20 *Loucks visit to French chemical weapons testing site in Algeria:* Major General Charles E. Loucks, USA, Retired, interviewed by Mr. Morris C. Johnson, 1984, U.S. Army Military History Institute, Senior Officer Military History Program, Project 84-8, transcript [MHI]; Papers of General Charles E. Loucks, *Desk Diary 1949* [MHI].

120 *British interest in Sarin production capability:* Chemical Defence Experimental Establishment, Porton Down, "Note on the Potential Value of Nerve Gases as C.W. Agents," September 6, 1947 [PRO, AIR 20/8730].

120 *British decision to build Sarin pilot plant:* (U.K.) Combined Chiefs of Staff, "Biological and Chemical Warfare Research and Development Policy (Draft)," undated [PRO WO 188/705].

120 *British Research Establishment at Sutton Oak:* G. B. Carter and Graham S. Pearson, "Past British Chemical Warfare Capabilities," *RUSI Journal,* February 1996, pp. 60, 62–63; United Kingdom, "Declaration of Past Activities Relating to Its Former Offensive Chemical Weapons Program," submitted to the Organisation for the Prohibition of Chemical Weapons, The Hague, May 27, 1997.

121 *Selection of location for U.K. Sarin pilot plant:* Author's interview with Ron Manley.

121 *Sarin pilot plant at Nancekuke:* G. B. Carter and Graham S. Pearson, "Past British Chemical Warfare Capabilities," *RUSI Journal,* February 1996, pp. 59–68; (UK) Chiefs of Staff Committee, Chemical Warfare Sub-Committee, "First Requirements for Nerve Gas Weapons: Memorandum by Ministry of Supply," CW(52) 13 (Secret), October 23, 1952 [PRO DEFE 41/157]; C. A. H. Pidcock, (U.K.) Ministry of Defence, "Estimation of Quantity of New Gases Required for Operational Purposes in 1957" (Top Secret), June 14, 1949 [PRO AIR 20/8731].

CHAPTER SEVEN: BUILDING THE STOCKPILE

122 *Cut in Chemical Corps budget, plan to move to Camp Siebert:* Jeffrey K. Smart, Command Historian, U.S. Army Chemical and Biological Defense Command,

Aberdeen Proving Ground, Md., "Oral History Interview: Elmer H. Engquist, February 26, 1993," transcript, pp. 4–5.

122 *Research at Edgewood on G-series agents:* U.S. Army Chemical Corps, Research and Development Project Card (New Projects), "Agents of the G Series" (Secret), A1.13, Project No. 4-08-03-05, March 26, 1948 [FOIA].

122–23 *Debate over Sarin versus Soman:* Wise, *Cassidy's Run,* pp. 65–66.

123 *Selection of Sarin as the standard agent:* U.S. Army Chemical Corps, Office of the Chief, Memorandum to the Chairman, Chemical Corps Technical Committee, Subcommittee Report "V," Subject: "Classification of Quick-Acting, Nonpersistent Agent, GB, as a Substitute Standard Type" (Secret), revised May 18, 1948, p. V5.

123 *Top secret assessment by Joint Chiefs (January 1949):* JCS, Joint Intelligence Group, Memorandum for the Joint Strategic Plans Group, Subject: "Estimate of Soviet Capabilities for Employing Biological and Chemical Weapons" (Top Secret), J.I.G. 297/3, January 27, 1949.

124 *President Roosevelt pledge of no first use:* U.S. Arms Control and Disarmament Agency, *Arms Control and Disarmament Agreements,* pp. 10–11.

124 *NSC meeting on February 1, 1950:* Fredericks, "The Evolution of Post–World War II United States Chemical Warfare Policy," p. 11–19.

124 *Memorandum by General Bradley:* Omar N. Bradley, Chairman, JCS, Memorandum for the Secretary of Defense, Subject: "Chemical Warfare Policy," January 18, 1950 [NSA].

124 *NSC-62:* "Chemical Warfare, NSC 62," in National Security Council, *Current Policies of the Government of the United States of America Relating to the National Security,* vol. 2: *Functional Policies,* Part XI: "Warfare, Chemical Warfare" (Top Secret) [NSA].

124 *Retention of Tabun as an emergency war reserve:* U.S. Army Chemical Corps, Office of the Chief, Chemical Corps Technical Committee, Army Chemical Center, Md., Subject: "Classification of GA as a Limited Standard Type" (Secret), September 30, 1949 [FOIA].

125 *General McAuliffe memorandum:* U.S. Army Chemical Corps, Office of the Chief (Major General A. C. McAuliffe), "Acceleration of G Agent Program" (Secret), April 14, 1950 [FOIA].

125 *McAuliffe speech to American Chemical Society:* "War of Nerves," *Time,* vol. 55, no. 18 (May 1, 1950), p. 44.

125 *May 1950 article in* Time *magazine:* Ibid.

125 *Outbreak of Korean War:* www.korean-war.com.

126 *Stevenson committee report:* "Letter of Transmittal," *Report of the Secretary of Defense's Ad Hoc Committee on Chemical, Biological, and Radiological Warfare,* June 30, 1950, pp. iii–iv [NSA].

127 *Joint Chiefs recommendation:* JCS, 1837/14, September 7, 1950 [NARA, RG 218], cited in Fredericks, "The Evolution of Post–World War II United States Chemical Warfare Policy," p. 11–25.

127 *R&D budget of Chemical Corps tripled in size:* Fredericks, "The Evolution of Post–World War II United States Chemical Warfare Policy," p. 11–26.

127 *Reactivation of Dugway Proving Ground:* Hersh, *Chemical and Biological Warfare,* pp. 137–143.

128 *Selection of DMHP process:* U.S. Army Chemical Corps, Research and Development Project Card, "GB Unit Plant Design" (Secret), December 1950 [FOIA].

128 *Secret allocation of $50 million:* Fredericks, "The Evolution of Post–World War II United States Chemical Warfare Policy," p. 11–26.

128 *Code name "Gibbett" for Sarin plant:* Tab C, Memorandum for the Secretary of Defense, Subject: "Chemical, Biological and Radiological Warfare," December 15, 1951 [NSA].

128 *Major Levy quote on technical challenges:* Major Stanley Levy, Chairman, Industrial Mobilization Review Committee, Memorandum to Commanding Officer, Chemical Corps Procurement Agency, Subject: "Proposed Contract for G Agent Manufacturing and Munitions Loading Plant" (Secret), CMLWD-PP, July 6, 1950 [FOIA].

128 *M34 Sarin cluster bomb:* U.S. Army Chemical Corps, Office of the Chief, Chemical Corps Technical Committee, Subject: "Military Characteristics for a Clusterable G-Series Bomb" (Secret), May 16, 1949 [FOIA].

129 *Selection of Site A and Site B:* Hylton, *The History of Chemical Warfare Plants and Facilities in the United States,* p. 2.

129 *National Intelligence Estimate:* DCI, National Intelligence Estimate no. 18, *The Probability of Soviet Employment of BW and CW in the Event of Attacks Upon the US* (Top Secret), December 15, 1950 [NARA].

130 *Construction of Site A:* "Muscle Shoals Makes G-Gas Basics," *Chemical and Engineering News,* vol. 32 (July 19, 1954).

130–32 *Technical problems at Site A:* Major Serge Tonetti, "Chemical Corps Phosphate Development Works," *Armed Forces Chemical Journal,* vol. 10, no. 5 (1956), pp. 32–33, 36; Hylton, *The History of Chemical Warfare Plants and Facilities in the United States,* pp. 4–58.

133 *Final two steps in Sarin production:* Stephen Black, Benoit Morel, and Peter Zapf, *Technical Aspects of Verification of the Chemical Weapons Convention,* Internal Technical Report, Program on International Peace and Security, Carnegie Mellon University, January 1991, pp. 39–41.

133 *Construction of Sarin plant at Site B:* "Vitro Builds Nerve Gas Plants," *Chemical and Engineering News,* vol. 32, no. 36 (September 5, 1954), pp. 3292–3493; "Army Permits Peak at Nerve Gas Facilities," *Chemical Engineering,* vol. 65, no. 19 (September 22, 1958), pp. 74–78; Hylton, *The History of Chemical Warfare Plants and Facilities in the United States,* pp. 59–75.

133 *Special materials used in Sarin plant:* Joint Chiefs of Staff, Memorandum for the Chairman, Munitions Board, Subject: "Priority for Chemical and Biological Warfare Facilities" (Top Secret), J.C.S. 1837/31, February 29, 1952 [NARA, RG 218].

133–34 *Delays in operating Site A:* JCS, Joint Strategic Plans Committee, Report to the Joint Chiefs of Staff on "Chemical (Toxic) and Biological Warfare Readiness" (Top Secret), J.S.P.C. 954/29, August 13, 1953 [NARA, RG 385].

134 *Delays in production of M34 bomb:* Memorandum from the Assistant Chief of Staff, G-4 Logistics, Department of the Army, to the Chief Chemical Officer, Subject: "Production of E101 Cluster, Bomb, GB-Filled" (Top Secret), G4/F3-392(SF), January 14, 1954 [FOIA].

134 *Purchase of dichlor from Shell Chemical Company:* Chemical Corps Advisory Council, "Ad Hoc Committee Meeting of Engineering & Production Committee, Muscle Shoals, Alabama, 5–6 April 1954" (Secret), AC-988, 1954, p. 7 [FOIA].

134 *APC process:* Black, Morel, and Zapf, *Technical Aspects of Verification of the Chemical Weapons Convention,* pp. 42–43.

134 *Ambros release, recruitment by Army Chemical Corps:* Hunt, *Secret Agenda,* p. 132.

134–35 *Deployment of U.S. chemical weapons on Okinawa:* Chief of Staff, U.S. Army, Memorandum for the Joint Chiefs of Staff, Subject: "Overseas Deployment of Toxic Chemical Agents," JCS 1837/46 (Top Secret), March 12, 1953 [NARA, RG 218].

135 *Request to use chemical weapons in Korean War:* McCarthy, *The Ultimate Folly,* p. 11.

135 *Chemical Corps public relations campaign:* J. H. Rothschild, "Germs and Gas: The Weapons Nobody Dares Talk About," *Harper's,* June 1959.

135 *Article on G agents by Cornelius Ryan:* Cornelius Ryan, "G-Gas: A New Weapon of Chilling Terror; We Have It—So Does Russia," *Collier's Magazine,* vol. 132, no. 13 (November 27, 1953), pp. 89–95.

136 *Transport of dichlor in railroad tank cars:* Author's interview with Sigmund R. Eckhaus.

136–37 *Ventilation system at Site B:* "Vitro Builds Nerve Gas Plants."

138 *Accidents in Sarin plant:* Hersh, *Chemical and Biological Warfare,* p. 107.

138–39 *Munitions filling line:* Author's interview with Sigmund R. Eckhaus.

139 *M5 detectors measured Sarin vapor in air:* "Vitro Builds Nerve Gas Plants."

139 *Pollution abatement system:* Author's interview with Sigmund R. Eckhaus.

139 *Cost overruns:* Hylton, *The History of Chemical Warfare Plants and Facilities in the United States,* pp. 95.

140 *Advisory Committee on New Agents:* U.S. Army Chemical Corps, Chemical and Radiological Laboratories, Army Chemical Center, Md., *Significant Accomplishments, Fiscal Year 1953* (Secret), CRLR 225, August 24, 1953, p. 5 [FOIA].

140 *Project Big Ben:* Author's interview with David M. Falk, Arlington, Va., April 21, 2004.

140–41 *Field trials of M34 cluster bomb:* U.S. Army Chemical Corps, "Minutes of the Meeting of the Chemical Corps Technical Committee" (Secret), Meeting No. 3, December 9, 1954, pp. 99–100, 108–109, 118–124 [NARA].

141 *Military antipathy to chemical weapons:* Gordon M. Burck, personal communication, June 27, 2004.

142 *JCS assessment of Soviet nerve-agent capability:* JCS, Joint Strategic Plans Com-

mittee, "Appendix to Enclosure J.S.P.C. 887/27: Chemical, Biological, and Radiological Warfare Annex" (Top Secret), May 1956 [NARA, RG 218].

CHAPTER EIGHT: CHEMICAL ARMS RACE

143 *Boris Libman biography:* Author's interview with Boris Libman.

144 *USSR Council of Ministers resolution (1952):* Derek Averre, "Chemical Weapons in Russia: After the CWC," *European Security,* vol. 8, no. 4 (Winter 1999), pp. 130–164.

144 *Soviet development of Sarin production process:* Author's interview with Boris Libman.

144 *Difficulty of obtaining special construction materials:* Boris Libman, "How Former Soviet Union Chemical Weapons Production Facilities Were Supplied with Chemical Equipment," *Communiqué No. 14* (prepared for the U.S. Army Chemical Corps), handwritten manuscript, undated.

145 *Sarin production plant in Stalingrad:* Sonia Ben Ouagrham, "Conversion of Russian Chemical Weapons Production Facilities: Conflicts with the CWC," *The Nonproliferation Review,* vol. 7, no. 2 (Summer 2000), p. 52.

145 *Recruitment of male engineers:* Boris Libman, "How Chemical Weapons Production Facilities Were Supplied with Raw Materials, Precursors and Hardware."

146 *Central Chemical Testing Site:* SIPRI, *The Rise of CB Weapons,* pp. 279–280, 285.

146 *Ghosh development of Amiton:* Ibid., pp. 74–75.

147 *Schrader's related patent:* Benjamin Garrett, "The CW Almanac," *ASA Newsletter,* no. 60, June 23, 1997, p. 15.

147 *Discovery of Tammelin esters:* Lars-Erik Tammelin, "Choline Esters: Substrates and Inhibitors of Cholinesterases," *Svensk Kemisk Tidskrift,* vol. 70, no. 4 (1958), pp. 157–181; Sten-Magnus Aquilonius, Torsten Fredriksson, and Anders Sundwall, "Studies on Phosphorylated Thiocholine and Choline Derivatives: I. General Toxicology and Pharmacology," *Toxicology and Applied Pharmacology,* vol. 6 (1964), pp. 269–279.

147 *Testing of nerve agents on volunteers at Porton:* United Kingdom, Ministry of Supply, Advisory Council on Scientific Research and Technical Development, Chemical Defense Advisory Board, "Annual Review of the Work of the Board for 1957" (Secret), A.C.14176 CDB.237, November 4, 1957 [NSA]; Rob Evans, "The Past Porton Down Can't Hide," *The Guardian,* May 6, 2004.

148–49 *Ronald Maddison:* David Williams and Tom Rawstorne, "British Nerve Gas Deaths 'Had Nazi Scientists,' " *Daily Mail* (London), October 25, 1999, p. 17; Sean Rayment, "Porton Nerve Gas Scientists Escape Criminal Charges," *Daily Telegraph* (London), June 8, 2003; BBC News, "Nerve Gas Death Was 'Unlawful,' " November 15, 2003.

148 *Maddison hoped to buy engagement ring:* Sally Pook, "Porton Down Unlawful Killing Verdict Opens Gates to Claims," *The Telegraph,* November 16, 2004.

149 *Twenty drops of 10 milligrams of Sarin:* Richard Smith, "Airman Used as Sarin Test Guinea Pig by MOD," *The Daily Mirror,* November 16, 2004.

149 *After twenty minutes, Maddison complained of feeling ill:* Rob Evans and Sandra Laville, "Porton Down Unlawfully Killed Airman in Sarin Tests," *The Guardian,* November 16, 2004.

149–50 *Alfred Thornhill story:* Antony Barnett, "Final Agony of RAF Volunteer Killed by Sarin—in Britain," *The Observer,* September 28, 2003.

151 *Memorandum by British Chiefs of Staff:* Chiefs of Staff, "Biological and Chemical Warfare Research and Development Policy: Annex (Draft)" [PRO, WO 188/705].

151 *British plan for war reserve of Sarin weapons:* U.K., Chiefs of Staff Committee, Chemical Warfare Sub-Committee, "First Requirements for Nerve Gas Weapons: Memorandum by Ministry of Supply" (Secret), CW(52) 13, October 23, 1952 [PRO, DEFE 41/157].

151 *Number of munitions needed to deliver one ton of Sarin:* Matthew Meselson, "The Myth of Chemical Superweapons," *Bulletin of the Atomic Scientists,* vol. 47, no. 3 (April 1991), p. 13.

151–52 *U.S. offer to sell Sarin stockpile to United Kingdom:* JCS, Joint Logistics Plans Committee, Directive, "Provision of Nerve Gas to the U.K." (Top Secret), J.L.P.C. 470/D, April 7, 1953 [NARA, RG 385]; United Kingdom, Ministry of Defence, "Text of U.S. Reply to British Aide-Memoire on the Production of Nerve Gas Weapons" (Top Secret), April 20, 1953 [NSA].

152 *Memorandum by Crawford:* K. N. Crawford, Chairman, Chemical Warfare Sub-Committee, Draft of letter to Lieutenant General Sir Nevill Brownjohn, Chief Staff Officer, Ministry of Defence (Top Secret), September 21, 1953 [PRO, WO 286/77].

153 *British decision to drop plan to buy U.S. weapons:* Royal Air Force, Vice Chief of the Air Staff, Memorandum to the Vice Chiefs of Staff, "Policy for Chemical Warfare" (Top Secret), D.D.Ops(B), April 23, 1953 [PRO, AIR 20/8734]; A. F. Boyd, Squadron Leader, Royal Air Force, Loose Minute, "Chemical Weapons," COS 1055/D.D.Ops.(B) (Top Secret), April 14, 1954 [PRO, AIR 20/8734].

153 *French chemical weapons program:* Author's interview with Olivier Lepick.

154 *Excessive toxicity of Amiton:* SIPRI, *The Rise of CB Weapons,* p. 75.

154 *Termed V agents for "venomous":* Harris and Paxman, *A Higher Form of Killing,* p. 184.

154 *Production halt at Nancekuke in 1955:* War Office, "CDE, Nancekuke (Process Research Division)" (Secret), DBCD/IL.1026/1687 A, undated [PRO, WO 32/21686]; Robert Mendick, "Nerve Gas Dump Cover-Up Exposed," *The Independent on Sunday,* February 6, 2000, p. 10.

154 *British decision to renounce active chemical stockpile:* Gradon Carter and Brian Balmer, "Chemical and Biological Warfare and Defence, 1945–90," in Bud and Gummett, *Cold War, Hot Science,* p. 298.

155 *NSC-5602/1:* "Basic National Security Policy," March 15, 1956, in Glennon, et al. *Foreign Relations of the United States, 1955–1957,* vol. XIX, *National Security Policy,* p. 246.

155 *Implementation of Stevenson committee report: Report of the Secretary of Defense's*

Ad Hoc Committee on Chemical, Biological, and Radiological Warfare, June 30, 1950, pp. 11–14; Office of the Secretary of Defense, Memorandum for the Secretary of Defense, Subject: "Chemical, Biological and Radiological Warfare" (Confidential), December 15, 1951.

155–56 *Miller report:* Hersh, *Chemical and Biological Warfare,* p. 23.

156 *Department of Defense directive (October 6, 1956):* Reuben B. Robertson, Jr., Deputy Secretary of Defense, Department of Defense Directive, Subject: "Chemical (Toxic) and Biological Warfare Readiness" (Top Secret), TS-3145.1, October 6, 1956 [NSA].

156 Field Manual on Law of Land Warfare: SIPRI, *CB Weapons Today,* pp. 195–196.

156 *Shutdown of Sarin production lines:* U.S. Army Chemical Corps Historical Office, Army Chemical Center, Maryland, *Summary of Major Events and Problems, U.S. Army Chemical Corps, Fiscal Year 1958* (Secret), March 1959, pp. 153, 155 [NSA].

156–57 *Environmental pollution at Rocky Mountain Arsenal:* Peter Pringle, "Lest We Forget the Days of Gas and Poison," *Independent,* April 13, 1989, p. 12; Penelope Purdy, "What Is Really Buried at the Arsenal? Sarin Saga Will Haunt Metro Denver," *Denver Post,* January 14, 2001, p. I-1.

157 *Underground injection of toxic wastes at Rocky Mountain Arsenal:* Jonathan Eberhart, "Geology: To Pump or Un-Pump," *Science News,* vol. 93, May 4, 1968, pp. 434–435; McCarthy, *The Ultimate Folly,* pp. 99–101.

CHAPTER NINE: AGENT VENOMOUS

158 *Synthesis of V-agents, selection of VX:* U.S. Army Chemical Corps Historical Office, Army Chemical Center, Md., *Summary of Major Events and Problems, United States Army Chemical Corps, Fiscal Year 1957* (Secret), October 1957, p. 94 [FOIA].

158–59 *Characteristics of VX:* U.S. Army Chemical Corps Historical Office, Army Chemical Center, Maryland, *Summary of Major Events and Problems, United States Army Chemical Corps, Fiscal Year 1958* (Secret), March 1959, p. 98 [FOIA].

159 *Development of transester process:* Author's interview with Sigmund R. Eckhaus.

159 *VX pilot plants at Edgewood:* U.S. Army Chemical Corps Historical Office, Army Chemical Center, Maryland, *Summary of Major Events and Problems, U.S. Army Chemical Corps, Fiscal Year 1958* (Secret), March 1959, p. 99 [NSA].

160 *V-Agent Team at Dugway Proving Ground:* U.S. Army Chemical Corps Historical Office, Army Chemical Center, Maryland, *Summary of Major Events and Problems, United States Army Chemical Corps, Fiscal Year 1957* (Secret), October 1957, p. 95 [FOIA].

160 *Concept of binary weapons:* SIPRI, *CB Weapons Today,* pp. 306–308.

160 *Attempt to produce binary explosive shell:* Brauch, *Der chemische Alptraum,* p. 121.

160 *Development of binary VX bomb:* Author's interview with William C. Dee.

161–62 *Nachmansohn research on cholinesterase from electric eels:* John Kobler, "The

Terrible Threat of Nerve Gas," *Saturday Evening Post,* vol. 230, no. 4 (July 27, 1957), pp. 28–29, 75–77.

162 *Synergistic effects of PAM and atropine:* U.K., Ministry of Supply, Chemical Defence Advisory Board, Advisory Council on Scientific Research and Technical Development, "Annual Review of the Work of the Board for 1957" (Secret), A.C. 14176, CDB.237, November 4, 1957 [NSA].

163 *Development of British "Autoject":* U.K. Ministry of Defence, *A Brief History of the Chemical Defence Establishment Porton,* March 1961 (Restricted), p. 36 [SPRU].

163 *Development of "Ace autoinjector":* Cornelius Ryan, "G-Gas: A New Weapon of Chilling Terror: We Have It—So Does Russia," *Collier's Magazine,* vol. 132, no. 13 (November 27, 1953), p. 92.

164 *Automatic field detectors for nerve agents:* U.K. Ministry of Defence, *A Brief History of the Chemical Defence Establishment Porton,* March 1961 (Restricted), p. 34; John Kobler, "The Terrible Threat of Nerve Gas," *Saturday Evening Post,* vol. 230, no. 4 (July 27, 1957), p. 76; U.S. Army Soldier and Biological Chemical Command, *History of Chemical and Biological Detectors, Alarms, and Warning Systems* (Aberdeen, Md.: SBCCOM, undated), p. 20.

164 *Tripartite research on V agents:* U.K., Ministry of Defence, CDEE Porton Down, Offensive Evaluation Committee, "11th Tripartite Conference on Toxicological Warfare, Conclusions and Recommendations" (Secret), Ptn./IT.4222/2099/57, May 15, 1957; U.K., Ministry of Defence, CDEE Porton Down, Offensive Evaluation Committee, "Conclusion and Recommendations of the 12th Tripartite Conference—BW/CW Applications" (Secret), Ptn./IT.4222/4141/57, October 1, 1957.

165 *British VX pilot plant at Nancekuke:* Author's interview with Ron Manley.

165 *Selection of Newport site for V-agent production:* Harris and Paxman, *A Higher Form of Killing,* p. 185.

165 *Solicitation of proposals from industry:* U.S. Army Chemical Corps Historical Office, Army Chemical Center, Maryland, *Summary of Major Events and Problems, United States Army Chemical Corps, Fiscal Year 1958* (Secret), March 1959, p. 158.

165–66 *Selection of FMC Corporation:* Hersh, *Chemical and Biological Warfare,* p. 102.

166 *Newport Army Chemical Plant:* Hylton, *The History of Chemical Warfare Plants and Facilities in the United States,* pp. 78–93.

166 *Development of M55 rocket:* Clarke, *The Silent Weapons,* p. 48; U.S. Army Chemical Corps Historical Office, Army Chemical Center, Maryland, *Summary of Major Events and Problems, United States Army Chemical Corps, Fiscal Year 1960* (Secret), April 1961, pp. 108, 154.

167 *Development of Sarin rockets:* James Baar, "Army Seeks Poison Gas Missiles," *Missiles and Rockets,* May 16, 1960, pp. 10–11.

167 *Honest John missile warhead:* U.S. Army Chemical Corps Historical Office, Army Chemical Center, Maryland, *Summary of Major Events and Problems,*

United States Army Chemical Corps, Fiscal Year 1960 (Secret), April 1961, p. 110 [NSA].

169 *U.S. release of VX information to France:* U.K., Minister of War, "Release by the United States of Information on Toxic Munitions to France" (Secret), R.H.S./428/62, July 5, 1962 [PRO, DEFE 13/440].

169 *French production and testing of Sarin and VX:* Lepick, *Les armes chimiques,* p. 98; author's interview with Olivier Lepick.

170 *Project 112 Task Group:* U.S. Department of Defense, "DOD Releases Project SHAD Fact Sheets," News Release 264-02, May 23, 2003.

170 *Tripling of Chemical Corps budget:* Harris and Paxman, *A Higher Form of Killing,* p. 186.

170–71 *Six recommendations in Project 112 report:* General L. L. Lemnitzer, Chairman, JCS, Memorandum for the Secretary of Defense, Subject: "Biological and Chemical Weapons and Defense Programs" (Top Secret), CM-551-62, February 14, 1962 [NARA, RG 218].

171 *General Taylor memorandum to Secretary McNamara:* General Maxwell D. Taylor, Chairman, JCS, Memorandum for the Secretary of Defense, Subject: "Negotiations for Storage Rights in USEUCOM" (Secret), December 17, 1962 [NARA, RG 218].

171 *Founding of Deseret Test Center:* Jeffrey K. Smart, "History of Chemical and Biological Warfare: An American Perspective," in Sidell, Takafuji, and Franz, eds., *Medical Aspects of Chemical and Biological Warfare,* p. 55.

171–72 *Other U.S. chemical weapons test sites:* McCarthy, *The Ultimate Folly,* pp. 36–38.

172 *First public mention of VX production:* "Lethal Nerve Gas Is Now Being Made at Newport Ordnance Plant," *Terre Haute Star,* May 16, 1952, p. 1.

172–77 *Description of VX production at Newport:* James R. Polk, "U.S. Assembly Line Turns Out Deadly Nerve Gas for Military," *Washington Post,* April 22, 1964, p. 1; Hersh, *Chemical and Biological Warfare,* pp. 102–104; author's interview with Sigmund R. Eckhaus.

177 *VX stored in tanks and one-ton containers:* Author's interview with Sigmund R. Eckhaus.

177–80 *Dee development of binary VX bomb:* Author's interview with William C. Dee.

180 *Marshal Zhukov warning:* Harris and Paxman, *A Higher Form of Killing,* p. 145.

181 *Soviet synthesis of novel agents:* Author's interview with Boris Libman; U.K., MI-3, "Soviet Tactical Chemical Warfare" (Secret), April 26, 1963 [PRO, DEFE 24/31]; U.S. Department of the Army, Office of the Assistant Chief of Staff for Intelligence, Washington, D.C., Intelligence Staff Study, "Soviet Research and Development Capabilities for New Toxic Agents" (Secret), Project No. A-1735, July 28, 1958, pp. 4–8 [MHI].

181 *Soviets obtained the secret formula for VX:* Author's interview with Vil S. Mirzayanov.

181 *Soviet synthesis of R-33:* Wise, *Cassidy's Run,* p. 193.

181–82 *Structural differences between VX and R-33:* Author's interview with Ron Manley.

182   *Soviet decision to mass-produce Soman:* Wise, *Cassidy's Run,* p. 66.

182   *Production method developed by Tomilov:* Author's interview with Vil S. Mirza-yanov.

182–83   *Problems with the production of pinacolyl alcohol:* Boris Libman, "How Former Soviet Union Chemical Weapons Production Facilities Were Supplied with Chemical Equipment," Communiqué no. 14 (prepared for U.S. Army Chemical Corps), handwritten manuscript, undated.

183   *Soviet manufacture of silver-lined reactors:* Ibid.

183   *Guskov sent to Volgograd:* Author's interview with Vil S. Mirzayanov.

183   *Soviet delivery systems for nerve agents:* Director of Central Intelligence, National Intelligence Estimate no. 11-10-63, *Soviet Capabilities and Intentions with Respect to Chemical Warfare* (Secret), December 27, 1963, pp. 8–10 [NARA].

183   *Soviet chemical weapons deployed in Warsaw Pact states:* Institute for Defense and Disarmament Studies, "Chemical Weapons, Chronology 1989: 18 October," *Arms Control Reporter,* December 1989, p. 704.B.408.

183   *Soviet chemical warfare doctrine:* U.K., Ministry of Defence, "Soviet Tactical Chemical Warfare" (Secret), MI-3 (d)/13/9, April 26, 1963 [PRO, DEFE 24/31].

184   *CBIC assessment process, Kerlin report:* Author's interview with Garrett Cochran; Director of Central Intelligence (DCI), National Intelligence Estimate No. 11-10-63, *Soviet Capabilities and Intentions with Respect to Chemical Warfare* (Secret), December 27, 1963 [NARA]; DCI, National Intelligence Estimate No. 11-11-69, *Soviet Chemical and Biological Warfare Capabilities* (Secret), February 13, 1969 [NARA].

185   *NORAD sabotage scenario:* Greeman and Roberts, *Der kälteste Krieg.*

185   *NORAD requirement for chemical detector:* Major General W. H. Hennig, Chief of Staff, Headquarters North American Air Defense Command, Memorandum to Chief of Staff, Department of the Army, Attn: Chief Chemical Officer, Subject: "Biological and Chemical Warfare" (Confidential), JCS 1837/118, August 22, 1960 [NARA, RG 218].

185   *Soviet covert chemical attack in Rothschild book:* Rothschild, *Tomorrow's Weapons,* p. 26.

185–86   *British chemical rearmament R&D program in 1963:* Gradon Carter and Brian Balmer, "Chemical and Biological Warfare and Defence, 1945–90," in Bud and Gummet, *Cold War, Hot Science,* p. 299.

186   *British chemical rearmament plans:* United Kingdom, Ministry of Defence, "CW—Retaliatory Capability" (Top Secret), 86/CHEMICAL/960, August 25, 1965 [PRO, DEFE 24/31]; Memorandum from the Chief of the Defence Staff to the Secretary of State, "Chemical and Biological Warfare" (Top Secret/UK Eyes Only), October 28, 1965 [PRO, DEFE 24/31]; Memorandum from DAEP to DCGS, Loose Minute, "Chemical Warfare" (Top Secret), November 29, 1965 [PRO, DEFE 24/31].

186   *Australia joined Tripartite Program:* G. B. Carter and Graham S. Pearson, "Past British Chemical Warfare Capabilities," *RUSI Journal,* February 1996, p. 63; Canadian Embassy, Washington, D.C., "Some Historical Comments and Back-

ground on TTCP," prepared for the 25th Anniversary Meeting of the NAMRAD Principals, Washington, D.C., October 12–13, 1983.

187 *Fish kill on Volga River, punishment of Libman:* Wise, *Cassidy's Run,* p. 67; author's interview with Vil S. Mirzayanov.

187–88 *French-Algerian agreement on B2-Namous:* Mayer, *L'arme chimique,* pp. 157–158.

188 *French testing with "maquettes":* Ibid., pp. 159, 408.

188–89 *Operation Shocker:* Wise, *Cassidy's Run.*

CHAPTER TEN: YEMEN AND AFTER

190 *Egyptian "Izlis" program:* Dany Shoham, "Chemical and Biological Weapons in Egypt," *The Nonproliferation Review,* vol. 5, no. 3 (Spring–Summer 1998), p. 48.

190 *Egyptian military officers trained in Moscow:* Ibid., p. 49.

190 *Recruitment of West German scientists:* Ibid.

190 *Outbreak of civil war in Yemen:* W. Andrew Terrill, "The Chemical Warfare Legacy of the Yemen War," *Comparative Strategy,* vol. 10 (1991), pp. 109–119.

191 *Egyptian attack at Al-Kawma:* Hersh, *Chemical and Biological Warfare,* p. 283.

191 *Saudia Arabia filed a complaint with the U.N. Secretary-General:* Terrill, "The Chemical Warfare Legacy of the Yemen War," p. 111.

191 *U.N. observer group asked to investigate:* "U.N. Will Weigh Gas-Bomb Charge: Yemen Team Told to Check Reports About U.A.R.," *New York Times,* July 10, 1963, p. 3.

191 *Article by journalist Marquis Childs:* Marquis W. Childs, "Egypt Stored Nerve Gas Before War," *St. Louis Post-Dispatch,* June 18, 1967, p. 1.

191 *Prime Minister Wilson statement:* Harris and Paxman, *A Higher Form of Killing,* p. 234.

191–92 *Classified cable from U.S. Embassy in Beirut:* Hersh, *Chemical and Biological Warfare,* p. 285.

192 *Saudi Arabia submitted medical reports to United Nations:* Hersh, *Chemical and Biological Warfare,* pp. 283–284.

192 *Analysis showed V agent in contaminated sand:* Childs, "Egypt Stored Nerve Gas Before War."

192 *International Committee of the Red Cross (ICRC) investigation:* Terrill, "The Chemical Warfare Legacy of the Yemen War," p. 114.

193 *Evidence Egyptian munitions furnished by Soviets:* "Yemen—A Testing Ground for Soviet Poison Gases," *Elseviers Weekblad* (Amsterdam), November 25, 1967; "How Nasser Used Poison Gas," *U.S. News & World Report,* July 3, 1967, p. 60.

193 *Military Plant No. 801 at Abu Za'abal:* Shoham, "Chemical and Biological Weapons in Egypt," p. 48.

193 *Meselson testimony:* Testimony of Matthew S. Meselson, U.S. Senate, Committee on Foreign Relations, Hearing, *Chemical and Biological Warfare,* 91st Congress, 1st session, April 30, 1969 (Washington, D.C.: U.S. Government Printing Office, 1969), p. 47.

193 *U.S. diplomatic protest:* Terrill, "The Chemical Warfare Legacy of the Yemen War," p. 114.

194 *Talking points for Johnson-Kosygin meeting:* Arms Control and Disarmament Agency (drafter: Herbert Scoville), "President's Meeting with Chairman Kosygin, June 1967, Background Paper and Contingency Talking Points: Use of Chemical Warfare Agents in Middle East" (Secret), AKV/B-11, June 16, 1967 [NARA].

194 *Statement by State Department spokesman:* Hersh, *Chemical and Biological Warfare,* p. 284.

194 *Unnamed U.S. official quoted by Hersh:* Ibid., p. 285.

194 *Johnson administration muted its criticism:* Terrill, "The Chemical Warfare Legacy of the Yemen War," p. 114.

195 *Egyptian troop buildup in Sinai Peninsula:* Israel Defense Forces, "The Six Day War," available online at www.idf.il/english/history/sixday.stm.

195 *Israeli reconnaissance mission found chemical shells:* Jack Anderson, "The Growing Chemical Club," *Washington Post,* August 26, 1984, p. C7.

195 *Frantic efforts by Israeli government to procure gas masks:* "Bonn to Sell Israel 20,000 Gas Masks," *New York Times,* June 2, 1967, p. 14. (*Note:* Gerhard Schröder was a Christian Democratic politician during the 1960s, not to be confused with the Social Democratic chancellor of the same name.)

196 *Israel returned the gas masks to West Germany unused:* "West Germans to Take Back Gas Masks Unused by Israel," *New York Times,* July 1, 1967, p. 5.

196 *Egypt's last use of chemical weapons in Yemen:* Terrill, "The Chemical Warfare Legacy of the Yemen War," p. 114.

196 *U.S. shortage of conventional munitions for Vietnam:* Gordon M. Burck, personal communication, June 27, 2004.

197 *Leakage of M55 rockets:* Army Materiel Command, Deputy Chief of Staff for Nuclear & Chemical Matters, M55 Functional Task Group, "M55 GB/VX Rocket: Stockpile Assessment Plan," March 18, 1985; U.S. Army Material Systems Analysis Activity, Aberdeen Proving Ground, Md., "Independent Evaluation/Assessment of Rocket, 115mm: Chemical Agent (GB or VX) M55," October 1985, p. 40.

197 *Lots of Sarin used in M55 rockets:* Author's interview with Sigmund R. Eckhaus.

197 *Recycling of Sarin from M34 bomblets:* Ibid.

198 *Open-pit burning of M55 rockets at Dugway:* Ibid.

199 *Operation CHASE 5:* E. J. Hogendoorn, "A Chemical Weapons Atlas," *Bulletin of the Atomic Scientists,* vol. 53, no. 5 (September–October 1997); Jeffrey K. Smart, "History of Chemical and Biological Warfare: An American Perspective," in Sidell, Takafuji, and Franz, eds., *Medical Aspects of Chemical and Biological Warfare,* pp. 62–63.

199 *Project 112 field trials:* U.S. General Accounting Office, "Chemical and Biological Defense: DOD Needs to Continue to Collect and Provide Information on Tests and Potentially Exposed Personnel," Report to the Senate and House Committees on Armed Services, GAO-04-410, May 2004.

199–200 *Project SHAD trials:* Thom Shanker and William J. Broad, "Sailors Sprayed with Nerve Gas in Cold War Test, Pentagon Says," *New York Times,* May 24, 2001, p. A1; U.S. Department of Defense, "DOD Expands SHAD Investigation," News Release No. 355-02, July 9, 2002; U.S. Department of Defense, "Briefing on Cold War–Era Chemical and Biological Warfare Tests, Presenter: Dr. William Winkenwerder, Jr.," News Transcript, October 9, 2002; Thom Shanker, "U.S. Troops Were Subjected to Wider Toxic Testing," *New York Times,* October 9, 2002, p. A18; Thom Shanker, "Defense Dept. Offers Details of Toxic Tests Done in Secret," *New York Times,* October 10, 2002, p. A32.

200 *Trials on Hawaii:* Thom Shanker, "U.S. Tested a Nerve Gas in Hawaii," *New York Times,* November 1, 2002, p. A18.

200 *Trials at Army Tropic Test Center in Panama:* Fellowship of Reconciliation, *Test Tube Republic: Chemical Weapons Tests in Panama and U.S. Responsibility* (San Francisco, July 31, 1998), pp. 5, 6, 13.

200 *Fort Greely Military Reservation, Alaska:* Korey Capozza, "Northern Exposure," *The Nation,* August 18–25, 2003, p. 32.

200 *Trials at Gerstle River Test Site:* Deborah Funk, "Were They Exposed to Dangerous Toxins?," *Army Times,* July 8, 2002.

201 *Blueberry Lake incident:* Michael Getler, "Nerve Gas Lost 3 Years," *Washington Post,* January 6, 1971, pp. A1, A8; Reuters, "Deadly Gas Lay Unnoticed for 2 Years in Lake," *Baltimore Sun,* January 7, 1971, p. A6; Alaska Community Action on Toxics, "Critical Cleanup Target: Gerstle River Test Site," online at www.akaction.net/pages/critical/gerstle.html.

202 *Mysterious death of fifty-three caribou:* Brauch, *Der chemische Alptraum,* p. 146.

CHAPTER ELEVEN: INCIDENT AT SKULL VALLEY

203 *Description of Dugway Proving Ground:* Jerry Steelman, "Dugway Proving Ground—A Unique Test Facility," *CML* (Army Chemical Review), July 1990, pp. 35–38.

203 *Warning sign at entrance:* Hersh, *Chemical and Biological Warfare,* p. 137.

203–04 *Description of life at Dugway:* Author's interviews with William C. Dee and Garrett Cochran.

204 *Description of test grids:* Steelman, "Dugway Proving Ground," pp. 36–38.

204 *Principle underlying safety of testing:* Hersh, *Chemical and Biological Warfare,* p. 139.

204 *Calls from Dugway officials to sheriff of Tooele County:* Philip M. Boffey, "Nerve Gas: Dugway Accident Linked to Utah Sheep Kill," *Science,* vol. 162, no. 3861 (December 27, 1968), p. 1463.

204 *A million pounds of nerve agent had been released:* Scott Simon, "Evidence of Radiation and Chemical Testing Disclosed," National Public Radio, Weekend Edition, May 27, 1995, transcript no. 1125-12.

204–06 *VX trial on March 13, 1968:* Virginia Brodine, Peter P. Gaspar, and Albert J. Pallmann, "The Wind from Dugway," *Environment Magazine,* vol. 2

(January–February 1969), reprinted as Appendix 2 in U.S. House, Committee on Government Operations, Subcommittee on Conservation and Natural Resources, Hearing, *Environmental Dangers of Open-Air Testing of Lethal Chemicals,* 91st Congress, 1st session, May 20–21, 1969 (Washington, D.C.: U.S. Government Printing Office, 1969), pp. 209–219; Boffey, "Nerve Gas: Dugway Accident Linked to Utah Sheep Kill," p. 1461.

206 *Sheep acted "crazy in the head":* Boffey, "Nerve Gas: Dugway Accident Linked to Utah Sheep Kill," p. 1461.

207 *Sheep deaths at White Rock in Skull Valley:* Department of Defense, Office of Public Affairs, "Status Report on Investigation of Sheep Deaths in Utah," April 18, 1968.

207 *Visit to Skull Valley by USDA experts:* Kent R. Van Kampen, et al., "Organic Phosphate Poisoning of Sheep in Skull Valley, Utah," *Journal of the American Veterinary Medical Association,* vol. 154, no. 6, March 15, 1969, pp. 623–630.

207 *Appointment of Dr. Osguthorpe to investigate outbreak:* Statement of D. A. Osguthorpe (D.V.M.), Veterinarian, in U.S. House, Committee on Government Operations, Subcommittee on Conservation and Natural Resources, *Environmental Dangers of Open-Air Testing of Lethal Chemicals,* p. 63.

208 *Affected flocks fell within a triangular area:* David J. Sencer, M.D., Director, CDC, "Investigation of Sheep Deaths—Skull Valley, Utah," Communicable Disease Center, Atlanta, Ga., April 29, 1968, Appendix 7, in U.S. House, Committee on Government Operations, Subcommittee on Conservation and Natural Resources, *Environmental Dangers of Open-Air Testing of Lethal Chemicals,* p. 247.

208 *Visits by Osguthorpe to Dugway and affected flocks:* Statement of D. A. Osguthorpe (D.V.M.), in U.S. House, Committee on Government Operations, Subcommittee on Conservation and Natural Resources, *Environmental Dangers of Open-Air Testing of Lethal Chemicals,* pp. 63–64.

208 *Army technical report provided to Senator Moss:* Responses to questions by Dr. Osguthorpe, in U.S. House, Committee on Government Operations, Subcommittee on Conservation and Natural Resources, *Environmental Dangers of Open-Air Testing of Lethal Chemicals,* p. 94.

209 *Governor Rampton called meeting at the Utah State Capitol:* Ibid., p. 93.

210 *CDC epidemiological study:* Boffey, "Nerve Gas: Dugway Incident Linked to Utah Sheep Kill," p. 1462.

210 *Study by USDA Poisonous Plant Research Laboratory:* Ibid.

211 *CDC comparison analysis of sample of VX:* Ibid., p. 1463.

211 *Sheep turned out to be unusually susceptible:* Ibid., p. 1462.

211–12 *Advisory Committee report:* Department of the Army, *Report of the Interagency Ad Hoc Advisory Committee for Review of Testing Safety at Dugway Proving Ground,* Washington, D.C., November 1968, Appendix 1 in U.S. House, Committee on Government Operations, Subcommittee on Conservation and Natural Resources, *Environmental Dangers of Open-Air Testing of Lethal Chemicals,* pp. 187–208.

212 *Financial compensation of ranchers:* Boffey, "Nerve Gas: Dugway Incident Linked to Utah Sheep Kill," p. 1464.

212 *Enduring legacy of distrust:* Benjamin C. Garrett, "Leading a Horse to Water: Investigation of Alleged CW Use, Dugway Proving Ground (USA), July 1976," Presentation at Pugwash Meeting no. 254, 13th Workshop of the Pugwash Study Group on the Implementation of the Chemical and Biological Weapons Conventions, Oegstgeest, The Netherlands, April 8–9, 2000, p. 3.

212 First Tuesday *broadcast and McCarthy quote:* McCarthy, *The Ultimate Folly,* p. 126.

213 *Army briefing for House members:* Ibid., pp. 127–128.

213 *Hearings by House subcommittee:* U.S. House, Committee on Government Operations, Subcommittee on Conservation and Natural Resources, Hearing, *Environmental Dangers of Open-Air Testing of Lethal Chemicals.*

214 *Disclosure of open-air tests at Fort McClellan and Edgewood:* Marjorie Hunter, "Nerve Gas Tested at Open-Air Sites," *New York Times,* July 12, 1969, pp. 1, 8; Steve Vogel, "Military Reveals Testing of Nerve Agents in Md.," *Washington Post,* October 10, 2002, p. A2; Frank D. Roylance, Ariel Sabar, and Tom Bowmann, "Nerve Agents Released in Md. During Open-Air Tests in '60s," *Baltimore Sun,* October 10, 2002.

214–15 *Accident involving Sarin bombs on Okinawa:* Robert Kentley, "Nerve Gas Accident: Okinawa Mishap Bares Overseas Deployment of Chemical Weapons," *Wall Street Journal,* July 18, 1969, p. 1; Neil Sheehan, "U.S. Said to Keep Nerve Gas Abroad at Major Bases," *New York Times,* July 19, 1969, pp. 1, 2; Kensei Yoshida, "Some Okinawans Knew of U.S. Chemical Activities," *New York Times,* July 19, 1969, p. 2.

215 *Response to* Wall Street Journal *story:* Takashi Oka, "Protest Is Strong in Japan, Okinawa," *New York Times,* July 19, 1969, p. 2; Takashi Oka, "Okinawa Report on Gas Provides Windfall for Opposition in Japan," *New York Times,* July 20, 1969, p. 4; Neil Sheehan, "U.S. Will Remove Nerve-Gas Arms at Okinawa Base," *New York Times,* July 23, 1969, pp. 1, 4.

215 *James Reston quote:* McCarthy, *The Ultimate Folly,* p. 141.

215–16 *Quote by Japanese official:* Ibid., p. 130.

216 *Congress passed Public Law 91-121:* Jeffrey K. Smart, "History of Chemical and Biological Warfare: An American Perspective," in Sidell, Takafuji, and Franz, eds., *Medical Aspects of Chemical and Biological Warfare,* p. 63.

216–17 *Nixon decision of November 25, 1969:* James M. Naughton, "Nixon Renounces Germ Weapons, Orders Destruction of Stocks; Restricts Use of Chemical Arms," *New York Times,* November 26, 1969, p. 1.

217 *Controversy over proposed CHASE mission:* "Statement of Congressman Richard D. McCarthy on Chemical and Biological Warfare Policies and Practices" (press release), April 21, 1969; McCarthy, *The Ultimate Folly,* pp. 103–108.

218 *Testimony by Eckhaus and Bass:* Author's interview with Sigmund R. Eckhaus.

218 *National Academy of Sciences recommendation:* Joel Primack and Frank von Hip-

pel, "Matthew Meselson and Federal Policy on Chemical and Biological Warfare," in Primack, *Advice and Dissent,* pp. 143–164.

218–20 *Operation CHASE 10:* Richard D. Lyons, "Army Will Transport Nerve Gas Across South for Disposal in Sea," *New York Times,* July 30, 1970, pp. 1, 11; Richard D. Lyons, "Nerve Gas Trains Will Cross 7 States," *New York Times,* July 31, 1970, p. 41; Associated Press, "Army to Stockpile Nerve Gas Antidote," *New York Times,* August 1, 1970, p. 26; Associated Press, "Nerve Gas Is Sunk off Florida Coast," *New York Times,* August 19, 1970, p. 1, 6.

220–21 *Removal of U.S. chemical weapons from Okinawa (Operation Red Hat):* Anonymous, "U.S. to Remove Gas from Okinawa Soon," *New York Times,* December 5, 1970, p. 9; Reuters, "Transfer of War Gas Completed," *New York Times,* September 22, 1971, p. 12.

222 *Conditions on Johnston Island:* Owen Wilkes, "Chemical Weapon Burnoff in Central Pacific," *Peacelink,* vol. 83 (July 1990), pp. 5–10.

CHAPTER TWELVE: NEW FEARS

224 *Army plans to eliminate Chemical Corps:* "Army Chemical Corps to Be Phased Out," *Chemical & Engineering News,* January 22, 1973, p. 3.

224 *Chemical Corps motives to acquire binary weapons:* Gordon M. Burck, "Decision-Making in the United States Nerve Gas Weapon Program," unpublished paper, 1985.

224 *Increasing share of Chemical Corps R&D budget devoted to binaries:* Julian Perry Robinson, *The United States Binary Nerve-Gas Programme: National and International Implications* (Falmer, U.K.: University of Sussex ISIO Monographs, First Series, no. 10, 1975), pp. 4, 10.

224–25 *Use of DF in binary Sarin artillery shell:* Lois Ember, "Binary Chemical Arms Study Fires House Debate," *Chemical and Engineering News,* May 7, 1984, p. 24.

225–26 *Correction of instability problem with binary artillery shell:* Author's interview with William C. Dee.

226 *Binary artillery shell entered engineering development:* SIPRI, *CB Weapons Today,* p. 308.

226 *One test involved live Sarin:* U.S. General Accounting Office, Report to the Congress by the Comptroller General of the United States, *U.S. Lethal Chemical Munitions Policy: Issues Facing the Congress* (Secret/NOFORN), 1990.

226 *Callaway announcement of binary program:* John Finney, "Army Will Spend $200 Million for Safer Nerve Gas," *New York Times,* December 10, 1973.

226–27 *Egyptian chemical warfare preparations:* William Beecher, "Egypt Deploying Nerve Gas Weapons," *Boston Globe,* June 6, 1976, p. 1.

227 *Egypt transferred chemical arsenal to Syria for $6 million:* Dany Shoham, "Chemical and Biological Weapons in Egypt," *Nonproliferation Review,* vol. 5, no. 3 (Spring–Summer 1998), p. 49.

227 *Israeli military intelligence lulled into complacency:* Malcolm Gladwell, "Connecting the Dots," *New Yorker,* March 10, 2003, p. 83.

227 *Description of Yom Kippur War:* Jewish Virtual Library, "The 1973 War," www.us-israel.org/jsource/History/1973toc.html.

227 *Israeli discovery of Arab armored vehicles with chemical protection:* Jeffrey K. Smart, "History of Chemical and Biological Warfare: An American Perspective," in Sidell, Takafuji, and Franz, eds., *Medical Aspects of Chemical and Biological Warfare,* pp. 64–65.

228 *Egyptian prisoners of war admitted under interrogation:* Beecher, "Egypt Deploying Nerve Gas Weapons."

228 *Israel ordered manufacture of 3 million gas masks:* Ibid.

228–29 *CIA estimates of Soviet chemical troops and vehicles:* David Binder, "U.S. Re-Emphasizing Chemical Warfare," *New York Times,* June 5, 1978, pp. A1, A6.

229 *Testimony of General Creighton W. Abrams:* General Creighton W. Abrams, Army Chief of Staff, Testimony before the House Defense Appropriations Subcommittee, March 5, 1974.

229 *Comment by Representative Robert Sikes:* Ibid.

229–30 *Egyptian TAB nerve-agent antidote kit, CIA Project Grand Plot:* Jack Anderson, "Flawed Antidote for Nerve Gas Pushed by CIA," *Washington Post,* October 4, 1984, p. 11.

230 *V-agent production at Novocheboksarsk:* Derek Averre, "Chemical Weapons in Russia: After the CWC," *European Security,* vol. 8, no. 4 (Winter 1999), pp. 130–164.

231 *Fire at Novocheboksarsk on April 28, 1974:* Ibid.

231 *Soviet "Foliant" program:* Bill Gertz, "Russia Dodges Chemical Arms Ban," *Washington Times,* February 4, 1997.

231 *"Series F" code name:* Author's interview with Vil S. Mirzayanov. According to Mirzayanov, the USSR Council of Ministers approved several top secret research programs on various aspects of chemical and biological warfare, all having code names beginning with the letter F. In addition to Foliant, these programs reportedly included Flora (military herbicides), Flute (toxin weapons for assassination), Fouette, Fagot, Flask, Ferment (genetically engineered pathogens), and Factor.

231–32 *Description of GosNIIOKhT:* Wise, *Cassidy's Run,* p. 63.

232 *Staff and branches of GosNIIOKhT:* Vil S. Mirzayanov, "Dismantling the Soviet/Russian Chemical Weapons Complex: An Insider's View," in Amy E. Smithson, ed., *Chemical Weapons Disarmament in Russia: Problems and Prospects* (Washington, D.C.: Henry L. Stimson Center, Report no. 17, October 1995), p. 22, footnote 4. See also Amy E. Smithson, *Toxic Archipelago: Preventing Proliferation from the Former Soviet Chemical and Biological Weapons Complexes* (Washington, D.C.: Henry L. Stimson Center, Report No. 32, December 1999), p. 11.

232 *Development of A-230, A-232:* Author's interview with Vil S. Mirzayanov.

233 *A-232 was a phosphate:* Mirzayanov, "Dismantling the Soviet/Russian Chemical Weapons Complex: An Insider's View," pp. 24–25.

233 *Martinov visit to GITOS:* Author's interview with Vil S. Mirzayanov.

233–34 *Testing of Foliant agents at Shikhany:* Author's interview with Ron Manley.

234 *Trials of nerve agents at Camp de Mourmelon:* Author's interview with Olivier Lepick.

234 *Extension of French use of B2-Namous:* Mayer, *L'arme chimique,* pp. 159–160.

235 *Nixon summit meeting with Brezhnev: Washington Post,* Superpower Summits, 1959–1995, "Joint Communiqué, Moscow, July 3, 1974," available online at www.washingtonpost.com/wp-srv/inatl/longterm/summit/archive/archive.htm.

235 *Senate ratification of Geneva Protocol:* ACDA, *Arms Control and Disarmament Agreements,* pp. 13–14.

235–36 *Chemical disarmament negotiations in Geneva:* Deborah Shapley, "Chemical Warfare: Binary Plan, Geneva Talks on a Collision Course," *Science,* vol. 184, no. 4143 (June 21, 1974), pp. 1267–1269; "Hearings Probe U.S. Chemical Warfare Stance," *Chemical & Engineering News,* vol. 52, no. 21 (May 27, 1974), pp. 10–11; "House Stirs Up Chemical Warfare Issue," *Chemical & Engineering News,* vol. 52, no. 33 (August 19, 1974), pp. 16, 19–20.

236 *Army decision not to abolish Chemical Corps:* Smart, "History of Chemical and Biological Warfare," p. 65.

236 *Countries of chemical weapons proliferation concern:* Rear Admiral Thomas A. Brooks, Director of Naval Intelligence, Statement on Intelligence Issues before the Seapower, Strategic, and Critical Materials Subcommittee of the House Armed Services Committee, March 7, 1991; "Navy Report Asserts Many Nations Seek or Have Poison Gas," *New York Times,* March 10, 1991, p. 10; R. Jeffrey Smith, "Confusing Data on Chemical Capability," *Washington Post,* March 10, 1991, p. A21; E. J. Hogendoorn, "A Chemical Weapons Atlas," *Bulletin of the Atomic Scientists,* vol. 53, no. 5 (September–October 1997); Colonel Richard Price, "Yugoslav Chemical Warfare Capability, Mostar's History of Chemical Weapon Research, Development, Production: What, When, Where, How Much?," *The ASA Newsletter,* no. 71 (February 1999); Human Rights Watch, *Clouds of War: Chemical Weapons in the Former Yugoslavia,* vol. 9, no. 5 (D) (March 1997).

236–37 *Origins of Iraqi chemical weapons program:* UNMOVIC, "Unresolved Disarmament Issues: Iraq's Proscribed Weapons Programmes" (Working Document), March 6, 2003, p. 139.

237 *Al-Hazen Institute:* Ibid.

237 *Saddam Hussein's rise to power:* Baram, *Building Toward Crisis,* p. 2.

237 *Arab Projects and Development:* Efraim Karsh, "Rational Ruthlessness: Non-Conventional and Missile Warfare in the Iran-Iraq War," in Karsh, Navias, and Sabin, eds., *Non-Conventional Weapons Proliferation in the Middle East,* pp. 32–33.

237–38 *Iraqi negotiations with Pfaudler:* Timmerman, *The Death Lobby,* pp. 36–38.

238 *ICI officials were suspicious of Iraqi request:* U.S. Department of State, cable from U.S. Information Service, London, to SecState, Washington D.C., "Observer Article Describes Iraqi Chemical Weapons Factory" (Unclassified), March 15, 1984.

238 *Iraqis turned to the Italian company Montedison:* Timmerman, *The Death Lobby,* p. 49.

238 *Plant built by Klöckner Industrie:* Ibid., p. 233.

238 *U.S.-Soviet chemical weapons talks resumed:* Will Lepkowski, "Chemical Warfare: One of the Dilemmas of the Arms Race," *Chemical & Engineering News,* vol. 56, no. 1 (January 2, 1978), pp. 16–21.

239 *M687 Sarin artillery shell ready for production:* Department of the Army, Headquarters, United States Army Armament Command, "Record of Decision/Action on Type Classification of Projectile, 155mm, Lethal Binary, GB2, XM687" (Confidential), January 7, 1977 [FOIA].

239 *Carter administration review of chemical warfare policy:* Zbigniew Brzezinski, Presidential Review Memorandum, NSC-27, Subject: "Chemical Warfare" (Secret), May 19, 1977 [NSA].

239 *Letter from Secretary of State Vance to Defense Secretary Brown:* Jack Anderson, "Poisonous Advantage," *Washington Post,* June 14, 1981, p. C7.

239 *Statement by General Lennon:* Lepkowski, "Chemical Warfare: One of the Dilemmas of the Arms Race," p. 17.

240 *Statement by Stephen Douglass, Jr.:* Ibid.

240 *Statement by Gen. Alexander Haig:* Harris and Paxman, *A Higher Form of Killing,* p. 226.

240 *"Wintex" chemical warfare scenario:* Ibid., pp. 230–232.

240 *Administration deleted Army's funding requests for binary weapons:* Lois Ember, "Chemical Weapons: Build Up or Disarm?," *Chemical & Engineering News,* December 15, 1980, p. 23.

241 *Chemical Corps failed to maintain the unitary stockpile:* Gordon M. Burck, "Decision-Making in the United States Nerve Gas Weapons Program," unpublished paper, 1985.

241–42 *Reports of Soviet chemical warfare in Southeast Asia and Afghanistan:* Director of Central Intelligence, *Use of Toxins and Other Lethal Chemicals in Southeast Asia and Afghanistan* (Secret), Special National Intelligence Estimate, Memorandum to Holders, March 2, 1983.

242 *U.S.-Soviet negotiations ended after twelve rounds:* "Todeswolken über Europa," *Der Spiegel,* no. 8 (February 22, 1982), p. 52.

242–43 *Ichord and Jackson amendments:* Wayne Biddle, "Restocking the Chemical Arsenal," *New York Times Magazine,* May 24, 1981, p. 36.

243 *Statement by Senator Carl Levin:* Senator Carl Levin, *Congressional Record—Senate,* September 16, 1980, p. S12643.

243–44 *Congress approved funding for Pine Bluff plant:* Biddle, "Restocking the Chemical Arsenal," p. 36; Jeffrey K. Smart, "History of Chemical and Biological Warfare: An American Perspective," in Sidell, Takafuji, and Franz, eds., *Medical Aspects of Chemical and Biological Warfare,* p. 71.

CHAPTER THIRTEEN: BINARY DEBATE

245 *Defense Science Board report:* Office of the Under Secretary of Defense for Acquisition and Technology, Defense Science Board, *Report of the 1980 DSB Summer Study Task Force on Chemical Warfare and Biological Defense* (Washington, D.C.: Department of Defense, December 1980).

245–46 *Reasons for opposition to binary weapons:* Author's interview with John Isaacs.

246 *Letter from twelve Senators to Chairman Tower:* Letter to the Honorable John Tower, Chairman, Senate Committee on Armed Services, signed by Nancy Kassebaum, Gary Hart, David Pryor, William Proxmire, Edward Kennedy, Thad Cochran, George Mitchell, Lowell Weicker, Donald Riegle, Paul Tsongas, Walter Huddleston, and Dave Durenberger, March 12, 1982.

246 *1982 congressional votes on binary weapons production:* John Isaacs, "Nervous About Nerve Gas," *Bulletin of the Atomic Scientists,* vol. 39, no. 10 (December 1983), pp. 7–8.

246 *Technical problems with Bigeye binary VX bomb:* Stephen Budiansky, "Qualified Approval for Binary Chemical Weapons," *Science,* vol. 234, no. 4779 (November 21, 1986), p. 930; Keith Morrison, "Modernizing Chemical Weapons Needless," *Defense News,* April 13, 1987, p. 31.

247 *Intermediate-volatility agent for MLRS:* Author's interview with Sigmund R. Eckhaus.

247 *XM135 Binary Chemical Warhead:* Jeffrey K. Smart, "History of Chemical and Biological Warfare: An American Perspective," in Sidell, Takafuji, and Franz, eds., *Medical Aspects of Chemical and Biological Warfare,* p. 71.

247–48 *Joint weapons testing by United States and France:* Author's interview with William C. Dee.

248 *Senator Pryor letter to Senator Tower:* "For the Record," *Washington Post,* February 18, 1983.

248 *1983 votes on binary weapons production:* "Senate Floor Action: Binary Weapons Debate," *1983 CQ Almanac* (Washington, D.C.: Congressional Quarterly, 1984), p. 11; "Conference Action: Chemical Weapons," *1983 CQ Almanac,* pp. 491–492.

249 *Dorothy Bush's criticism of Vice President's votes: Washington Post,* November 11, 1983; John Isaacs, "Mother Knows Best . . ." *Bulletin of the Atomic Scientists,* April 1989, pp. 3–4.

249 *Iranian revolution:* www.bbc.co.uk/persian/revolution/.

249 *Iranian "human wave" tactics:* Rolf Ekéus, "Iraq's Real Weapons Threat," *Washington Post,* June 29, 2003, p. B7.

250 *Project 922 at Al-Rashad:* UNMOVIC, "Unresolved Disarmament Issues: Iraq's Proscribed Weapons Programmes" (Working Document), March 6, 2003, p.

141; Bob Drogin, "Iraqi Weapons Expert Insists Search Is Futile," *Los Angeles Times,* June 4, 2003, p. 1.

250 *Iraqi strategic cooperation with Egypt:* Dany Shoham, "Chemical and Biological Weapons in Egypt," *The Nonproliferation Review,* vol. 5, no. 3 (Spring–Summer 1998), p. 51; Associated Press, "CIA: Egyptian Help Key to Iraqi Chemical Weapons," March 13, 2005.

250 *Construction of SEPP:* UNMOVIC, "Unresolved Disarmament Issues," p. 141.

250 *Iraqi use of tear gas and mustard:* W. Andrew Terrill, "The Chemical Warfare Legacy of the Yemen War," *Comparative Strategy,* vol. 10 (1991), p. 117.

250–51 *More than thirty Western firms supplied materials and equipment:* Philip Shenon, "Declaration Lists Companies That Sold Chemicals to Iraq," *New York Times,* December 21, 2002; U.S. Department of State, cable from U.S. Information Service, London, to SecState, Washington, D.C., "Chemical Weapon Materials Sold to Iran and Iraq" (Unclassified), April 6, 1984; Michael Dobbs, "U.S. Had Key Role in Iraqi Buildup," *Washington Post,* December 30, 2002, p. A1.

251 *Karl Kolb GmbH:* "Middle East Chemical Weapons," *MEDNEWS: Middle East Defense News,* vol. 2, no. 2–3 (October 24, 1988), p. 4; Timmerman, *The Death Lobby,* pp. 109–112.

251 *Iraqi purchase of munitions:* UNMOVIC, "Unresolved Disarmament Issues," p. 141.

251 *Scientists employed at Muthanna:* Christophe Ayad, "Armes chimiques: Un témoignage," *Libération,* April 17, 2003.

252 *Memorandum to Secretary of State Shultz:* U.S. Department of State, Memorandum from Jonathan T. Howe, Bureau of Politico-Military Affairs, to the Secretary of State, Subject: "Iraq Use of Chemical Weapons," November 1, 1983 [NSA].

252 *U.S. démarche to Iraqi Foreign Ministry:* U.S. Department of State, Briefing Paper, "Iraqi Illegal Use of Chemical Weapons" (Secret), drafted by PM/TMP, November 16, 1984 [NSA].

252 *Reagan dispatched Rumsfeld to Baghdad:* U.S. Department of State, cable from U.S. Embassy London to Secretary of State, "Rumsfeld Mission: December 20 Meeting with Iraqi President Saddam Hussein" (Secret), December 21, 1983 [NSA].

253 *Iraq's use of chemical weapons became more effective:* Timothy V. McCarthy and Jonathan B. Tucker, "Saddam's Toxic Arsenal: Chemical and Biological Weapons in the Gulf Wars," in Lavoy, Sagan, and Wirtz, *Planning the Unthinkable,* p. 64.

253 *1983 Soviet decree to develop Novichok agents:* Vil S. Mirzayanov, "Dismantling the Soviet/Russian Chemical Weapons Complex: An Insider's View," in Amy E. Smithson, ed., *Chemical Weapons Disarmament in Russia: Problems and Prospects* (Washington, D.C.: Henry L. Stimson Center, Report no. 17, October 1995), p. 23.

253–54 *Development of Novichok-5:* Ibid., p. 24; author's interview with Vil S. Mirzayanov.

254 *Production of Novichok agents:* Mirzayanov, "Dismantling the Soviet/Russian

Chemical Weapons Complex: An Insider's View," p. 25; James Ring Adams, "Russia's Toxic Threat," *Wall Street Journal,* April 20, 1996, p. A14.

254 *Pavlodar Chemical Plant:* Gulbarshyn Bozheyeva, "The Pavlodar Chemical Weapons Plant in Kazakhstan: History and Legacy," *The Nonproliferation Review,* vol. 7, no. 2 (Summer 2000), pp. 136–145.

254–55 *Czech development of binary agent:* Jiří Matousek and Ivan Masek, "On the New Potential Supertoxic Lethal Organophosphorus Chemical Warfare Agents with Intermediary Volatility," *ASA Newsletter,* 94-5 (October 7, 1994), pp. 1, 10–11; J. Bajgar, J. Fusek, and J. Vachek, "Treatment and Prophylaxis Against Nerve Agent Poisoning," *ASA Newsletter,* 94-4, pp. 10–11.

255 *Matousek statement:* Matousek and Masek, "On the New Potential Supertoxic Lethal Organophosphorus Chemical Warfare Agents," p. 11.

255 *Iranian offensive north of Basra:* Julian Perry Robinson and Jozef Goldblat, "Chemical Warfare in the Iran-Iraq War," *SIPRI Fact Sheet on Chemical Weapons,* no. 1 (Stockholm: Stockholm International Peace Research Institute, May 1984); Karsh, "Rational Ruthlessness," p. 37.

255–56 *Iraqi broadcast warning about use of poison gas:* Department of State, confidential cable from U.S. Interests Section Baghdad to Secretary of State, Washington, D.C., "Iraqi Warning re Iranian Offensive," February 22, 1984.

256 *Description of Majnoon marshes:* Major General Edward Fursdon, "Iraqis Dig In to Secure Oil-Rich Majnoon Marshes," *Daily Telegraph,* March 26, 1984, p. 4.

256 *Iranian capture of Majnoon Islands:* Ibid.

256–57 *Iraqi counterattack on Majnoon Islands:* Javed Ali, "Chemical Weapons and the Iran-Iraq War: A Case Study in Noncompliance," *The Nonproliferation Review,* vol. 8, no. 1 (Spring 2001), pp. 43–58.

257 *Execution of Saddam order for chemical attack:* McCarthy and Tucker, "Saddam's Toxic Arsenal: Chemical and Biological Weapons in the Gulf Wars," in Lavoy, Sagan, and Wirtz, *Planning the Unthinkable,* pp. 63–64.

257 *Iraqi use of Tabun at Majnoon Islands:* Hooshang Kadivar and Stephen C. Adams, "Treatment of Chemical and Biological Warfare Injuries: Insights Derived from the 1984 Iraqi Attack on Majnoon Island," *Military Medicine,* vol. 156, no. 4 (April 1991), pp. 171–177.

257–58 *Evacuation of Iranian casualties to Ahwaz:* Author's interview with Colonel Jonathan Newmark.

258 *White House statement of March 5, 1984:* The White House, Office of the Press Secretary, "Press Statement: Iraq's Use of Chemical Weapons," March 5, 1984.

258 *U.N. investigation of Iraqi chemical attack:* U.N. Secretary-General, "Report of the Specialists Appointed by the Secretary-General to Investigate Allegations by the Islamic Republic of Iran Concerning the Use of Chemical Weapons," U.N. Security Council Document S/16433, March 26, 1984.

258 *Experts examined bomb fragments, casualties:* Richard M. Preece, "Iran-Iraq War: Implications for U.S. Policy," *CRS Issue Brief* (Washington, D.C.: Congressional Research Service), October 14, 1987, p. 8.

258 *Results of chemical analyses:* U.N. Secretary-General, "Report of the Specialists

Appointed by the Secretary-General," p. 9; Eliot Marshall, "Iraq's Chemical Warfare: Case Proved," *Science*, vol. 224 (April 13, 1984), pp. 130–131.

259 *Nerve-agent casualties not "photogenic" for propaganda purposes:* Author's interview with Colonel Jonathan Newmark.

259 *Tepid Western response to Iraqi chemical warfare:* R. Jeffrey Smith, "Relying on Chemical Arms: Early Use Was Central to Recent Iraqi Battle Tactics," *Washington Post*, August 10, 1990, p. A25.

259 *Iran decision to acquire chemical weapons:* Cordesman, *Weapons of Mass Destruction in the Middle East*, p. 83; Paula DeSutter, "Deterring Iranian NBC Use," *Strategic Forum*, no. 110 (Washington, D.C.: National Defense University, April 1997), p. 3.

259 *U.S. interdiction of shipment of potassium fluoride:* U.S. Department of State, Action Memorandum to the Secretary of State, Subject: "Iraq's Use of Chemical Weapons: Control on U.S. Ingredients" (Confidential), March 14, 1984.

260 *West German Chancellor Kohl intervened to stop deliveries:* "Chemical Weapons in the Middle East," *MEDNEWS: Middle East Defense News*, vol. 2, no. 2–3 (October 24, 1988), p. 4.

260 *Adelman testimony:* Kenneth Adelman, Director, U.S. Arms Control and Disarmament Agency, quoted in Ronald D. Stricklett, "Chemical Proliferation: The Changing Environment," U.S. Army Dugway Proving Ground, Document no. DPG-S-TA-85-07, June 1985, p. 1.

260 *Reagan administration's "two-track policy":* Office of the Chief of Public Affairs, Headquarters, U.S. Army Training and Doctrine Command, "The Chemical Strategy—Negotiation and Modernization," *Army Chemical Review*, January 1988, pp. 23–24; Caspar W. Weinberger, "The Deterrence of Chemical Warfare," in Secretary of Defense, *Annual Report to Congress, Fiscal Year 1983*, section III, pp. 143–147.

260 *Quote by General Bernard Rogers:* "Todeswolken über Europa," *Der Spiegel*, no. 8 (February 22, 1982), p. 34.

261 *Vice President Bush presented first draft of CWC:* Michael Krepon, "Verification of a Chemical Weapons Convention," in Roberts, ed., *Chemical Disarmament and US Security*, p. 81.

261 *1984 votes on binary weapons production:* "House Floor Action: Nerve Gas Curb," *1984 CQ Almanac* (Washington, D.C.: Congressional Quarterly, 1985), pp. 42–43; Council for a Livable World, "Why Did the House Vote Overwhelmingly Against Nerve Gas in 1984?," press release, May 23, 1984.

262 *Quality control problems with Iraqi Tabun:* UNMOVIC, "Unresolved Disarmament Issues," p. 142.

262 *Description of Muthanna production complex:* Timmerman, *The Death Lobby*, pp. 111–112.

262 *Iraqi production of Sarin at Muthanna:* Ibid., p. 189.

262–63 *Founding of Australia Group:* See Australia Group Web site, www.australia-group.net; U.S. Department of State, Bureau of Nonproliferation, "Fact Sheet:

Australia Group," January 6, 2004, available online at www.state.gov/t/np/rls/fs/27800.htm.

263 *"Three strikes and you're out":* Author's interview with John Isaacs.

263 *Administration launched a full-court press:* Lois R. Ember, "Pentagon Pressing Hard for Binary Chemical Arms Funds," *Chemical & Engineering News,* February 25, 1985, pp. 26–28; "Reagan Administration Mounts Push for Binary Chemical Weapons," *National Guard,* May 1985, pp. 8–9.

263 *Testimony by Pentagon officials:* Author's interview with William C. Dee.

263 *Kroesen study:* U.S. Senate, Committee on Armed Services, Hearing, *Department of Defense Authorization for Appropriations for Fiscal Year 1986: Binary Chemical Modernization,* February 28, 1985, pp. 1455–1483.

263 *Declassified version of Kroesen report:* Walter Pincus, "Ex-Officers Pushed for Chemical Arms," *Washington Post,* January 22, 1986, p. 4.

264 *Stoessel commission report:* Chemical Warfare Review Commission, *Report of the Chemical Warfare Review Commission* (Washington, D.C.: U.S. Government Printing Office, 1985).

264 *Critics challenged credibility of the Stoessel commission:* "Chemical Weapons Panel Draws Fire," *Congressional Quarterly,* May 4, 1985, p. 861.

264 *Representative Porter arranged for CIA briefings on Soviet threat:* Walter Pincus, "2 Agencies at Odds on a Soviet Threat," *Washington Post,* January 23, 1986, p. 19.

264–65 *Opinion piece by Representatives Porter and Fascell:* Dante B. Fascell and John E. Porter, "New Nerve-Gas Weapons That We Don't Need," *Washington Post,* June 17, 1985, p. 11.

265–66 *1985 votes on binary weapons production:* "Defense Authorization: Chemical Weapons," *1985 CQ Almanac* (Washington, D.C.: Congressional Quarterly, 1986), p. 149; John Isaacs, "November—Critical Month for Arms Control," *Bulletin of the Atomic Scientists,* September 1985, pp. 3–4.

266 *Congressional conditions on binary production:* "Binary Chemical Weapons Legislation: Title 1, Section 119 of the Department of Defense Authorization Act for FY 1986—Conditions on Spending Funds for Binary Chemical Munitions, Approved by the House on June 19, 1985," in U.S. House, Committee on Foreign Affairs, Subcommittee on Arms Control, International Security and Science, *Binary Chemical Weapons: Selected Documents,* 99th Congress, 2nd session (Washington, D.C.: U.S. Government Printing Office, 1986), p. 1.

266 *Representative Porter prediction:* "Defense Authorization: Nerve Gas," *1985 CQ Almanac,* p. 157.

CHAPTER FOURTEEN: SILENT SPREAD

267 *NATO approval of U.S. binary production program:* Bill Keller, "New Nerve Gas Plan Leaves Tests and Storage Undecided," *New York Times,* August 12, 1985, p. 1; David Dickson, "Approval Seen for New U.S. Chemical Weapons," *Science,*

vol. 232, no. 4750 (May 2, 1986), p. 567; R. Jeffrey Smith, "Binary Deployment Remains Controversial," *Science*, vol. 232, no. 4750 (May 2, 1986), p. 568.

267–68 *Reagan ordered production of binary projectile:* John H. Cushman, "Reagan Announces U.S. Is Resuming Production of Chemical Weapons," *International Herald Tribune*, July 31, 1986, p. 1; The White House, "Statement by Principal Deputy Press Secretary Speakes on the Binary Chemical Munitions Program," July 29, 1986.

268 *Failed effort to block production of binary projectile:* John Isaacs, "Using Summitry to Thwart Congress," *Bulletin of the Atomic Scientists*, December 1986, pp. 4–5.

268 *Airlift of binary weapons in wartime:* Stephen Budiansky, "Qualified Approval for Binary Chemical Weapons," *Science*, vol. 234, no. 4779 (November 21, 1986), p. 932.

268 *Conservative critics said administration had "shot itself in the foot":* Thomas F. O'Boyle, "White House Has Shot Itself in the Foot with West German Nerve-Gas Accord," *Wall Street Journal*, July 29, 1986, p. 30.

269 *Representative Fascell made public GAO report on Bigeye bomb:* Walter Pincus, "Gas-Shell Production Ordered: Long Moratorium Is About to End," *Washington Post*, July 30, 1986, p. 6.

269 *French ACACIA binary weapons program:* Meyer, *L'arme chimique*, pp. 109–110; Lepick, *Les armes chimiques*, p. 98; David Dickson, "France to Produce Binary Weapons?," *Science*, vol. 234, November 28, 1986, p. 1070; Laurent Mossu, "Roland Dumas à Genève: La France ne fabriquera pas d'armes chimiques," *Le Figaro*, February 8, 1989, p. 3.

269–70 *Saddam delegated authority for chemical weapons use:* McCarthy and Tucker, "Saddam's Toxic Arsenal: Chemical and Biological Weapons in the Gulf Wars," in Lavoy, Sagan, and Wirtz, eds., *Planning the Unthinkable*, p. 65.

270 *Defense Intelligence Agency shared intelligence with Iraqi military:* Patrick E. Tyler, "Officers Say U.S. Aided Iraq in War Despite Use of Gas," *New York Times*, August 18, 2002, p. 1; Michael Dobbs, "U.S. Had Key Role in Iraqi Buildup," *Washington Post*, December 30, 2002, p. A1.

270 *Poor quality of Iraqi Sarin:* UNMOVIC, "Unresolved Disarmament Issues: Iraq's Proscribed Weapons Programmes" (Working Document), March 6, 2003, p. 73.

271 *Unit 2100 facility at Al-Haditha:* Paul Salopek, "Saddam Tested Deadly Weapons on Humans, Accounts Say," *Chicago Tribune*, July 16, 2003.

271 *Mass execution of Iraqi political prisoners with nerve gas:* Ibid.

271 *Construction of chemical plants at Fallujah:* Timmerman, *The Death Lobby*, pp. 232–233.

271 *Iraqi production of Cyclosarin:* UNMOVIC, "Unresolved Disarmament Issues," p. 143.

272 *Iraqi development of VX:* Ibid., pp. 80–82.

272 *Letter from General Nazar al-Khazarji:* Ibid., p. 143.

272 *Poor quality of Iraqi VX:* Author's interview with Ron Manley.

272–73 *Iran's production and use of chemical weapons:* Gregory F. Giles, "The Islamic Republic of Iran and Nuclear, Biological, and Chemical Weapons," in Lavoy, Sagan, and Wirtz, *Planning the Unthinkable,* pp. 79–103.

273 *Chemical Research Institute at Nukus:* Judith Miller, "U.S. and Uzbeks Agree on Chemical Arms Plant Cleanup," *New York Times,* May 25, 1999, p. A3.

273–74 *Accidental poisoning of Zheleznyakov:* Wise, *Cassidy's Run,* pp. 191–192; Will Englund, "Russia Still Doing Secret Work on Chemical Arms," *Baltimore Sun,* October 18, 1992, p. 1A; author's interview with Vil S. Mirzayanov.

275 *Gorbachev's act of "diplomatic jujitsu":* Michael Krepon, "Verification of a Chemical Weapons Convention," in Roberts, ed., *Chemical Disarmament and US Security,* p. 81.

275–76 *"Open house" at Shikhany Central Chemical Weapons Testing Site:* A. Gorokhov and A. Serbin, "Trust: Report from the Soviet Shikhany Military Installation," *Pravda* (in Russian), October 4, 1987, p. 6 [translated in FBIS-SOV-87-196, October 9, 1987, pp. 5–7].

277 *"The visit was very carefully orchestrated":* "Soviets Reveal CW Capabilities," *International Defense Review,* vol. 20 (November 1987), p. 1453.

277 *Statement of General Pilakov:* Gorokhov and Serbin, "Trust: Report from the Soviet Shikhany Military Installation."

277 *Soviets concealed several aspects of chemical arsenal:* Gary Thatcher, "Soviet Chemical Arsenal: How Superior Is It?," *Christian Science Monitor,* December 14, 1988, pp. B1–B9; Boris Libman, "Lack of Confidence Between the American and Russian Parties," *Communiqué no. 36* (report prepared for the U.S. Army Chemical Corps, handwritten manuscript, undated).

277–78 *Faked demonstration of nerve-agent neutralization:* Author's interview with Vil S. Mirzayanov.

278 *Start of production of M687 binary artillery shell:* U.S. Congress, General Accounting Office, "Status of the Army's M687 Binary Program," Report No. GAO/NSIAD-90-295 (September 1990), p. 3.

278–79 *No commercial company was willing to supply dichlor:* Lois Ember, "Army Seeks Firms to Make Nerve Gas," *Chemical and Engineering News,* August 23, 1982, pp. 32, 34.

280 *Anfal campaign against the Iraqi Kurds:* Physicians for Human Rights, "Winds of Death: Iraq's Use of Poison Gas Against Its Kurdish Population," February 1989; James Bruce and Tony Banks, "Growing Concern over Iraqi Use of CW," *Jane's Defence Weekly,* September 24, 1988, p. 715.

280 *Ali Hassan al-Majid statement, "I will kill them all.":* Christopher Dickey and Evan Thomas, "How Saddam Happened," *Newsweek,* September 23, 2002, p. 37.

280 *Description of Iraqi Kurdish town of Halabja:* Pamela Constable, "Cloud over Halabja Begins to Dissipate," *Washington Post,* August 7, 2003, p. A10.

280 *Saddam ordered a "special strike":* Ghassan Shirbil, "Opening His Books, Al-

Khazarji: 'Saddam Ordered Bombing Halabja with Chemical Weapons After Receiving a False Report,' " *Al-Hayah* (London), in Arabic, November 28, 2002, p. 10; translated in FBIS, document no. GMP20021128000095.

281 *Chemical attack against Halabja:* C. J. Chivers, "Still Suffering from '88 Gas Attack, a Village Distrusts Iraq's Arms Report," *New York Times,* December 11, 2002, p. A22; Christine Gosden and Mike Amitay, "Lesson of Iraq's Mass Murder," *Washington Post,* June 2, 2002, p. B7; "The Trial of 'Chemical Ali,' " *Washington Times,* December 20, 2004, p. 20.

281 *Description in Human Rights Watch report:* Human Rights Watch, *Genocide in Iraq: The Anfal Campaign Against the Kurds* (New York: Human Rights Watch, July 1993), available online at http://hrw.org/reports/1993/iraqanfal/ANFAL3.htm.

281–82 *Flight of refugees from Halabja:* Len Kelly, Dale Dewar, and Bill Curry, "Experiencing Chemical Warfare: Two Physicians Tell Their Story of Halabja in Northern Iraq," *Canadian Journal of Rural Medicine,* vol. 9, no. 3 (Summer 2004), pp. 178–182.

282 *Casualties from Halabja attack:* U.S. Department of State, Bureau of Public Affairs, "Saddam's Chemical Weapons Campaign: Halabja, March 16, 1988," March 14, 2003.

282 *U.S. intelligence intercepted Iraqi radio messages:* James Bruce and Tony Banks, "Growing Concern over Iraqi Use of CW," *Jane's Defence Weekly,* September 24, 1988, p. 715.

282–83 *Account by British correspondent:* David Hirst, "Iran Puts Dead on Show After Gas Raid," *The Guardian,* March 22, 1988, p. 1.

283–84 *Type of nerve agent used in Halabja attack:* Author's interview with Ron Manley; author's interview with Julian Perry Robinson.

284 *Iraq denied request by U.N. Secretary-General to investigate:* Bruce and Banks, "Growing Concern over Iraqi Use of CW."

284 *U.S. government accused Iran as well as Iraq:* Joost R. Hiltermann, "Iran's Nuclear Posture and the Scars of War," Middle East Report Online, January 18, 2005, www.merip.org/mero/mero011805.html.

284 *U.N. Security Council Resolution 598:* United Nations, "Resolution 598 (1987) of 20 July 1987," *Resolutions and Decisions of the Security Council 1987* (New York: United Nations, 1988), pp. 5–6.

284 *Iraqi use of nerve agents at Al-Fao:* Lee Waters, "Chemical Weapons in the Iran/Iraq War," *Military Review,* vol. 70 (October 1990), pp. 57–63.

284 *Battlefield tour by Lieutenant Colonel Francona:* R. Jeffrey Smith, "Relying on Chemical Arms: Early Use Was Central to Recent Iraqi Battle Tactics," *Washington Post,* August 10, 1990, pp. A25, A27.

285 *War of the cities:* Ephraim Karsh, "Rational Ruthlessness: Non-Conventional and Missile Warfare in the Iran-Iraq War," in Karsh, Navias, and Sabin, eds., *Non-Conventional Weapons Proliferation in the Middle East,* p. 41.

285 *Iraqi plan for chemical attack on Tehran:* Amatzia Baram, "An Analysis of Iraqi

WMD Strategy," *The Nonproliferation Review,* vol. 8, no. 2 (Summer 2001), p. 29.

285 *Mass evacuation of Tehran:* Chaim Herzog, "A Military-Strategic Overview," in Karsh, ed., *The Iran-Iraq War: Impact and Implications,* p. 263.

285 *Psychological impact of Iraqi chemical threat:* Thomas L. McNaugher, "Ballistic Missiles and Chemical Weapons: The Legacy of the Gulf War," *International Security,* vol. 15 (Fall 1990), p. 22.

285 *Iranian chemical casualties:* Giles, "The Islamic Republic of Iran and Nuclear, Biological, and Chemical Weapons," p. 83; Scott Peterson, "Lessons from Iran on Facing Chemical War," *Christian Science Monitor,* November 19, 2002.

286 *Physicians for Human Rights investigation:* Physicians for Human Rights, *Winds of Death: Iraq's Use of Poison Gas Against Its Kurdish Population, Report of a Medical Mission to Turkish Kurdistan* (Cambridge, Mass.: PHR, February 1989).

286 *Analysis of soil samples from bomb craters:* Lois Ember, "Chemical Weapons: Residues Verify Iraqi Use on Kurds," *Chemical and Engineering News,* vol. 711 (May 3, 1993), p. 8.

286 *Flora Lewis warning:* Flora Lewis, "Move to Stop Iraq," *New York Times,* September 14, 1988, p. A31.

286 *Statement by Robert Gates:* Robert M. Gates, Deputy Director, Central Intelligence Agency, "The CIA and the University," address before the Association of Former Intelligence Officers, October 10, 1987, *Periscope,* vol. 12, no. 4 (1987), pp. 17–19.

286–87 *Statement by Ali Akbar Hashemi Rafsanjani:* Islamic Republic News Agency (IRNA), Tehran, English broadcast, transcribed in FBIS Daily Report: Near East and South Asia, October 19, 1988, pp. 55–56.

287 *Iran contracted with the Swiss company Krebs AG:* Lois Ember, "U.S. Vexed by Spread of Chemical Weapons," *Chemical & Engineering News,* March 27, 1989, p. 23.

287 *Krebs plant at Abu Za'abal, Egypt:* Michael R. Gordon and Stephen Engelberg, "Egypt Accused of Big Advance in Gas for War," *New York Times,* March 10, 1989, pp. A1, A2; Christopher Walker, "Egypt Denies Claim That It Is Building Poison Gas Factory," *The Times* (London), March 11, 1989, p. 8; Michael R. Gordon and Stephen Engelberg, "Poison Gas Fears Lead U.S. to Plan New Export Curbs," *New York Times,* March 26, 1989, p. A1; Dany Shoham, "Chemical and Biological Weapons in Egypt," *The Nonproliferation Review,* vol. 5, no. 3 (Spring–Summer 1998), p. 50.

287 *Syrian chemical weapons program:* M. Zuhair Diab, "Syria's Chemical and Biological Weapons: Assessing Capabilities and Motivations," *The Nonproliferation Review,* vol. 5, no. 1 (Fall 1997), pp. 104–111; Ahmed S. Hashim, *Chemical and Biological Weapons and Deterrence, Case Study 1: Syria* (Washington, D.C.: Chemical and Biological Arms Control Institute, 1998), pp. 5–9; Magnus Normark et al., *Syria and WMD: Incentives and Capabilities* (Umeå: Swedish Defence Research Agency, FOI-R-1290-SE, June 2004), pp. 34–37.

287 *Syria sought chemical weapons as a deterrent:* Walter Pincus, "Syria Built Arsenal as 'Equalizer,' " *Washington Post,* April 17, 2003, p. A30.

287 *Syrian acquisition of Western chemical technology:* Gazit, ed., *The Middle East Military Balance, 1993–1994,* p. 229; "Middle East Chemical Weapons," *MED-NEWS: Middle East Defense News,* vol. 2, no. 2–3 (October 24, 1988), p. 5; Dany Shoham, "Guile, Gas and Germs: Syria's Ultimate Weapons," *Middle East Quarterly,* vol. 9, no. 3 (Summer 2002), available online at www.meforum.org/article/493.

287 *Three Syrian chemical weapons production facilities:* "Syria's Secret Weapons," *Jane's Intelligence Digest,* May 2, 2003.

287–88 *Syrian Scud missiles with chemical warheads:* Arieh O'Sullivan, "Does Syria Really Have a War Option?," *Jerusalem Post,* January 11, 2000; Ze'ev Schiff, "Syrian Scud Fired with Chemical Warhead," *Ha'aretz,* July 13, 2001; Dany Shoham, "Poisoned Missiles: Syria's Doomsday Deterrent," *Middle East Quarterly,* vol. 9, no. 4 (Fall 2002), available online at www.meforum.org/article/510.

288 *Integration of nerve-agent warheads with ballistic missiles:* David B. Ottaway, "Middle East Weapons Proliferate: Concern Heightened by Chemical Arms, Missile Capabilities," *Washington Post,* December 19, 1988, p. A1.

CHAPTER FIFTEEN: PEACE AND WAR

289 *Director-general of Muthanna wrote to senior Iraqi officials:* UNMOVIC, "Unresolved Disarmament Issues: Iraq's Proscribed Weapons Programmes" (Working Document), March 6, 2003, p. 143.

289 *Iraqi development of field-mixed chemical weapon:* Ibid., pp. 71–72, 144.

290 *Iraqi development of "true" binary artillery shell:* Ibid., p. 72.

290 *Further development of manufacturing process for VX:* UNSCOM, "Note for the File" [interrogation of General Hussein Kamel in Amman, Jordan, August 22, 1995], pp. 12–13.

290 *Iraq produced a VX intermediate known as "dibis":* UNMOVIC, "Unresolved Disarmament Issues," pp. 80–82.

290–91 *Controversy over the Army's access to thionyl chloride:* Tony Capaccio, "Army Pressing Defiant Firms for Nerve Gas Ingredient," *Defense Week,* vol. 11, no. 14 (March 26, 1990), p. 15; R. Jeffrey Smith, "Suppliers Reject Poison Gas Program; U.S. May Act," *Washington Post,* March 27, 1990, p. A5; Lois Ember, "Chemical Weapons: Firms Deny Sale of Chemical to Army," *Chemical & Engineering News,* April 2, 1990, p. 4.

292 *Marquardt Corporation linked to "Ill Wind" scandal:* John M. Broder, "Cheney Prepares to Halt All Chemical Weapons Output," *Los Angeles Times,* July 12, 1990, p. A22.

292 *Trial inspection at DuPont plant:* Federation of American Scientists, "Chemical Weapons Convention Chronology," available online at www.fas.org/nuke/control/cwc/chron.htm.

292 *Baker and Shevardnadze signed Wyoming MOU:* Federation of American Scien-

tists, "U.S.-Russian Wyoming Memorandum of Understanding on Chemical Weapons," available online at www.fas.org/nuke/control/cwc/news/cwmou. htm.

293 *President Bush speech to U.N. General Assembly:* The White House, Office of the Press Secretary, "The President's Chemical Weapons Initiative," September 25, 1989.

293 *United States would retain option to keep producing binary weapons:* Michael R. Gordon, "Bush Keeps Option to Make Poison Gas," *New York Times,* October 15, 1989, p. 8.

294 *Criticism of U.S. proposal to retain a security stockpile:* John Isaacs, "Where Does Bush Stand on Chemical Weapons?" *Bulletin of the Atomic Scientists,* December 1989, p. 3; "Banning Chemical Weapons," *Technology Review,* October 1990, pp. 33–40.

294 *Representative Fascell statement:* Gordon, "Bush Keeps Option to Make Poison Gas."

294 *Soviets knew they could not reject U.S. proposal:* Author's interview with Serguei Batsanov.

294 *Negotiation of Bilateral Destruction Agreement (BDA):* Don Oberdorfer, "U.S. Ready to Set Date to End Chemical Arms Production," *Washington Post,* May 9, 1990; Michael R. Gordon, "In a Switch, Bush Offers to Stop Producing Chemical Weapons: Soviet Concessions Toward Deep Cuts Are Sought," *New York Times,* May 9, 1990, pp. A1, A8.

295 *Signing of BDA on June 1, 1990:* "Documentation: Agreement Between the United States of America and the Union of Soviet Socialist Republics on Destruction and Non-Production of Chemical Weapons and on Measures to Facilitate the Multilateral Convention on Banning Chemical Weapons," *Bulletin of Peace Proposals,* vol. 21 (1990), pp. 363–369.

295 *Key provisions of the BDA:* Spiers, *Chemical and Biological Weapons,* p. 94.

295 *Cheney told Army that binary production would end:* John M. Broder, "Cheney Prepares to Halt All Chemical Weapons Output," *Los Angeles Times,* July 12, 1990, p. A22.

295–97 *Removal of U.S. chemical weapons from West Germany (Operation Steel Box):* Mattias Plügge, "CW in the FRG—A Hazardous Withdrawal?," *International Defense Review,* vol. 23, no. 2 (February 1990), pp. 123–124; Ian Murray, "Transport of US Gas Weapons Stirs German Alarm," *The Times* (London), July 26, 1990, p. 7; Marc Fisher, "U.S. Starts Removing Nerve Gas," *International Herald Tribune,* July 27, 1990, p. 1; Ian Murray, "Chemical Weapons Begin the Voyage to Oblivion," *The Times* (London), July 27, 1990, p. 9; Tom Rhodes, "A Farewell to Arms," *The European,* July 27–29, 1990, p. 1.

297 *Completion of Operation Steel Box:* John Deniston and Terri Ferguson, "Chemical Transport: Moving Chemical Munitions from Germany was a Complex Mission," *EurArmy,* November 1990, p. 28; Major General John C. Heldstab, Deputy Chief of Staff, Operations, Headquarters, U.S. Army Europe, Memo-

randum for CINCEUR, Subject: "STEEL BOX European Phase After-Action Report," December 9, 1990 [SPRU].

297 *Storage of U.S. chemical weapons on Johnston Island:* Owen Wilkes, "Chemical Weapon Burnoff in Central Pacific," *Peacelink,* no. 83 (July 1990), p. 10.

297 *Chemical weapons destruction technologies:* U.S. Congress, Office of Technology Assessment, *Disposal of Chemical Weapons: Alternative Technologies—Background Paper,* OTA-BP-O-95 (Washington, D.C.: U.S. Government Printing Office, June 1992), p. 3.

298 *Saddam Hussein speech on April 1, 1990:* Speech by Saddam Hussein on 1 April 1990, translated from Arabic in "President Warns Israel, Criticizes U.S.," FBIS-NES-90-064, 3 April 1990, pp. 32–36.

298 *Iraqi development of R-400 bomb and Al-Hussein warhead:* UNMOVIC, "Unresolved Disarmament Issues," pp. 146–147.

298 *Flight tests of chemical warhead for Al-Hussein missile:* Ibid., p. 147.

298 *Thunderstrike project:* Timothy V. McCarthy and Jonathan B. Tucker, "Saddam's Toxic Arsenal: Chemical and Biological Weapons in the Gulf Wars," in Lavoy, Sagan, and Wirtz, *Planning the Unthinkable,* pp. 65–67.

298–99 *Saddam Hussein statement on April 12, 1990:* Transcript of the meeting with a delegation of U.S. senators, FBIS-NES-90-076, April 17, 1990, p. 7.

299 *Iraqi invasion of Kuwait, Bush response:* Atkinson, *Crusade,* p. 4.

299 *Soviet testing on Ustyurt Plateau:* Judith Miller, "U.S. and Uzbeks Agree on Chemical Arms Plant Cleanup," *New York Times,* May 25, 1999, p. A3.

299–300 *Mirzayanov biography:* Author's interview with Vil S. Mirzayanov.

301 *Cover-up of contamination at Volgograd:* Ibid.

302 *End of binary agent production at Pine Bluff:* Institute for Defense and Disarmament Studies, "US Chemical Weapon Chronology 1991," *Arms Control Reporter,* May 1992, p. 704.E-1.17.

302 *Size of U.S. chemical weapons stockpile:* Office of the Assistant Secretary of Defense for Public Affairs, "U.S. Chemical Weapons Stockpile Information Declassified," News Release no. 024-96, January 22, 1996.

302 *U.S.-Soviet verification measures for BDA:* Gordon, "In a Switch, Bush Offers to Stop Producing Chemical Weapons."

302 *Chapayevsk destruction facility never opened:* Plügge, "CW in the FRG—A Hazardous Withdrawal?," p. 123.

302–03 *United Nations ultimatum to Iraq:* U.S. News and World Report, *Triumph Without Victory,* pp. 181–182.

303 *Muthanna churned out Sarin and Cyclosarin:* UNMOVIC, "Unresolved Disarmament Issues," p. 72.

303 *General Kamel asked for thirty-one trailers to transport munitions:* McCarthy and Tucker, "Saddam's Toxic Arsenal," p. 70.

303 *Iraqi dispersal of chemical munitions:* UNMOVIC, "Unresolved Disarmament Issues," p. 148.

303 *Special mobile missile unit under SSO control:* McCarthy and Tucker, "Saddam's Toxic Arsenal," pp. 74–78.

304 *CIA estimated Iraq had more than 1,000 tons of agents:* Atkinson, *Crusade,* p. 86.

304 *Reprogramming of CAMs to detect Cyclosarin:* Author's interview with Ron Manley.

304 *U.S. troops given pyridostigmine bromide (PB):* Beatrice Alexandra Golomb, *A Review of the Scientific Literature as It Pertains to Gulf War Illnesses,* vol. 2: *Pyridostigmine Bromide* (Santa Monica, Calif.: RAND National Defense Research Institute, 1999).

305 *Porton Down organized emergency response teams:* Author's interview with Ron Manley.

305 *Baker meeting with Tariq Aziz on January 9, 1991:* Baker, *The Politics of Diplomacy,* p. 359.

305 *U.S. Air Force used combination of munitions:* Atkinson, *Crusade,* p. 89.

305–06 *Czech chemical detachment detected nerve agent:* U.S. Department of Defense, Assistant Secretary of Defense for Public Affairs, News Briefing, "Czechoslovakian Chemical Report," November 10, 1993, p. 2; Philip Shenon, "Czechs Say They Warned U.S. of Chemical Weapons in Gulf," *New York Times,* October 19, 1996.

306 *General Schwarzkopf quote:* Schwarzkopf, *It Doesn't Take a Hero,* p. 509.

306 *FOX vehicle:* "The Fox Vehicle," www.gulflink.osd.mil/camp_mont2/tabe.htm.

307 *Mission Oriented Protective Posture (MOPP):* GulfLink, "Summary Paper on MOPP Procedures," available online at www.gulflink.osd.mil/mopp/mopp_so1.htm.

308 *Israeli chemical defense preparations during Gulf War:* Ariel Levite, "Israel Intensifying Preparations to Counter Chemical Attack," *Armed Forces Journal International,* May 1990, p. 60; Joel Brinkley, "Israelis' Fear of a Poison Gas Attack Is Growing," *New York Times,* August 24, 1990, p. A8; Cole, *The Eleventh Plague,* pp. 103–121.

308 *Isaac Stern rehearsal with Israel Philharmonic:* Cole, *The Eleventh Plague,* p. 114.

309 *Israeli casualties from misuse of gas masks and atropine:* E. Karsenty, J. Shemer, I. Alshech, et al., "Medical Aspects of the Iraqi Missile Attacks on Israel," *Israeli Journal of Medical Science,* vol. 27 (1991), pp. 603–607.

309 *Reasons for Iraqi nonuse of chemical weapons:* W. Andrew Terrill, "Chemical Warfare and 'Desert Storm': The Disaster That Never Came," *Small Wars and Insurgencies,* vol. 4, no. 2 (Autumn 1993), pp. 263–279.

309 *To saturate a square kilometer of territory with Sarin:* Atkinson, *Crusade,* p. 87.

309 *Demolition of bunkers at Khamisiyah:* Philip Shenon, "Study Sharply Raises Estimate of Troops Exposed," *New York Times,* July 24, 1997, p. A18.

310 *U.N. Security Council Resolution 687:* Amin Saikal, "The Coercive Disarmament of Iraq," in Wright, *Biological Warfare and Disarmament,* p. 267.

310 *Iraq initially declared 10,000 chemical munitions:* Author's interview with Ron Manley.

311 *Bush policy decisions:* Federation of American Scientists, "Chemical Weapons Convention Chronology."

311–12 *Manley meeting with Graham Pearson:* Author's interview with Ron Manley.

312 *Manley meeting in New York with Barrass and Gee:* Ibid.

312 *UNSCOM Destruction Advisory Panel:* Ibid.

312 *UNSCOM Chemical Destruction Group (CDG):* Ibid.; Amin Saikal, "The Coercive Disarmament of Iraq," in Wright, ed., *Biological Warfare and Disarmament,* p. 273.

313–14 *Explosive destruction of 122 mm rockets:* Author interview with Ron Manley.

314 *Total number of Iraqi chemical weapons destroyed:* UNMOVIC, "Unresolved Disarmament Issues," p. 148.

314 *Iraq's unilateral destruction of chemical weapons:* Ibid.

CHAPTER SIXTEEN: WHISTLE-BLOWER

315 *Award of Lenin Prize to Petrunin, Kuntsevich, and Yevstavyev (April 1991):* Vil S. Mirzayanov, "Dismantling the Soviet/Russian Chemical Weapons Complex: An Insider's View," in Amy E. Smithson, ed., *Chemical Weapons Disarmament in Russia: Problems and Prospects* (Washington, D.C.: Henry L. Stimson Center, Report no. 17, October 1995), p. 24.

315 *Article by Mirzayanov in* Kuranty: Ibid., p. 26.

316 *Article by Mirzayanov and Fedorov:* Oleg Vishnyakov, "Binary Bomb Exploded" [interview with Vil Mirzayanov and Lev Fedorov], *Novoye Vremya,* no. 44 (October 1992), pp. 4–9 (translated in FBIS, JPRS-TAC-92-033, November 14, 1992, pp. 44–49); Igor Ryabov, " 'Chemical War' Against an Invisible Enemy," *Novoye Vremya,* no. 5 (February 1994), pp. 4–6 (translated in FBIS, JPRS-TAC-94-008-L, July 27, 1994).

316 *Mirzayanov interview with Englund:* Will Englund, "Chemical Weapons Shadow Moscow," *Baltimore Sun,* March 19, 1992, p. 14A; Will Englund, "Ex-Soviet Scientist Says Gorbachev's Regime Created New Nerve Gas in '91," *Baltimore Sun,* September 15, 1992, p. 3A; Will Englund, "Russia Still Doing Secret Work on Chemical Arms," *Baltimore Sun,* October 18, 1992, p. 1A.

316–17 *FSB arrest and imprisonment of Mirzayanov:* Fred Hiatt, "Russia Jails Scientist over State Secrets," *Washington Post,* October 27, 1992, pp. A21, A27; Serge Schmemann, "K.G.B.'s Successor Charges Scientist," *New York Times,* November 1, 1992, p. 4.

317 *Martinov vowed that Mirzayanov would be convicted:* Mirzayanov, "Dismantling the Soviet/Russian Chemical Weapons Complex," p. 27.

318 *Endgame phase of CWC negotiations:* U.S. Arms Control and Disarmament Agency, "Fact Sheet: Chemical Weapons Negotiations at the Conference on Disarmament," August 13, 1992; Hassan Mashhadi, "How the Negotiations Ended," *Chemical Weapons Convention Bulletin,* no. 17 (September 1992), pp. 1, 28–30.

318 *Development of Australian "model treaty" and chairman's text:* Bernauer, *Chemistry of Regime Formation,* p. 29.

319 *CWC opened for signature in Paris:* Secretary of State Lawrence Eagleburger, "Remarks upon Signing the Chemical Weapons Convention, Paris, France,

January 13, 1993," *U.S. Department of State Dispatch*, vol. 4, no. 3 (January 18, 1993), p. 26.

320 *Uglev article in* Novoye Vremya: Oleg Vishnyakov, "Interview with a Noose Around the Neck" [interview with Vladimir Uglev], *Novoye Vremya*, No. 6, February 4, 1993, pp. 40–41 (translated in FBIS, JPRS-TAC-93-007, April 13, 1993, pp. 39–42).

320–21 *Russian government crackdown on journalists:* Will Englund, "2 Russian Papers Investigated After New Disclosures on Chemical Arms," *Baltimore Sun*, June 11, 1993, p. 21A.

321 *Drozd development of Novichok-7:* Mirzayanov, "Dismantling the Soviet/Russian Chemical Weapons Complex: An Insider's View," p. 28.

321 *Protests by scientific and human rights organizations:* Frank von Hippel, "Russian Whistleblower Faces Jail," *Bulletin of the Atomic Scientists*, vol. 49, no. 2 (March 1993) pp. 7–8; Gale Colby, "Fabricating Guilt," *Bulletin of the Atomic Scientists*, vol. 49, no. 8 (October 1993), pp. 12–13; Gale Colby and Irene Goldman, "When Will Russia Abandon Its Secret Chemical Weapons Program?," *Demokratizatsiya* (Winter 1993–94), pp. 148–154.

321 *Mirzayanov trial:* Sonni Efron, "Russian Scientist Faces Trial for Chemical-Arms Report," *Los Angeles Times*, January 5, 1994, p. A6; Fred Hiatt, "Russian Court Opens Unprecedented Secrets Trial," *Washington Post*, January 25, 1994, p. A1.

321 *Mirzayanov case dismissed for lack of evidence:* Mirzayanov, "Dismantling the Soviet/Russian Chemical Weapons Complex," p. 28.

322 *Mirzayanov received award at 1995 AAAS annual meeting:* American Association for the Advancement of Science, Scientific Freedom and Responsibility Award, 1995, http://archives.aaas.org/people.php?p_id=318.

322 *Mirzayanov underwent CIA polygraph test:* Author's interview with Vil S. Mirzayanov.

322 *Letter from Union of Khimprom Workers:* "News Chronology: 4 February 1995," *Chemical Weapons Convention Bulletin*, no. 28 (June 1995), p. 13.

322 *Behind-the-scenes discussions between U.S. and Russia:* Elisa D. Harris, "Outlawing Chemical and Biological Weapons," presentation at the Paul C. Warnke Conference on the Past, Present and Future of Arms Control, Georgetown University, Washington, D.C., January 28, 2004.

323 *Russian noncompliance with Wyoming MOU:* Vladimir Gusar, "Third-Generation Chemical Weapons Are Being Produced and Tested as Before," *VEK*, no. 12, March 26–April 1, 1993, p. 2 (translated in FBIS, JPRS-TAC-93-007, April 13, 1993, p. 43); Will Englund, "Russia Still Doing Secret Work on Chemical Arms," *Baltimore Sun*, October 18, 1992, p. 1A.

323 *Kirpichev working at secret institute in Shikhany:* Author's interview with Vil S. Mirzayanov.

324 *General Kuntsevich indictment and firing:* Mirzayanov, "Dismantling the Soviet/Russian Chemical Weapons Complex," p. 28.

324 *Chemical Destruction Group finished its work:* Author's interview with Ron Manley.

325 *Phenomenon of "secondary" proliferation:* Carl W. Ford, Jr., Assistant Secretary of State for Intelligence and Research, Testimony before the Senate Committee on Foreign Relations, Hearing, *Reducing the Threat of Chemical and Biological Weapons,* March 19, 2002, p. 11.

325 *Statement by Mamdouh Ateya:* Reuters, "Arabs 'Need Chemical Weapons,' " *The Independent,* July 28, 1988, p. 8.

325–26 *Robinson quote:* "Middle East Chemical Weapons," *Mednews: Middle East Defense News,* vol. 2, no. 3 (October 24, 1988), p. 3.

326 Yin He *incident of August 1933:* Rone Tempest, "China Demands U.S. Apology; Search of Ship Fails to Find Warfare Chemicals," *Chicago Sun-Times,* September 6, 1993, p. 10; "Statement by the Ministry of Foreign Affairs of the People's Republic of China on the "Yin He" Incident, September 4, 1993, available online at www.nti.org/db/china/engdocs/ynhe0993.htm.

326–28 *Chizuo Matsumoto, origins of Aum Shinrikyo cult:* David E. Kaplan, "Aum Shinrikyo (1995)," in Tucker, *Toxic Terror,* p. 297; Kaplan and Marshall, *The Cult at the End of the World,* pp. 8–11.

328 *Compound in Kamikuishiki:* Kaplan and Marshall, *The Cult at the End of the World,* p. 21.

328 *Tokyo government recognized Aum as a religious organization:* "Matsumoto's Aum Cult Grew Rapidly in Late 80s," *The Daily Yomiuri* (English-language Web site of *Yomiuri Shimbun*), February 16, 2004, transcribed in FBIS, document no. JPP20040216000107.

328–29 *Aum pursued a variety of moneymaking operations:* Kaplan and Marshall, *The Cult at the End of the World,* pp. 21–22.

329 *"Seidaishi" and "Seigoshi":* Shoko Egawa, "From the Other Witness Stand: Following the Aum Case" (serialized article no. 61), *Shukan Yomiuri,* March 28, 1999, pp. 46–47, translated in FBIS, document no. FTS19990717000186.

329 *Aum practitioners wore battery-powered caps:* Miwa Suzuki, "Children Take Charge of Japan's Lethal Doomsday Cult," *Agence France Presse,* August 12, 1999.

329 *Apocalyptic ideology of Aum Shinrikyo:* Ian Reader, "Spectres and Shadows: Aum Shinrikyo and the Road to Megiddo," *Terrorism and Political Violence,* vol. 14, no. 1 (2002), pp. 147–186.

329 *Asahara moved the date of doomsday:* Shoko Egawa, "From the Other Witness Stand: Following the Aum Case" (serialized article no. 35), *Shukan Yomiuri,* September 20, 1998, pp. 122–123, translated in FBIS, document no. FTS19990517000505.

329 *Asahara developed religious concepts to rationalize murder:* Kensaku Tokiu, "Aum Officer Warns of Revival," *Shukan Gendai,* April 17, 1999, pp. 234–236, translated in FBIS, document no. FTS19990507001911.

330 *Aum candidates ran for Parliament in 1990 election:* Anthony T. Tu, "Anatomy of Aum Shinrikyo's Organization and Terrorist Attacks with Chemical and Biological Weapons," *Archives of Toxicology, Kinetics and Xenobiotic Metabolism,* vol. 7, no. 3 (Autumn 1999), p. 46.

330 *Aum organized itself into twenty-two "ministries":* Ibid., p. 47.

330 *Biographies of cult leaders:* "Police Versus Aum—Officials of Cult Being Arrested One After the Other," *Sande Mainichi,* April 30, 1995, pp. 28–33, translated in FBIS, document no. FTS19950430000039.

330 *Construction of factory to produce AK-47 rifles:* Kaplan, "Aum Shinrikyo (1995)," in Tucker, ed., *Toxic Terror,* p. 212.

331 *Failed effort to produce biological weapons:* William J. Broad, "How Japan Germ Terror Alerted World," *New York Times,* May 26, 1998, pp. A1, A10.

331 *Aum ties with Oleg Lobov:* Kaplan and Marshall, *The Cult at the End of the World,* p. 72.

331 *Asahara trip to Moscow:* Shoko Egawa, "From the Other Witness Stand: Following the Aum Case" (serialized article no. 39), *Shukan Yomiuri,* October 18, 1998, pp. 118–119, translated in FBIS, document no. FTS19990515000045.

332 *Lobov provided access to Soviet expertise:* Author's interview with Alexander Pikayev.

332 *Masami Tsuchiya, selection of Sarin:* Brackett, *Holy Terror,* p. 114.

CHAPTER SEVENTEEN: THE TOKYO SUBWAY

333 *Construction of Sarin plant in Satian 7:* Shoko Egawa, "From the Other Witness Stand: Following the Aum Case" (serialized article no. 41), *Shukan Yomiuri,* November 1, 1998, pp. 46–47, translated in FBIS, document no. FTS19950515001098.

333 *Description of interior of Satian 7:* Kaplan and Marshall, *The Cult at the End of the World,* pp. 119–120.

333 *Equipment ordered through front companies:* Japan, National Police Agency, *Police White Paper 1996* (Tokyo: Printing Bureau, Ministry of Finance, August 30, 1996), unofficial translation by Robert Mauksch, Monterey Institute of International Studies.

335 *Bar containing wine for senior cult leaders:* Shoko Egawa: "From the Other Witness Stand: Following the Aum Case" (serialized article no. 42), *Shukan Yomiuri,* November 1, 1998, pp. 46–47, translated in FBIS, document no. FTS19990515001098.

335 *Initial production of Sarin intermediates, foul odors:* Anthony T. Tu, "Anatomy of Aum Shinrikyo's Organization and Terrorist Attacks with Chemical and Biological Weapons," *Archives of Toxicology, Kinetics and Xenobiotic Metabolism,* vol. 7, no. 3 (Autumn 1999), p. 51.

335 *Aum purchase of Russian military helicopter:* Kaplan and Marshall, *The Cult at the End of the World,* p. 193.

335 *Attempted assassination of Ikeda:* Shoko Egawa, "From the Other Witness Stand: Following the Aum Case" (serialized article no. 54), *Shukan Yomiuri,* February 7, 1999, pp. 46–47, translated in FBIS, document no. FTS19990701000162.

335–36 *Sarin trials at sheep station in Australia:* Kaplan and Marshall, *The Cult at the End of the World,* pp. 126–134.

336 *Aum plan for Matsumoto attack:* Japan, National Police Agency, *Police White Paper 1996.*

336–38 *Description of Matsumoto attack:* "Matsumoto: A Dry Run for Tokyo," in U.S. Senate, Committee on Government Affairs, Permanent Subcommittee on Investigations, Staff Statement, *Global Proliferation of Weapons of Mass Destruction: A Case Study on the Aum Shinrikyo,* October 31, 1995; Kaplan and Marshall, *The Cult at the End of the World,* pp. 137–141; A. Oppenheimer, "Aum Shinrikyo: Lessons to be Learnt," *Jane's Terrorism & Security Monitor,* March 1, 2004.

338 *Kono telephoned Matsumoto Emergency Services:* Kaplan and Marshall, *The Cult at the End of the World,* pp. 142–145.

338 *Casualties of Matsumoto attack:* "Five Long Years Have Passed, Yet Fear of Another Matsumoto Sarin Incident Is Not Fading," *Mainichi Shimbun,* June 27, 1999, p. 25, translated in FBIS, document no. FTS19990824000226; Kaplan and Marshall, *The Cult at the End of the World,* pp. 142–146.

338 *Police found dead animals under grove of trees:* Y. Seto, et al., "Toxicological Analysis of Victims' Blood and Crime Scene Evidence Samples in the Sarin Gas Attack Caused by the Aum Shinrikyo Cult," in Tu and Gaffield, eds., *Natural and Selected Synthetic Toxins,* p. 319.

338–39 *Police pressured Mr. Kono for confession:* Kaplan and Marshall, *The Cult at the End of the World,* p. 145.

339 *"Song of Sarin":* Brackett, *Holy Terror,* p. 119; Staff Statement, *Global Proliferation of Weapons of Mass Destruction: A Case Study on the Aum Shinrikyo,* October 31, 1995, pp. 60–61.

340 *Asahara set Sarin production target of seventy tons:* Tu, "Anatomy of Aum Shinrikyo's Organization and Terrorist Attacks with Chemical and Biological Weapons," p. 55.

340 *Technical problems with Sarin plant:* Shoko Egawa: "From the Other Witness Stand: Following the Aum Case" (serialized article no. 42), *Shukan Yomiuri,* November 1, 1998, pp. 46–47, translated in FBIS, document no. FTS19990515001098.

340 *Leakage of DMMP from Satian 7:* Tu, "Anatomy of Aum Shinrikyo's Organization and Terrorist Attacks with Chemical and Biological Weapons," p. 52.

340 *Failed attempt to recruit Russian scientists:* Kyle B. Olson, "Aum Shinrikyo: Once and Future Threat?," *Emerging Infectious Diseases,* vol. 5, no. 4 (July–August 1999), pp. 513–516.

340 *Decision to shut down Sarin plant:* Brackett, *Holy Terror,* p. 117.

341 *Tsuchiya synthesis of VX, use for assassinations:* Monterey Institute of International Studies, Center for Nonproliferation Studies, "Chronology of Aum Shinrikyo's CBW Activities," 2001.

341 *Yomiuri Shimbun article:* Tu, "Anatomy of Aum Shinrikyo's Organization and Terrorist Attacks with Chemical and Biological Weapons," p. 52.

341 *Remodeling of Satian 7:* Brackett, *Holy Terror,* pp. 117–118; Kaplan and Marshall, *The Cult at the End of the World,* pp. 215–216.

342 *Nakagawa secretly buried supply of DF:* Tu, "Anatomy of Aum Shinrikyo's Organization and Terrorist Attacks with Chemical and Biological Weapons," p. 55.

342 *Press visit to Satian 7, attorney's allegations:* Kaplan and Marshall, *The Cult at the End of the World,* pp. 216–217.

342 *Two army sergeants tipped off the cult:* Brackett, *Holy Terror,* p. 124.

342 *Crisis meeting of Aum leaders:* Tatshuhito Ida, "Subway Sarin Attack Case: No Progress Seen in the 'Matsumoto Trial,' " *Yomiuri Shimbun,* April 6, 1999, p. 25, translated in FBIS, document no. FTS19990517000879.

342 *Five million passengers ride Tokyo subway daily:* Robyn Pangi, "Consequence Management in the 1995 Sarin Attacks on the Japanese Subway System," BCSIA Discussion Paper 2002–4, John F. Kennedy School of Government, Harvard University, February 2002, p. 8.

343 *Doubt in Asahara seen to reflect shallowness of faith:* Shoko Egawa, "From the Other Witness Stand: Following the Aum Case" (serialized article no. 49), *Shukan Yomiuri,* October 25, 1998, pp. 118–119, translated in FBIS, document no. FTS1999-154001879.

343–44 *Synthesis of Sarin by Endo and Nakagawa:* Anthony T. Tu, "Overview of Sarin Terrorist Attacks in Japan," in Tu and Gaffield, eds., *Natural and Selected Synthetic Toxins,* p. 306.

344 *Acetonitrile added to jump-start evaporation of Sarin:* Kyle B. Olson, "Overview: Recent Incidents and Responder Implications," in U.S. Public Health Service, *Proceedings of the Seminar on Responding to the Consequences of Chemical and Biological Terrorism,* p. 2-40.

344 *Final preparations for Tokyo subway attack:* Brackett, *Holy Terror,* pp. 126–130; Kaplan and Marshall, *The Cult at the End of the World,* pp. 242–243.

345 *Hirose thoughts during Sarin attack:* Shoko Egawa, "From the Other Witness Stand: Following the Aum Case" (serialized article no. 33), *Shukan Yomiuri,* September 6, 1998, pp. 48–49, translated in FBIS, document no. FTS19990516000861.

345 *Cultists burned clothes and threw umbrellas in water:* Kaplan and Marshall, *The Cult at the End of the World,* p. 251.

346 *Emergency response to Sarin attack:* Amy E. Smithson and Leslie-Anne Levy, *Ataxia: The Chemical and Biological Terrorism Threat and the US Response,* Report No. 35 (Washington, D.C.: Henry L. Stimson Center, October 2000), pp. 91–101; Per Kulling, "The Terrorist Attack with Sarin in Tokyo on 20 March 1995" (Stockholm: Socialstyrelsen), November 19, 1998, available online at www.sos.se/SOS/PUBL/REFERENG/9803020.htm.

346 *Number of ambulances:* Tetsu Okumura, et al., "The Tokyo Subway Sarin Attack: Disaster Management, Part 1: Community Emergency Response," *Academic Emergency Medicine,* vol. 5. no. 6 (June 1998), pp. 613–617.

346 *Casualties from Sarin attack:* Nicholas D. Kristoff, "Hundreds in Japan Hunt Gas Attackers After 8 Die: Police Tighten Security Steps at Stations," *New York Times,* March 21, 1995, p. A1.

346 *Police confirmed Sarin:* Smithson and Levy, *Ataxia,* p. 94.

347 *Victims divided into three categories:* Sadayoshi Ohbu, et al., "Sarin Poisoning on Tokyo Subway," *Southern Medical Journal,* June 1997, available online at www.sma.org/smj/97june3.htm; Fred Sidell, U.S. Army Medical Research Institute of Chemical Defense, "U.S. Medical Team Briefing," in U.S. Public Health Service, *Proceedings of the Seminar on Responding to the Consequences of Chemical and Biological Terrorism,* p. 2-32.

347 *About 3,700 casualties were "worried well":* Sidell, "U.S. Medical Team Briefing," p. 2-33.

347 *Secondary exposures from off-gassing of Sarin:* Okumura, et al., "The Tokyo Subway Sarin Attack, Disaster Management, Part 1," p. 615.

347–48 *Decontamination of victims' clothes:* Jun Sato, "Sarin Cleanup Remembered," *The Daily Yomiuri* (English-language Web site of *Yomiuri Shumbun*), March 16, 2003.

348 *Police raid on Aum facilities:* Japan, National Police Agency, *Police White Paper 1996.*

348 *Police used caged canaries as Sarin detectors:* Tu, "Aum Shinrikyo's Chemical and Biological Weapons," p. 66.

348 *Murder of Murai by* yakuza: Olson, "Aum Shinrikyo: Once and Future Threat?," p. 515.

348 *Arrest of Asahara:* Kaplan and Marshall, *The Cult at the End of the World,* p. 281.

350 *Senator Helms blocked ratification of CWC:* Michael Krepon, Amy E. Smithson, and John Parachini, *The Battle to Obtain US Ratification of the Chemical Weapons Convention* (Washington, D.C.: Henry L. Stimson Center, Occasional Paper no. 35, July 1997).

CHAPTER EIGHTEEN: THE EMERGING THREAT

351 *UNSCOM retrieval of Iraqi documents from Muthanna:* Krasno and Sutterlin, *The United Nations and Iraq,* pp. 67–68.

352 *Pentagon admitted nerve-agent release at Khamisiyah:* Suzanne Gamboa, "VA Orders More Study of Deaths After Gulf War; Destroyed Iraqi Nerve Gases May Have Affected Soldiers," *Washington Post,* March 4, 2002, p. A17; U.S. General Accounting Office, *Gulf War Illnesses: DOD's Conclusions about U.S. Troops' Exposure Cannot Be Adequately Supported,* GAO-04-159, June 2004, p. 7.

353 *Research by Robert W. Haley, M.D.:* R. W. Haley, J. Hom, P. S. Roland, et al., "Evaluation of Neurologic Function in Gulf War Veterans: A Blinded Case-Control Study," *Journal of the American Medical Association,* vol. 277 (1997), pp. 223–230; R. W. Haley, W. W. Marshall, G. G. McDonald, M. A. Daugherty, F. Petty, and J. L. Fleckenstein, "Brain Abnormalities in Gulf War Syndrome: Evaluation with 1H MR Spectroscopy," *Radiology,* vol. 215 (2000), pp. 807–817; Robert W. Haley, Scott Billecke, and Bert N. La Du, "Association of Low PON1 Type Q (Type A) Arylesterase Activity with Neurologic Symptom Complexes in

Gulf War Veterans," *Toxicology and Applied Pharmacology*, vol. 157 (1999), pp. 227–233.

354 *Negotiation of conditions in Senate resolution of ratification:* "Senate Advice and Consent Subject to Conditions," *Congressional Record—Senate,* April 24, 1997, pp. S3651–S3657.

354–55 *Ratification vote in Senate:* Michael Krepon, Amy E. Smithson, and John Parachini, *The Battle to Obtain US Ratification of the Chemical Weapons Convention* (Washington, D.C.: Henry L. Stimson Center, Occasional Paper No. 35, July 1997).

356–57 *Russian chemical weapons destruction:* Jonathan B. Tucker, "Russia's New Plan for Chemical Weapons Destruction," *Arms Control Today,* vol. 31, no. 6 (July–August 2001), pp. 9–13.

357 *Analysis of wipe samples from Iraqi "special" warheads:* Krasno and Sutterlin, *The United Nations and Iraq,* pp. 68–69.

357 *UNSCOM Technical Evaluation Meeting:* UNSCOM, "Report of the Group of International Experts on VX," October 23, 1998, available online at www.cns.miis.edu/research/iraq/vxreprt.htm.

357–58 *Israeli debate over CWC ratification:* Aluf Benn, "Terms for Chemical Nonproliferation Treaty," *Ha'aretz* (in Hebrew), July 16, 1992, pp. A1, A8, translated in FBIS, JPRS-TND-92-024, July 21, 1992, pp. 12–13; Steve Rodan, "Bitter Choices: Israel's Chemical Dilemma," *Jerusalem Post,* August 18, 1997; Stephanie Nebehay, "Israel Not Ready to Ratify Chemical Weapons Pact," Reuters, September 4, 1997; Lieutenant Colonel David Eshel, "Israel Grapples with CWC Ratification," *Armed Forces Journal International,* September 1997, p. 24; Yair Evron, *Weapons of Mass Destruction in the Middle East,* Occasional Paper no. 39 (Washington, D.C.: Henry L. Stimson Center, March 1998), pp. 36–37; Emily Landau and Tamar Malz, "Israel's Arms Control Agenda," *Strategic Assessment* (Jaffee Center for Strategic Studies, Tel Aviv University), vol. 2, no. 4 (February 2000), pp. 21–25; Aluf Benn, "A Difficult Choice: the Chemical Weapons Convention," *Ha'aretz,* November 29, 2000; Gerald M. Steinberg, "Israeli Policy on the CWC," *OPCW Synthesis,* November 2002, pp. 29–31.

358 *Israeli chemical weapons program:* Avner Cohen, "Israel: Reconstructing a Black Box," in Wright, ed., *Biological Warfare and Disarmament,* pp. 181–212. See also, Avner Cohen, "Israel and Chemical/Biological Weapons: History, Deterrence, and Arms Control," *The Nonproliferation Review,* vol. 8, no. 3 (Fall–Winter 2001), pp. 27–53.

358 *Israeli scientists visited French chemical test site:* Hersh, *The Samson Option,* pp. 63–64.

359 *After Six-Day War, Israel expanded production of CW:* Ian Black, "Israel Tries to Counter Arab Nerve Gas Threat," *The Guardian,* December 12, 1986, p. 9.

359 *Israel Institute for Biological Research (IIBR) at Nes Ziona:* Avner Cohen, "Israel: Reconstructing a Black Box," in Wright, ed., *Biological Warfare and Disarmament: New Problems/New Perspectives,* pp. 181–212; Uzi Mahmaimi, "Israeli Jets

Equipped for Chemical Warfare," *Sunday Times* (London), October 4, 1998; "Israeli Plant May Produce Biological Arms," *Salt Lake Tribune,* October 25, 1998; "Residents Near Israeli Institute Fear Nerve Gas Leak," *Fort Worth Star-Telegram,* November 1, 1998.

359 *IIBR's Organic Chemistry Department:* See IIBR Web site, www.iibr.gov.il.

359 *IIBR paper on V agents:* Julian Perry Robinson, "Behind the VX Disclosure," *New Scientist,* January 9, 1975, p. 50.

359 *MALMAB responsible for security at IIBR:* Yossi Melman, "Foreign Sources Say . . ." *Ha'aretz* (Internet edition), December 31, 2004.

359 *Crash of El Al cargo plane in Amsterdam in 1992:* Mouin Rabbani, "El Al Flight 1862 and Israel's Chemical Secrets," *Middle East International,* October 16, 1998, pp. 20–22.

360 *Allegations of cover-up:* Marlise Simons, "6 Years After Crash, Talk of Cover-Up: Health Problems Linked to 1992 Crash of El Al Cargo Plane in Amsterdam," *New York Times,* February 9, 1999, p. A18.

360 NRC Handelsblad *reported on its investigation:* Rabbani, "El Al Flight 1862," p. 20.

360 *189 liters of DMMP supplied by Solkatronic:* Ibid.

361 *"We fly sugar" quote by El Al spokesman:* Joel Greenberg, "Nerve-Gas Element Was in El Al Plane Lost in 1992 Crash," *New York Times,* October 2, 1998, p. A1.

361 *Statement by Ze'ev Schiff:* Ze'ev Schiff, "The 'Chemical Clock,' " *Ha'aretz* (in English), July 2, 1999.

362 *Bin Laden and Sudan:* Clarke, *Against All Enemies,* pp. 141–145.

363 *Intelligence reports implicating Al-Shifa:* James Risen, "To Bomb Sudan Plant, or Not: A Year Later, Debates Rankle," *New York Times,* October 27, 1999.

363 *Ani link to Al-Shifa Factory:* Associated Press, "Top Iraqi Scientist Surrenders to U.S.," *Washington Post,* April 19, 2003, p. A20.

364 *Meeting of "Small Group":* Michael Barletta, "Chemical Weapons in the Sudan: Allegations and Evidence," *The Nonproliferation Review,* vol. 6, no. 1 (Fall 1998), p. 116.

365 *Sandy Berger quote:* Benjamin and Simon, *The Age of Sacred Terror,* p. 260.

365–66 *U.S. cruise missile attack on Al-Shifa Factory:* Seymour M. Hersh, "The Missiles of August," *New Yorker,* October 12, 1998, p. 40; Clarke, *Against All Enemies,* p. 188.

367 *Criticisms of Al-Shifa attack:* Barletta, "Chemical Weapons in the Sudan: Allegations and Evidence," pp. 119; Eric Croddy, "Dealing with Al Shifa: Intelligence and Counterproliferation," *International Journal of Intelligence and Counterintelligence,* vol. 15, no. 1 (2002), pp. 52–60.

367 *Al-Fadl testimony:* Daniel Benjamin and Steven Simon, "A Failure of Intelligence?" *New York Review of Books,* vol. 48, no. 20 (December 20, 2001), p. 76; Stephen F. Hayes, "Connecting the Dots in 1998, but Not in 2003," *The Weekly Standard,* vol. 9, no. 16, December 29, 2003.

367 *Clarke testimony:* Testimony of Richard A. Clarke, National Coordinator for Security, Infrastructure Protection, and Counterterrorism, before a joint hearing of the Senate Select Committee on Intelligence and the House Permanent Select Committee on Intelligence, *Joint Inquiry Briefing by Staff on U.S. Government Counterterrorism Organizations (Before September 11, 2001) and on the Evolution of the Terrorist Threat and U.S. Response, 1986–2001,* June 11, 2002, pp. 52–53.

368 *George W. Bush State of the Union address:* Office of the Press Secretary, The White House, "President Delivers State of the Union Address," January 29, 2002.

368 *CNN broadcast Al-Qaeda videotapes of experiments on dogs:* Cable News Network, "Tapes Shed New Light on bin Laden's Network," August 19, 2002, available online at www.cnn.com/2002/US/08/18/terror.tape.main/.

369 *National Security Strategy:* The White House, "The National Security Strategy of the United States of America," September 2002, p. 15, available online at www.whitehouse.gov/nsc/nss.pdf.

369 *Events leading up to Iraq War:* Bryan Burrough, Evgenia Peretz, David Rose, and David Wise, "The Path to War," *Vanity Fair,* May 2004, pp. 228–294.

369 *October 2002 National Intelligence Estimate:* "Key Judgments from the National Intelligence Estimate on Iraq's Continuing Programs for Weapons of Mass Destruction, October 2002," in Cirincione, Mathews, and Perkovich, *WMD in Iraq: Evidence and Implications,* Appendix 1, pp. 63–65.

369 *"Dusty" agents:* National Intelligence Council, *Iraq's Chemical Warfare Capabilities: Potential for Dusty and Fourth-Generation Agents,* Memorandum to Holders of NIE 2002-16HC, November 2002.

369–70 *UNMOVIC inspections in Iraq:* Blix, *Disarming Iraq.*

370 *Secretary Powell speech to U.N. Security Council:* "U.S. Secretary of State Colin Powell's Address to the UN Security Council, February 5, 2003," in Cirincione, Mathews, and Perkovich, *WMD in Iraq: Evidence and Implications,* Appendix 4, p. 88.

370–71 *President Bush ordered invasion of Iraq:* George W. Bush, "Address to the Nation on War with Iraq," Remarks in Washington, D.C., March 17, 2003.

372 *Preliminary findings of Iraq Survey Group:* Fred Kaplan, "The Iraq Sanctions Worked," *Slate,* October 7, 2003; author's interview with David Kay.

372 *ISG final report:* Iraq Survey Group, *Comprehensive Report of the Special Advisor to the DCI on Iraq's WMD (Weapons of Mass Destruction),* September 30, 2004.

373 *Post-war looting:* James Glanz and William J. Broad, "Looting at Weapons Plants Was Systematic, Iraqi Says," *New York Times,* March 13, 2005, p. 1.

373 *Effectiveness of U.N. inspections:* Walter Pincus, "Former U.N. Inspectors Cite New Report as Validation," *Washington Post,* October 8, 2004, p. A30.

373 *Blix metaphor* BEWARE THE DOG: Associated Press, "Hans Blix Says Iraq Probably Destroyed Most WMDs Long Ago," September 17, 2003.

373 *Saddam Hussein's balancing act:* David Johnston, "Saddam Hussein Sowed Confusion About Iraq's Arsenal as a Tactic of War," *New York Times,* October 7,

2004, p. A22; Dana Priest, "Hussein's Aims, Capabilities Often Differed," *Washington Post,* October 8, 2004, p. A7.

374 *Reopening of Maddison case:* BBC News, " 'Tricked' into Nerve Gas Tests," November 15, 2004.

EPILOGUE: TOWARD ABOLITION

375–76 *Dismantling of binary plant at Pine Bluff:* Amy Riggin, "Arsenal Sees End of Era," *Pine Bluff Commercial,* January 6, 2005, available online at www.pbcommercial.com/articles/2005/01/06/news/news1.txt.

376 *Vast majority of nations have signed and ratified the CWC:* See the Web site of the Organization for the Prohibition of Chemical Weapons at www.opcw.org.

376 *Libyan accession to the CWC:* Joby Warrick and Peter Slevin, "Libya's Disclosures Put Weapons in New Light," *Washington Post,* March 2, 2004, p. A1; Associated Press, "Libya Details Its Chemical Weapons Hoard," *Washington Post,* March 6, 2004, p. A14; Organization for the Prohibition of Chemical Weapons, "OPCW Team Visits Libya," Press Release No. 4, February 5, 2004.

377 *Statement by Syrian President Bashar alAssad:* Benedict Brogan, "We Won't Scrap WMD Stockpile Unless Israel Does, Says Assad," *Daily Telegraph* [London], January 6, 2004, p. 1.

377 *Iranian chemical weapons program:* Khalil Hasan, "Iran Said to Be Making Chemical Arms," *Daily Times* (Pakistan), January 31, 2005.

377 *North Korean chemical weapons program:* Associated Press, "Chem. Weapons Remain Threat to S. Korea," *New York Times,* February 27, 2003; "DPRK's Biological, Chemical Arms Plants Located Along Yalu River," *Yomiuri Weekly* (Tokyo), December 1, 2002, pp. 14–16, translated in FBIS, document no. JPP20021124000044; Ruriko Kubota, "Former North Korean Chemist Reveals DPRK Carried Out Human Experimentation with Poison Gas at Prison Camp Beginning 25 Years Ago," *Sankei Shimbun* (Tokyo), March 22, 2004, translated in FBIS, document no. JPP20040322000060.

377 *Brad Roberts quote:* Brad Roberts, "Implementing the Biological Weapons Convention: Looking Beyond the Verification Issue," in Oliver Thraenert, ed., *The Verification of the Biological Weapons Convention: Problems and Perspectives* (Bonn: Friedrich Ebert Stiftung, 1992), p. 104.

377 *Al-Abud network:* Charles J. Hanley, "Iraq Insurgents Fail to Brew Chemical Arms," Associated Press, April 12, 2005.

378 *Proliferation Security Initiative:* Andrew C. Winner, "The Proliferation Security Initiative: The New Face of Interdiction," *The Washington Quarterly,* vol. 28, no. 2 (Spring 2005), pp. 129–143.

380 *Weakening of CWC verification regime:* Jonathan B. Tucker, "The Chemical Weapons Convention: Has It Enhanced U.S. Security?," *Arms Control Today,* vol. 31, no. 3 (April 2001), pp. 8–12.

380 *Limitations on chemical sampling and analysis:* Author's interview with Ron Manley.

382 *Problems of Russian chemical weapons destruction:* Susan B. Glasser, "Cloud over Russia's Poison Gas Disposal," *Washington Post,* August 24, 2002, pp. A1, A17; Jonathan B. Tucker, "Russia's New Plan for Chemical Weapons Destruction," *Arms Control Today,* vol. 31, no. 6 (July–August 2001), pp. 9–13; Sam Nunn, "We're Failing to Meet the Real Nuclear Threat," *London Sunday Times,* March 20, 2005.

382–83 *Problems of U.S. chemical weapons destruction:* U.S. General Accounting Office, Report to the Chairman, House Committee on Armed Services, "Delays in Implementing the Chemical Weapons Convention Raise Concerns About Proliferation," GAO-04-361, March 2004.

384 *Nonlethal chemical weapons, Moscow theater incident:* William M. Arkin, "Pulling Punches: Big Plans for Futuristic, Nonlethal Weapons Are Afoot, But Their Use Would Raise Troubling Questions," *Los Angeles Times,* January 4, 2004, p. M1; "Non-Lethal Weapons, the CWC and the BWC," *The CBW Conventions Bulletin,* no. 61 (September 2003), pp. 1–2; Robin Coupland, "Incapacitating Chemical Weapons: A Year After the Moscow Theater Siege," *The Lancet,* vol. 362, no. 9393 (October 25, 2003), p. 1346.

385 *Impact on CWC of advances in chemical technology:* International Union of Pure and Applied Chemistry (IUPAC), "Impact of Scientific Developments on the Chemical Weapons Convention" (special issue), *Pure and Applied Chemistry,* vol. 74, no. 12 (December 2002).

BIBLIOGRAPHY

BOOKS

Atkinson, Rick. *Crusade: The Untold Story of the Persian Gulf War.* Boston: Houghton Mifflin, 1993.

Baker, James A. *The Politics of Diplomacy: Revolution, War and Peace, 1989–1992.* New York: G. P. Putnam, 1995.

Baram, Amatzia. *Building Toward Crisis: Saddam Husayn's Strategy for Survival.* Washington, D.C.: Washington Institute for Near East Policy, 1998.

Benjamin, Daniel, and Steven Simon. *The Age of Sacred Terror: Radical Islam's War Against America.* New York: Random House, 2002.

Bernauer, Thomas. *The Chemistry of Regime Formation.* Brookfield, Vt.: Dartmouth Publishing Co. for the United Nations Institute of Disarmament Research, 1993.

Blix, Hans. *Disarming Iraq.* New York: Pantheon Books, 2004.

Borkin, Joseph. *The Crime and Punishment of I. G. Farben.* New York: Free Press, 1978.

Bower, Tom. *The Paperclip Conspiracy: The Hunt for the Nazi Scientists.* Boston: Little, Brown, 1987.

Brackett, D. W. *Holy Terror: Armageddon in Tokyo.* New York: Weatherhill, 1996.

Bradley, Omar. *A Soldier's Story.* New York: Henry Holt and Co., 1951.

Brauch, Hans Günter. *Der chemische Alptraum.* Berlin: Verlag J. H. W. Dietz Nachf., 1982.

——— and Rolf-Dieter Müller. *Chemische Kriegführung—Chemische Abrüstung: Dokumente und Kommentare,* Teil I: *Dokumente aus deutschen und amerikanischen Archiven.* West Berlin: Berlin Verlag, 1985.

Brown, Frederic J. *Chemical Warfare: A Study in Restraints.* Princeton, N.J.: Princeton University Press, 1968.

Bud, Robert, and Philip Gummet. *Cold War, Hot Science: Applied Research in Britain's Defence Laboratories, 1945–90.* New York: Harwood Academic Publishers, 1999.

Bullock, Allan. *Hitler: A Study in Tyranny.* New York: Harper & Row, abridged ed., 1971.

Bibliography

Carter, G. B. *Chemical and Biological Defence at Porton Down, 1916–2000.* London: Her Majesty's Stationery Office, 2000.

Charles, Daniel. *Master Mind: The Rise and Fall of Fritz Haber, the Nobel Laureate Who Launched the Age of Chemical Warfare.* New York: Ecco, 2005.

Cirincione, Joseph, Jessica T. Mathews, and George Perkovich. *WMD in Iraq: Evidence and Implications.* Washington, D.C.: Carnegie Endowment for International Peace, 2004.

Clarke, Richard A. *Against All Enemies: Inside America's War on Terror.* New York: Free Press, 2004.

Clarke, Robin. *The Silent Weapons.* New York: David McKay Co., 1968.

Cole, Leonard A. *The Eleventh Plague: The Politics of Biological and Chemical Warfare.* New York: W. H. Freeman, 1997.

Cordesman, Anthony. *Weapons of Mass Destruction in the Middle East.* London: Brassey's, 1991.

Diest, Wilhelm, ed. *The German Military in the Age of Total War.* Warwickshire, U.K.: Berg Publishers, 1985.

Drell, Sidney D., Abraham D. Sofaer, and George D. Wilson. *The New Terror: Facing the Threat of Biological and Chemical Weapons.* Stanford, Calif.: Hoover Institution Press, 1999.

DuBois, Josiah E. *The Devil's Chemists: 24 Conspirators of the International Farben Cartel Who Manufacture Wars.* Boston: Beacon Press, 1952.

Duffy, Christopher. *Red Storm on the Reich.* New York: Athaneum, 1991.

Eggleston, Wilfred. *Scientists at War.* London: Oxford University Press, 1950.

Fredericks, William Curtis. *The Evolution of Post–World War II United States Chemical Warfare Policy,* master's thesis, Oxford University, Oxford, U.K., January 15, 1988.

Friedman, Leon. *The Law of War: A Documentary History.* New York: Random House, 1972.

Fries, Amos A., and Clarence J. West. *Chemical Warfare.* New York: McGraw-Hill, 1921.

Fussell, Paul. *The Great War and Modern Memory.* Oxford, U.K.: Oxford University Press, 1975.

Gazit, Schomo, ed. The *Middle East Military Balance, 1993–1994.* Boulder, Colo.: Westview Press for the Jaffee Center for Strategic Studies, Tel Aviv University, 1994.

Gellermann, Guenther W. *Der Krieg, der nicht stattfand.* Bonn: Bernard & Graefe Verlag, 1986.

Glennon, John P., William K. Klingaman, David S. Patterson, and Ilana Stern, eds. *Foreign Relations of the United States, 1955–1957,* Vol. XIX, *National Security Policy.* Washington, D.C.: U.S. Government Printing Office, 1990.

Goudsmit, Samuel A. *ALSOS.* New York: Harry Schuman, 1947.

Greeman, Clive, and Gwynne Roberts. *Der kälteste Krieg: Professor Frucht und das Kampfstoff-Geheimnis.* Berlin: Ulstein, 1982.

Groehler, Olaf. *Der lautlose Tod.* East Berlin: Verlag der Nation, 1984, p. 327.

Bibliography

Haber, Ludwig Fritz. *The Poisonous Cloud: Chemical Warfare in the First World War.* Oxford: Clarendon Press, 1986.

Hahn, Otto. *Mein Leben.* München: Verlag F. Bruckmann, 1968.

Harris, Robert, and Jeremy Paxman. *A Higher Form of Killing: The Secret Story of Chemical and Biological Warfare.* New York: Hill and Wang, 1982.

Hastings, Max. *Overlord: D-Day and the Battle for Normandy.* London: Michael Joseph, 1984.

Hayes, Peter. *Industry and Ideology: IG Farben in the Nazi Era.* Cambridge, U.K.: Cambridge University Press, 2nd ed., 2001.

Heine, Jens Ulrich. *Verstand & Schicksal: Die Männer der I.G. Farbenindustrie A.G. (1925–1945) in 161 Kurzbiographien.* Weinheim: VCH Verlag, 1990.

Hersh, Seymour M. *Chemical and Biological Warfare: America's Hidden Arsenal.* Indianapolis: Bobbs-Merrill Co., 1968.

———. *The Samson Option: Israel, America and the Bomb.* London: Faber and Faber, 1991.

Hirsch, Walter. *Soviet BW and CW Preparations and Capabilities.* trans. Intelligence Branch, Office of the Chief, U.S. Army Chemical Corps, c. 1947 [MHI].

Hunt, Irmgard A. *On Hitler's Mountain: Overcoming the Legacy of a Nazi Childhood.* New York: HarperCollins, 2005.

Hunt, Linda. *Secret Agenda: The United States Government, Nazi Scientists, and Project Paperclip, 1945 to 1990.* New York: St. Martin's Press, 1991.

Hylton, A. R. *The History of Chemical Warfare Plants and Facilities in the United States, Studies on the Technical and Control Aspects of Chemical and Biological Weapons,* vol. 4. Kansas City, Kans.: Midwest Research Institute for the U.S. Arms Control and Disarmament Agency, November 1972.

Johnston, Harold. *A Bridge Not Attacked: Chemical Warfare Civilian Research During World War II.* Singapore: World Publishing Co., 2003.

Kaplan, David E., and Andrew Marshall. *The Cult at the End of the World: The Terrifying Story of the Aum Doomsday Cult, from the Subways of Tokyo to the Nuclear Arsenals of Russia.* New York: Crown Publishers, 1996.

Karsh, Efraim, ed. *The Iran-Iraq War: Impact and Implications.* New York: St. Martin's Press, 1989.

———, Martin S. Navias, and Philip Sabin, eds., *Non-Conventional Weapons Proliferation in the Middle East.* Oxford, U.K.: Clarendon Press, 1993.

Kleber, Brooks E., and Dale Birdsell. *The United States Army in World War II—The Technical Services—The Chemical Warfare Service: Chemicals in Combat.* Washington, D.C.: Office of the Chief of Military History, U.S. Army, 1966.

Köhler, Otto. *Und heute die ganze Welt: Die Geschichte der IG Farben, BAYER, BASF und HOECHST.* Cologne, Germany: PapyRossa Verlag, 1990.

Krasno, Jean E., and James S. Sutterlin. *The United Nations and Iraq: Defanging the Viper.* Westport, Conn.: Praeger, 2003.

Krause, Joachim, and Charles K. Mallory. *Chemical Weapons in Soviet Military Doctrine: Military and Historical Experience, 1915–1991.* Boulder, Colo.: Westview Press, 1992.

Bibliography

Lavoy, Peter R., Scott D. Sagan, and James J. Wirtz, eds. *Planning the Unthinkable: How New Powers Will Use Nuclear, Biological, and Chemical Weapons.* Ithaca, N.Y.: Cornell University Press, 2000.

Lepick, Olivier. *Les armes chimiques.* Paris: Presses Universitaires de France, Series Que Sais-Je, 1999.

McCarthy, Richard M. *The Ultimate Folly: War by Pestilence, Asphyxiation and Defoliation.* New York: Alfred A. Knopf, 1970.

Meyer, Claude. *L'arme chimique.* Paris: Ellipses Édition, 2001.

Morel, Benoit, and Kyle Olson, eds., *Shadows and Substance: The Chemical Weapons Convention.* Boulder, Colo.: Westview Press, 1993.

Murakami, Haruki. *Underground: The Tokyo Gas Attack and the Japanese Psyche.* New York: Vintage Books, 2001.

Ochsner, Hermann. *History of German Chemical Warfare in World War Two,* Part I: *The Military Aspect.* Washington, D.C.: Historical Office, Office of the Chief of the Chemical Corps, 1949.

O'Donnell, James P. *The Bunker: The History of the Reich Chancellery Group.* Boston: Houghton-Mifflin, 1978.

Owen, Wilfred. *Collected Poems of Wilfred Owen.* New York: New Directions, 1965.

Pfingsten, Otto. *Dr. Gerhard Schrader: Der Erfinder des Schädlingsbekämpfungsmittels E605.* Wendeburg, Germany: Verlag Uwe Krebs, 2003.

Plumpe, Gottfried. *Die I.G. Farbenindustrie AG: Wirtschaft, Technik und Politik, 1904–1945.* Berlin: Duncker & Humblot, 1990.

Price, Richard M. *The Chemical Weapons Taboo.* Ithaca, N.Y.: Cornell University Press, 1997.

Primack, Joel R. *Advice and Dissent: Scientists in the Political Arena.* New York: Basic Books, 1974.

Roberts, Brad, ed. *Chemical Disarmament and U.S. Security* Boulder, Colo.: Westview Press, 1992.

Rothschild, J. H. *Tomorrow's Weapons: Chemical and Biological.* New York: McGraw-Hill, 1964.

Russell, Edmund. *War and Nature: Fighting Humans and Insects with Chemicals from World War I to Silent Spring.* Cambridge, U.K.: Cambridge University Press, 2001.

Saunders, Bernard Charles. *Some Aspects of the Chemistry and Toxic Action of Organic Compounds Containing Phosphorus and Fluorine.* Cambridge, U.K.: Cambridge University Press, 1957.

Schwarzkopf, General H. Norman. *It Doesn't Take a Hero.* New York: Bantam, 1993.

Shirer, William L. *The Rise and Fall of the Third Reich.* New York: Simon & Schuster, 1959.

Sidell, Frederick R., Ernest T. Takafuji, and David R. Franz, eds. *Medical Aspects of Chemical and Biological Warfare.* Washington, D.C.: Borden Institute, Walter Reed Army Medical Center, 1997.

Simon, Leslie E. *German Research in World War II: An Analysis of the Conduct of Research.* New York: John Wiley & Sons., 1947.

Sloan, Roy. *The Tale of Tabun: Nazi Chemical Weapons in North Wales.* Lhanrwst: Gwasg Carreg Gwalch, 1998.

Speer, Albert. *Inside the Third Reich.* New York: Simon & Schuster, reissued 1997.

Spiers, Edward M. *Chemical and Biological Weapons: A Study of Proliferation.* New York: Macmillan, 1994.

Stock, Thomas, and Karlheinz Los, eds. *The Challenge of Old Chemical Munitions and Toxic Armament Wastes.* London: Oxford University Press for the Stockholm International Peace Research Institute, 1997.

Stockholm International Peace Research Institute (SIPRI). *The Problem of Chemical and Biological Warfare,* vol. 1: *The Rise of CB Weapons.* New York: Humanities Press, 1971.

———. *The Problem of Chemical and Biological Warfare,* vol. 2: *CB Weapons Today.* New York: Humanities Press, 1973.

Timmerman, Kenneth R. *The Death Lobby: How the West Armed Iraq.* Boston: Houghton Mifflin, 1991.

Toy, Arthur D. F. *Phosphorus Chemistry in Everyday Living.* Washington, D.C.: American Chemical Society, 1976.

Trevor-Roper, H. R. *The Last Days of Hitler.* London: Pan Books, 1952.

Tromp, Hylke, ed. *Non-Nuclear War in Europe: Alternatives for Nuclear Defence.* Gronigen, Netherlands: Gronigen University Press, 1987.

Tu, Anthony T., and William Gaffield, eds. *Natural and Selected Synthetic Toxins: Biological Implications.* Washington, D.C.: American Chemical Society, 2000.

Tuchman, Barbara W. *The Guns of August.* New York: Macmillan, 1962.

Tucker, Jonathan B., ed. *Toxic Terror: Assessing Terrorist Use of Chemical and Biological Weapons.* Cambridge, Mass.: MIT Press, 2000.

U.S. Arms Control and Disarmament Agency (ACDA). *Arms Control and Disarmament Agreements: Texts and Histories of the Negotiations.* Washington, D.C.: Arms Control and Disarmament Agency, 1990.

U.S. Congress, Office of Technology Assessment. *Technologies Underlying Weapons of Mass Destruction,* OTA-BP-ISC-115. Washington, D.C.: U.S. Government Printing Office, December 1993.

U.S. House of Representatives, Committee on Government Operations, Subcommittee on Conservation and Natural Resources. Hearing, *Environmental Dangers of Open-Air Testing of Lethal Chemicals,* 91st Congress, 1st session, May 20–21, 1969. Washington, D.C.: U.S. Government Printing Office, 1969.

U.S. News and World Report, *Triumph Without Victory: The Unreported History of the Persian Gulf War.* New York: Times Books, 1992.

U.S. Public Health Service, Office of Emergency Preparedness. *Proceedings of the Seminar on Responding to the Consequences of Chemical and Biological Terrorism, July 11–14, 1995.* Washington, D.C.: Department of Health and Human Services, 1995.

Wise, David. *Cassidy's Run: The Secret Spy War over Nerve Gas.* New York: Random House, 2000.

Wiseman, Lieutenant Colonel D. J. C. *The Second World War, 1939–1945, Army,*

Bibliography

Special Weapons and Types of Warfare, vol. 1: *Gas Warfare.* London: The War Office, 1951.

Wright, Susan, ed. *Biological Warfare and Disarmament: New Problems/New Perspectives.* Lanham, Md.: Rowman & Littlefield, 2002.

ARCHIVAL SOURCES

CBW Archive, Science Policy Research Unit, University of Sussex (Brighton, U.K.).

Historical Office, Edgewood Area, Aberdeen Proving Ground (Aberdeen, Md.).

Military History Institute, U.S. Army War College (Carlisle Barracks, Pa.).

National Security Archive, George Washington University (Washington, D.C.).

Public Record Office (Kew, U.K.).

U.S. National Archives and Records Administration (College Park, Md.).

AUTHOR INTERVIEWS

Serguei Batsanov (former Soviet CWC negotiator), The Hague, Netherlands, August 21, 2003.

Don Clagett (director of industry verification, OPCW), The Hague, Netherlands, August 20, 2003.

Garrett Cochran (former CIA chemical weapons analyst), Washington, D.C., October 17, 2003.

William C. Dee (former developer of binary chemical weapons), Abingdon, Md., July 18, 2003.

Sigmund R. Eckhaus (former chemical weapons process engineer), Washington, D.C., May 21, 2001.

Elisa D. Harris (former National Security Council staff member specializing in chemical arms control), Washington, D.C., December 11, 2003.

John Isaacs (President, Council for a Livable World), Washington, D.C., July 25, 2003.

David Kay (former director, Iraq Survey Group), Washington, D.C., May 20, 2004.

Olivier Lepick (French chemical weapons expert), Paris, September 21, 2004.

Boris Libman (former Soviet chemical weapons process engineer), Philadelphia, Pa., May 7, 2003.

Ron Manley (former process engineer at Nancekuke), Christchurch, England, August 25, 2003.

Vil S. Mirzayanov (former Soviet chemical weapons specialist), Princeton, N.J., June 10, 2003.

Colonel Jonathan Newmark (physician with the U.S. Army Medical Research Institute of Chemical Defense), Aberdeen, Md., October 29, 2003.

Alexander Pikayev (Russian defense policy analyst), Washington, D.C., December 12, 2003.

Julian Perry Robinson (British chemical weapons expert), Washington, D.C., February 22, 2004.

INDEX

Note: Page numbers in italics refer to illustrations

Index

Printed in the United States
by Baker & Taylor Publisher Services